The final half-dozen years of steam operation saw the South Western dominated by rebuilt Bulleid Pacifics and BR Standard 4-6-0's. This reliance on a small number of classes was achieved by the rebuilding of the original designs which allowed the withdrawal of several older classes. West Country 34048 'Crediton', seen running round a train of empty stock which it has brought in from Bournemouth West, was an import from the Central (LBSCR) system, having spent much of its time prior to being rebuilt as a Brighton-based engine.

"...and the scene is now transported, gentles, to Southampton..."

This, the third in the SR series, follows the style set by its companions in concentrating not upon the history of the line in the conventional sense - dates of construction and opening have been repeated often enough - but upon the train services as they were during the post-war days of steam operation.

The train service was the very substance of history since the schedules were almost as immovable as the line itself and a study of the timetable for the 1950's is a study of the line at almost any time in the post-grouping period.

The times of trains were an important element in railway history and it mattered greatly that one was able to catch the same train - the engine and coaches might have developed during the interim - as one's parents and grandparents. To board a Dorchester train at half-past four in the evening and to know that one was enacting the same scene that one's predecessors had experienced gave in a strange sort of a way, a curious degree of security.

Similarly to turn up at Waterloo in the early evening, just in time to see the tail lamp of the last Wessex day express vanish into the night was irritating but one's irritation was that of previous generations who had fallen into the same trap and spent long nights in the Dorchester Mail.

How one regarded Liverpool Street - who

had recently replaced their ancient dearth of fast trains to Norwich with an express service that operated at half-past every other hour - as a set of radicals. The timetable, like Shakespeare, existed to be perpetuated and not to be played with. Those who interfere with the higher forms of literary expression - the timeta-

Unless stated otherwise all references to engine workings and train services are for the period Winter 1953/4 although changes during the following decade were minimal.

ble - for grubby commercial needs deserve nothing but contempt.

Whilst the LSWR timetable remained a fix-

Thanks for assistance with text and illustrations are due to P. Foster, Michael Miller (Toronto), John Edgington, C. Caddy, M. Bentley, N. Kendall, W. Becket, C. Bentley, J. Cat, P. Webb, J. House, J. Gilks and the staff of the Fareham District.

ture, so did many of the engine and carriage workings and, indeed, so much of the day to day working relied upon the memory of individuals - one had to know that the 06.35 Bournemouth - Waterloo changed engines at Basingstoke without having to look it up - that to change the plan in any way invoked a serious threat of chaos. To understand that an engine or set of coaches performed the same gyrations that they had in Lloyd George's day was also part of the historical continuum.

Not everything had stayed static and in its final days the Southern had taken the questionable step of replacing its well-tried King Arthur 4-6-0's with a rather mercurial fleet of Pacifics whose principal virtue seemed to be the unique stamp they placed upon the Southern in visual terms. The new engines were numerically greatest on the more important London trains but elsewhere the system, like its timetable, remained changeless with M7 0-4-4T's and T9 4-4-0's working branch and main line services in plentiful numbers whilst a significant number of goods trains remained in the charge of venerable 700 0-6-0's.

All these treasures and more are recorded in this book which will not only allow the reader to trace the trains that once ran south of Worting Junction but will tell him what sort of engines worked them.

THE ROUTE IN GENERAL

GW engines worked into the Southampton area whilst Eastleigh had a small number of Brighton-built LM engines on its books. The full quota of regional visitors was only completed when the Bulleid Pacifics had to be withdrawn for a short time in 1953 and a number of V2 2-6-2's and B1 4-6-0's drafted in for the duration. One of the Xpress publishing team recalls superintending the loan of V2's at Nine Elms and clearly recalls the surprise with which the South Western crews discovered the power of a V2. 60896 of Doncaster stands in Waterloo with the 16.34 Pullman from Bournemouth West on May 22nd, 1953.

For variety in all its forms, it was hard to find a main line anywhere in the country with a diversity of trains or motive power as rich as that of the Wessex main line of the South Western. Only two elements were missing – sleeping cars and mineral trains – but what remained more than made up for their absence and the ninety miles between Basingstoke and Weymouth were as varied as the trains which served them.

The huge port of Southampton guaranteed a continuous procession of goods and passenger trains – many of which were not timetabled – whilst the once insignificant market town of Eastleigh rivalled its neighbour in operational terms; the latter containing most of the motive power facilities for Southampton's needs as well as extensive marshalling yards for the connections with the Salisbury – Portsmouth route.

Rather strangely, Southampton and Eastleigh were the operational opposites of what each other should have been. The former was a large urban area with nearly two hundred thousand inhabitants yet its Central station had no more operating facilities than an ordinary London suburban station, consisting of an up and down main each with a passenger loop where slower trains could be overtaken by ex-presses.

With neither motive power nor crewing facilities, the Central was no place to start or terminate services – the last thing it needed was a rake of coaches blocking one of the lines waiting for an engine or traincrew to turn up – and therefore almost all originating trains for the Bournemouth direction had to start back at Eastleigh in order to keep traffic movements fluid at Southampton Central. The Terminus station, which possessed most conventional operating facilities, provided an alternative to the Central but suffered the disadvantage of being, as its name suggested, a dead-end which meant that every movement had to be accompanied by a reversal and two engines. Most of the services based on the Terminus were of a purely local nature although occasionally, because of the vagaries of stock movements, the odd Bournemouth-line service from there as opposed to Eastleigh.

The chief importance of Eastleigh lay in the fact it was the meeting point between the London and Weymouth and Salisbury to Portsmouth routes and because if this contained almost every feature to be found in a main line location. The passenger station, even though the best trains passed through non-stop, was far more substantial than Southampton Central whilst a considerable volume of freight traffic was dealt with in the large centralised marshalling yard. The number of goods trains that passed through Eastleigh without calling were few and far between.

West of Southampton the line skirted the New Forest coastline and served a dozen intermediate stations which were of sufficient importance to warrant a reasonably regular service of through stopping trains. Of these, the most important was Brockenhurst; a three-way junction with branches leading off to Lymington and to Bournemouth via West Moors and important enough to warrant stops by all but three services from Waterloo.

Almost all the branch services from Brockenhurst were self-contained M7 workings although through trains from London were a regular feature of the summer Saturday workings. The West Moors line had been the original main line to Hamworthy Junction and Dorchester and at times of pressure acted as a Bournemouth avoiding line.

Operations at Bournemouth were complicated by the fact the town had two stations neither of which was more commercially important than the other; both having to be served

The statement made by one writer that the Lord Nelson 4-6-0's dominated the Bournemouth line until 1962 is rather wide of the mark and in fact appearances, especially on scheduled expresses, tended to be widely spaced. Only five departures from Waterloo were booked to the class (the 05.40 and 11.30 services to Bournemouth, the 21.00 and 22.30 for Southampton Terminus and a stopping train to Salisbury), the long distance tally being completed by the 10.23 York - Bournemouth which one of the class worked from Oxford. The remainder of their duties revolved around stopping trains between Eastleigh and Bournemouth or Southampton boat trains. Opinion on the class was divided: Eastleigh men in the boat train link thought them the finest engines ever built whilst Bournemouth and Nine Elms drivers tended to be sparing in their praise. An unidentified member of the class swings the 15.50 Weymouth - Waterloo off Battledown flyover and across to the up main at Worting Junction. This was an Eastleigh diagram which returned west with the 22.30 Waterloo - Dorchester mail as far as Southampton Terminus.

so far as it was possible by the same trains. Thus Weymouth services from London would detached a Bournemouth West portion at the Central during the change of engines whilst terminating trains called at both stations; finishing their journeys in the West.

Through engine-working on the Weymouth trains was by no means as widespread as it was on the Waterloo – Exeter services which, with an engine shed at each end of the line, had been

relatively easy to arrange. Bournemouth was the main shed for the country end of the line and this meant that engines for Weymouth – London services had to start their duties with a local train from Bournemouth to Weymouth, work a return trip to London and then complete the diagram by working back to Bournemouth. Where it was possible this was done – the Royal Wessex engine, for example, achieved a very respectable 355 miles per day – but in

the majority of cases the timetable was not sufficiently elastic to allow through running to become universal and a number of Weymouth services continued to change engines at Bournemouth Central.

Although having to cut an engine off short was irritating to the diagrammer, to the observer it simply added to the pageantry of the South Western with Pacifics giving way to Lord Nelsons, King Arthurs or – in one case – a U

Engine	MN	LN	MN	WC	H15	MN	LN	MN	WC	WC	WC	WC	WC	WC	WC	LN	LN
WATERLOO	02.40	05.40	08.30	09.30	09.54	10.30	11.30	12.30	13.30	15.20	15.30	16.35	17.30	18.30	19.30	21.00	22.30
Southampton	04C58	07C49	10C26	11C31	12C40	11C59	13C19	13C58	15C20	16C53	17C26	18C08	19C23	20C01	21C26	22D47	00T53
Engine		WC	U			MN								LN			N15
Bournemouth	05C54	08C53	11C14	12W43		12C43	14W34	14W52	16C20	17C40	18W40	18C55	20W39	20W41	22W38		02C14
Dorchester		09.53	12.09			13.34			17.15	18.48		19.21		21.30			03.16
WEYMOUTH		10.07	12.24			13.47			17.33	19.06		19.55		21.43			

Engine	LN	LN	WC	WC	WC	WC	N15	WC	WC	MN	WC	MN	LN	T9	MN	WC	WC
WEYMOUTH					07.34		09.20		11.30		13.25			15.50	17.35	18.30	21.55
Dorchester					07.50		09.40		11.48		13.44			16.10	17.52	18.48	22.20
Bournemouth			06W35	07W20	08C40	08W35	10C33	11W05	12C41	13W05	14C40	15W05		17.18	18C41	21C52	23C32
Engine							LN								LN		N15/LN
Southampton	06T04	07E22	07C48	08C26	09C17	09C56	11C20	12C20	13C20	14C20	15C20	16C22	17D30	18C20	19C21	21C09	01T10
WATERLOO	08.24	09.01	10.20	10.00	10.50	11.55	12.50	14.20	14.49	16.14	16.50	18.29	19.19	20.23	20.50	22.56	03.53

LONDON - BOURNEMOUTH/WEYMOUTH SERVICE SUMMARY : 1954

Although outnumbered by an impressive array of local trains, the London expresses were the backbone of the timetable and are extracted above from the rest of the workings for the sake of convenience. It will be noted that despite their numerical superiority, Bulleid Pacifics did not have a monopoly of haulage and it was relatively easy to travel from Weymouth to London behind pre-1948 motive power; one service actually being diagrammed for a T9 and a Lord Nelson. During the period under review diesel locomotives 10000/1, 10201/2 were at work on the Southern; their Wessex duty being the 05.40 Waterloo - Weymouth, 11.30 Weymouth - Waterloo, 16.35 (Royal Wessex) Waterloo - Weymouth and the 21.55 Weymouth - Waterloo. In theory one of the four engines maintained the working for a week at a time before being changed for another diesel but in reality the steam diagrams had to be adhered to - the diesels were not the most robust of performers.

BR Standard 73065 of Sheffield Millhouses, seen on an up express at Southampton Central, was a very rare visitor to the area although several batches of the class had been at work from Nine Elms and Stewarts Lane from the mid-1950's as replacements for N15 King Arthur 4-6-0's.

class 2-6-0. Where the through workings were concerned, considerable ingenuity had to be exercised in order to provide engines for the Bournemouth West sections and where one service might command the might of a Merchant Navy Pacific, another would have a T9 4-4-0 or a 700 0-6-0. The variety seemed endless yet at Bournemouth West it was extended even further by the presence of the Som-

Train	Arr	Engine	Shed	Dep	Destination
MICHELDEVER : 1954					
22.30 WATERLOO		LN 4-6-0	EL 253	00/07	DORCHESTER
23.15 Eastleigh		H15 4-6-0	EL 310	00/14	Basingtoke
22.20 Feltham		S15 4-6-0	FEL 109	00/27	Southampton Docks
22.38 Nine Elms		H15 4-6-0	NE 73	00/45	Weymouth
19.35 Kensington (Pds)		BR4 2-6-0	EL 273	00/54	Poole
19.50 Dorchester		H15 4-6-0	NE 74	01/10	Nine Elms
22.45 Woking		S15 4-6-0	FEL 110	01/16	Eastleigh
00.45 Eastleigh		H15 4-6-0	NE 64	01/37	Feltham
23.25 Nine Elms		LN 4-6-0	EL 251	01/45	Southampton Docks
21.55 WEYMOUTH		LN 4-6-0	NE 31	02/03	WATERLOO
22.45 Feltham		S15 4-6-0	FEL 112	02/25	Bournemouth Central
00.50 Fratton		S15 4-6-0	FEL 101	03/00	Feltham
02.10 Clapham Jcn (Fish)		N15 4-6-0	EL 265	03/26	Fratton
00.45 Nine Elms		H15 4-6-0	NE 79	03/45	Eastleigh
02.40 WATERLOO		MN 4-6-2	NE 30	04/04	BOURNEMOUTH CENTRAL
03.10 Bevois Park (S'ton)		S15 4-6-0	FEL 108	04/09	Woking
00.55 Feltham		S15 4-6-0	FEL 102	05/10	Southampton Docks
04.40 Feltham		H15 4-6-0	NE 66	05/44	Feltham
03.20 Bournemouth Central		S15 4-6-0	FEL 110	06/20	Feltham
04.07 Feltham		S15 4-6-0	FEL 103	06/50	Southampton Docks
06.03 Southampton Terminus	06.53		EL 251	06.54	Waterloo
05.40 WATERLOO	07.09	LN 4-6-0	NE 31	07.10	WEYMOUTH
06.50 Reading	07.52	HALL 4-6-0	RDG(GW) 66	07.53	Southampton Terminus
07.22 Eastleigh	07.53	LN 4-6-0	EL 252	07.54	Waterloo
07.31 Woking	08.34	U 2-6-0	GUI 181	08.35	Southampton Terminus
06.35 BOURNEMOUTH CENTRAL	08.40	WC 4-6-2	BM 380	08.41	Waterloo
07.20 BOURNEMOUTH WEST		WC 4-6-0	NE 38	09/03	WATERLOO
08.20 Eastleigh	09.12	BR4 2-6-0	EL 273	09.13	Reading
07.34 WEYMOUTH		WC 4-6-2	BM 381	09/53	Waterloo
08.30 WATERLOO		MN 4-6-2	NE 32	09/54	WEYMOUTH
09.12 Reading	10.11	HALL 4-6-0	RDG(GW) 50	10.12	Portsmouth
07.43 Clapham Jcn (Pds)	10.26	H15 4-6-0	NE 65	10.27	Eastleigh
08.35 Bournemouth West	10.41	WC 4-6-2	BM 382	10.42	Waterloo
09.30 WATERLOO		WC 4-6-2	NE 33	10/54	BOURNEMOUTH WEST
10.13 Southampton Terminus	11.02	HALL 4-6-0	RDG(GW) 66	11.03	Reading
09.20 BOURNEMOUTH WEST		N15 4-6-0	BM 399	11/14	BIRKENHEAD
01.30 Hoo Junction		H15 4-6-0	EL 310	11/17	Eastleigh
10.30 WATERLOO		MN 4-6-2	NE 34	11/35	WEYMOUTH
09.54 WEYMOUTH		H15 4-6-0	NE 68	11/50	Southampton Central
09.20 WEYMOUTH		LN 4-6-0	NE 31	11/51	WATERLOO
11.59 Eastleigh	12.38	S15 4-6-0	FEL 103		
11.30 WATERLOO		LN 4-6-0	EL 252	12/43	BOURNEMOUTH WEST
11.05 BOURNEMOUTH WEST		WC 4-6-2	BM 383	13/05	WATERLOO
12.15 Reading	13.08	BR4 2-6-0	EL 273	13.09	Portsmouth
11.16 BOURNEMOUTH WEST		H15 4-6-0	BM 395	13/10	YORK
12.31 Southampton Terminus	13.21	U 2-6-0	GUI 181	13.22	Reading
12.30 WATERLOO (PULLMAN)		MN 4-6-2	NE 35	13/34	BOURNEMOUTH WEST
11.30 WEYMOUTH		WC 4-6-2	BM 386	13/51	WATERLOO
09.45 Feltham		S15 4-6-0	FEL 105	13/56	Eastleigh
13.15 Eastleigh (Pds)		H15 4-6-0	EL 311	14/10	Clapham Jcn
13.20 Southampton Docks		S15 4-6-0	FEL 104	14/30	Feltham
11.36 Bournemouth West	14.37	H15 4-6-0	NE 79	14.39	Woking
13.30 WATERLOO		WC 4-6-2	BM 381	14/43	WEYMOUTH
13.48 Reading	14.50	HALL 4-6-0	OXLEY 208	14.51	Portsmouth
13.05 BOURNEMOUTH WEST		MN 4-6-2	NE 32	15/02	WATERLOO
		S15 4-6-0	FEL 103	15.04	Winchester
03.20 Ashford		N15 4-6-0	EL 264	15/14	Eastleigh
09.30 BIRKENHEAD		N15 4-6-0	BM 399	15/42	BOURNEMOUTH WEST
13.25 WEYMOUTH		WC 4-6-2	BM 385	15/51	WATERLOO
14.45 Portsmouth	16.10	HALL 4-6-0	RDG(GW) 50	16.11	Reading
15.20 WATERLOO		WC 4-6-2	BM 382	16/22	WEYMOUTH
15.30 WATERLOO	16.50	WC 4-6-2	NE 37	16.51	BOURNEMOUTH WEST
10.23 YORK		LN 4-6-0	BM 395	16/58	BOURNEMOUTH WEST
15.05 BOURNEMOUTH WEST	17.08	MN 4-6-2	NE 34	17.10	WATERLOO
16.20 Southampton Docks		S15 4-6-0	FEL 102	17/30	Feltham
16.35 WATERLOO		WC 4-6-2	BM 383	17/38	WEYMOUTH
16.34 BOURNEMOUTH WEST		MN 4-6-2	NE 35	17/53	WATERLOO (PULLMAN)
17.10 Reading	18.01	U 2-6-0	GUI 181	18.02	Southampton Terminus
17.35 SOUTHAMPTON DOCKS		LN 4-6-0	EL 254	18/08	WATERLOO
17.18 Winchester		S15 4-6-0	FEL 103	18/15	Basingstoke
14.45 Feltham		S15 4-6-0	FEL 107	18/15	Southampton Docks
17.30 WATERLOO	18.46	WC 4-6-2	BM 380	18.47	BOURNEMOUTH WEST
17.17 Portsmouth	18.48	BR4 2-6-0	EL 273	18.49	Reading
17.05 BOURNEMOUTH WEST		LN 4-6-0	EL 253	19/03	WATERLOO
19.00 Basingstoke (Pds)		S15 4-6-0	FEL 106	19/16	Eastleigh
17.50 Bevois Park		H15 4-6-0	NE 68	19/20	Basingstoke
18.30 WATERLOO		WC 4-6-2	BM 386	19/34	WEYMOUTH
19.00 Reading	19.50	N15X 4-6-0	BSK 236	19.51	Southampton Terminus
17.35 WEYMOUTH		MN 4-6-2	NE 30	19/52	WATERLOO
18.55 Southampton Docks		N15 4-6-0	EL 264	20/13	Nine Elms
19.45 Eastleigh (Parcels)		H15 4-6-0	NE 72	20/28	Waterloo
19.30 WATERLOO	20.45	WC 4-6-2	BM 385	20.46	BOURNEMOUTH WEST
18.35 Bournemouth West	20.56	N15 4-6-0	EL 266	20.57	Reading
18.42 Waterloo (Pds)		H15 4-6-0	NE 64	21/37	Eastleigh
18.30 WEYMOUTH		WC 4-6-2	NE 33	21/40	WATERLOO
21.15 Eastleigh		S15 4-6-0	FEL 111	22/07	Feltham
21.00 WATERLOO		LN 4-6-0	EL 251	22/13	SOUTHAMPTON DOCKS
21.45 Eastleigh (Parcels)		HALL 4-6-0	OXLEY 208	22/19	Crewe
19.25 Nine Elms		H15 4-6-0	EL 311	22/25	Southampton Docks
19.55 Bournemouth West (Pds)		N15X 4-6-0	BSK 236	22/50	Waterloo (Parcels)
19.48 Weymouth	23.22	N15 4-6-0	EL 263	23.23	Reading
16.55 Worcester Pax		N15 4-6-0	EL 266	23/26	Eastleigh
21.56 Bevois Park (S'ton)		S15 4-6-0	FEL 106	23/35	Feltham
22.30 Southampton Docks		U 2-6-0	GUI 181	23/47	Feltham

Not only was Eastleigh the centre for all locomotive running matters but the workshops - which were quite separate from operating affairs - guaranteed a regular supply of foreign engines as other districts and divisions sent their engines to be attended to. Returning engine to their home sheds was sometimes a drawn-out affair; the engine concerned being used locally - such as Bricklayers Arm's Schools 4-4-0's being used between Salisbury and Waterloo - until an exchange could be arranged. Perhaps the strangest of all foreigners at Eastleigh were the ex-Plymouth, Devonport & South Western 0-6-2T's tanks although in Spring 1956 the two survivors of the class found themselves transferred to Eastleigh. 30757 'Earl of Mount Edgcumbe' glistens after a visit to the works.

ALTON : 1954

Train	Arr	Engine	Shed	Dep	Destination	Train	Arr	Engine	Shed	Dep	Destination
03.30 North Camp (Goods)	05.53	700 0-6-0	GUI 217			13.27 Waterloo	14.47	EMU		14.54	Waterloo
05.48 Farnham ECS	06.10	EMU				12.23 Fareham (Goods)	15.05	700 0-6-0	GUI 217		
05.05 Woking (Goods)	06.26	U 2-6-0	GUI 204	06.40	Light to Farnham	13.57 Waterloo	15.17	EMU			
05.35 Woking (Pds)	06.41	T9 4-4-0	GUI 187			14.15 Eastleigh	15.18	M7 0-4-4T	EL 302		
		EMU		06.54	Waterloo			EMU		15.24	Waterloo
05.25 Waterloo	06.56	EMU			(Att to 07.17)	14.27 Waterloo	15.47	EMU			
05.55 Waterloo	07.17	EMU		07.24	Waterloo	14.48 Fareham	15.47	M7 0-4-4T	GOS 387		
07.12 Farnham ECS	07.26	EMU						EMU		15.54	Waterloo
06.35 Eastleigh	07.34	M7 0-4-4T	EL 303					M7 0-4-4T	EL 302	16.10	S. Terminus
		700 0-6-0	GUI 217	07.38	Fareham	14.57 Waterloo	16.17	EMU		16.24	Waterloo
		EMU		07.44	Waterloo			M7 0-4-4T	GOS 387	16.30	Fareham
06.25 Waterloo	07.47	EMU				15.27 Waterloo	16.47	EMU		16.54	Waterloo
		M7 0-4-4T	EL 303	07.53	Eastleigh	15.59 Eastleigh	17.00	M7 0-4-4T	EL 298		
		EMU		07.54	Waterloo			700 0-6-0	GUI 217	17.02	Aldershot (Goods)
07.46 Farnham ECS	08.00	EMU		08.10	Waterloo			M7 0-4-4T	EL 298	17.05	Eastleigh
06.55 Waterloo	08.17	EMU		08.24	Waterloo	15.57 Waterloo	17.17	EMU		17.24	Waterloo
08.10 Farnham ECS	08.40	EMU				16.27 Waterloo	17.47	EMU			
07.40 Eastleigh	08.41	M7 0-4-4T	EL 299			16.55 Eastleigh	17.52	M7 0-4-4T	EL 299		
		EMU		08.46	Waterloo			EMU		17.54	Waterloo
07.25 Waterloo	08.48	EMU				16.47 Waterloo	17.59	EMU			
07.42 Gosport	08.51	M7 0-4-4T	GOS 387					M7 0-4-4T	EL 299	18.02	S. Terminus
		M7 0-4-4T	EL 299	08.55	Eastleigh			EMU		18.04	Farnham
		EMU		08.56	Waterloo	16.57 Waterloo	18.17	EMU		18.24	Waterloo
		M7 0-4-4T	GOS 387	09.05	Gosport			EMU			
07.55 Waterloo	09.17			09.24	Waterloo	17.18 Waterloo	18.31	EMU		18.40	Farnham
08.12 Aldershot (Pds)	09.36	700 0-6-0	GUI 218			17.27 Waterloo	18.47	EMU			
08.25 Waterloo	09.47	EMU		09.54	Waterloo	17.23 S. Terminus	18.49	M7 0-4-4T	EL 297		
08.57 Waterloo	10.17	EMU						EMU		18.54	Waterloo
		T9 4-4-0	GUI 187	10.20	Fareham (Goods)	17.47 Waterloo	19.01	EMU		19.15	Farnham
		EMU		10.24	Waterloo	17.57 Waterloo	19.17	EMU		19.24	Waterloo
		700 0-6-0	GUI 218	10.32	Light to Farnham			M7 0-4-4T	EL 297	19.25	Eastleigh
09.27 Waterloo	10.47	EMU		10.54	Waterloo	18.48 Fareham	19.46	T9 4-4-0	GUI 187		
10.16 Eastleigh	11.16	M7 0-4-4T	EL 299			18.27 Waterloo	19.47	EMU			
09.57 Waterloo	11.17	EMU		11.24	Waterloo			EMU		19.54	Waterloo
10.27 Waterloo	11.47	EMU						T9 4-4-0	GUI 187	20.05	Light to Guildford
08.33 Eastleigh (Goods)	11.53	700 0-6-0	EL 327			18.57 Waterloo	20.17	EMU			
		EMU		11.54	Waterloo	18.58 S. Terminus	20.18	LM2 2-6-2T	AJN 247		
		M7 0-4-4T	EL 299	12.05	Eastleigh			EMU		20.24	Waterloo
10.57 Waterloo	12.17	EMU		12.24	Waterloo			LM2 2-6-2T	AJN 247	20.38	Eastleigh
11.27 Waterloo	12.47	EMU				19.27 Waterloo	20.47	EMU		20.54	Waterloo
11.30 Gosport	12.49	M7 0-4-4T	GOS 387			19.57 Waterloo	21.17	EMU		21.24	Waterloo
		EMU		12.54	Waterloo	20.27 Waterloo	21.47	EMU		21.54	Waterloo
11.57 Waterloo	13.17	EMU		13.24	Waterloo	20.57 Waterloo	22.17	EMU		22.24	Waterloo
		M7 0-4-4T	GOS 387	13.30	Fareham	21.27 Waterloo	22.47	EMU		22.54	Waterloo
		700 0-6-0	EL 327	13.45	Eastleigh (Goods)	21.57 Waterloo	23.17	EMU		23.25	Farnham ECS
12.27 Waterloo	13.47	EMU		13.54	Waterloo	22.25 Waterloo	23.47	EMU		23.55	Woking ECS
12.57 Waterloo	14.17	EMU		14.24	Waterloo	22.57 Waterloo	00.17	EMU		00.25	Farnham ECS

The route via Alton was never a serious alternative to the line via Basingstoke although prior to the electrification of 1937 half a dozen through trains operated between London and Southampton with a running time of around three and a quarter hours. After 1937 all through running ceased although the Alton - Eastleigh Push and Pull trains continued to be shown in the timetable as through services 'via Alton'.

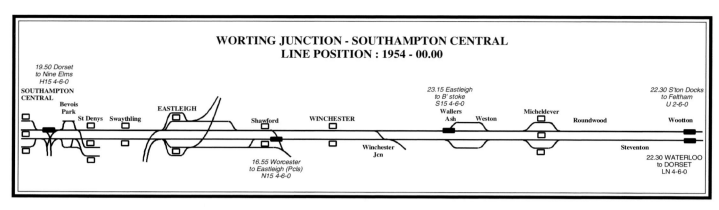

erset & Dorset whose trains arrived behind an extraordinary collection of engines ancient and modern.

Beyond Bournemouth the chief points of interest were to be found at Wareham and Dorchester. The former was the junction for Swanage, a coastal resort of sufficient note to warrant the provision of through coaches to and from Waterloo three times a day whilst the latter was, in theory, the end of the line since the final six miles of the route belonged to the Great Western who, up to the 1960's, continued to run several through expresses between Paddington and Weymouth.

It should not be thought that a trifle such as a transport act was sufficient to weld the two parties serving Weymouth into any sort of co-operative arrangement and for years after nationalisation the South Western retained full-blown locomotive facilities at Dorchester in order to provide engines and men for trains which Bournemouth was unable to cover. The shed at Weymouth looked after GWR workings to Westbury and Paddington and did nothing for Southern engines beyond turning and watering them.

Any thoughts that the Bulleid revolution was especially evident below Southampton can be discarded and in fact in the down direction at, for example, Brockenhurst one had to wait until almost midday until a West Country Pacific made an appearance: the Brighton – Bournemouth through train being the first service booked to a Pacific since the passage of the 02.40 ex Waterloo. Enthusiasts of the day were careful not to stake too much money on the Brighton actually turning up behind a Pacific since the availability of the Brighton-based 4-6-2's plummeted almost unbelievable depths at times leaving LBSCR Atlantics or BR 2-6-4 tanks to fill the gap.

Watching trains in the New Forest was a timeless activity and of the 120-odd arrivals and departures at Brockenhurst, no less than forty-nine were worked by M7 0-4-4 tanks; not all of which were confined to Lymington or West Moors locals.

The backbone of the service was repre-

COMBINED WORKING TIME TABLE AND ENGINE WORKINGS : 1954

Train	21.20			21.20	22.30	21.20	22.40	22.30		03.52		22.20	22.38	19.35
From	ELGH	Gds	Gds	ELGH	W'loo	ELGH	Bristol	W'loo	Gds	S'disbury		Feltham	N. Elms	Kens'tn
Class	Gds			Gds		Gds	Gds	Gds				Gds	Gds	Pcls
Engine	Q0-6-0	H15 4-6-0	Q0-6-0	Q0-6-0	LN 4-6-0	Q0-6-0	Hall 4-6-0	N15 4-6-0	Q0-6-0	WC 4-6-2	U 2-6-0	S15	H15 4-6-0	BR4 2-6-0
Shed	EL 324	EL 313	EL 317	EL 324	EL 253	EL 324	SPM	EL 263	BM 416	BM 381	DOR 426	FEL 109	NE 73	EL 273
WATERLOO					22.30									
Woking														
Basingstoke					23.53									00.37
Worting Jcn					23/59							00/13	00/30	00.45
Micheldever														
Wallers Ash														
WINCHESTER					00.23							00/44	01/01	01.27
Shawford														
Eastleigh Yard		23.35										01.06		
EASTLEIGH					00.33								01.23	01.40
EASTLEIGH		23/42			00.43								01.35	
Swaythling														
St Denys		23/53			00/50								01/45	
Bevois Park Yard														
Northam				00.36										
Southampton (T)					00.53			01.18						
SOUTHAMPTON CENT														
SOUTHAMPTON CENT		00/04		00/44				01/22				02/02		
Millbrook														
Redbridge		00/13										02/10		
Totton														
Lyndhurst Rd														
Beaulieu Rd														
BROCKENHURST	23.59	00/39		01/20				01.47				02.37		
Lymington Jcn	00/03													
Lymington Town														
Lymington Pier														
Sway														
New Milton														
Hinton Admirl														
Christchurch								02.06						
Pokesdown														
Boscombe														
Central Goods										03.30				
BOURNEMOUTH CENT								02.14						
BOURNEMOUTH CENT	00/40							02.28	03/34					
BOURNEMOUTH WEST														
Branksome										03.44				
Parkstone														
POOLE	01.01			01.30				02.41	03.55					
Hamworthy Jcn											05/13			
Holton Heath											05.18			
WAREHAM				01.50		02.05		02.57			05.29			
Worgret Jcn						02/08		03/01			05/31			
Corfe Castle														
SWANAGE														
Wool											05.37			
Moreton											05.46			
DORCHESTER						02.47		03.16			05.58	06.45		
Dorchester Jcn							03/03				06/00			
Wishing Well												06.53		
Upwey												06.57		
WEYMOUTH							03.30				06.09	07.02		

BR 5MT 4-6-0 73119 speeds through New Milton with a down Waterloo - Bournemouth holiday express. In 1960 twenty of these engines were given names - although not the original nameplates - from withdrawn N15 4-6-0's (73119 assumed the name 'Elaine' from 30747) in an attempt to rekindle an interest in railways in the way the Southern Region had done before the war. The scheme was not accompanied by much in the way of publicity and although the engines carried nameplates for a few years, the 5MT's - which for the most part were confined to outer-suburban workings - never came to be regarded as a rejuvenation of the N15's.

WEYMOUTH - BASINGSTOKE : WORKING TIMETABLE 1954

Train	22.26	19.50			22.26	00.50	22.26						03.20		03.20
From	H.Jcn	Dorset			H.Jcn	Fratton	H.Jcn						BM		BM
Class	Gds	Gds	Gds	Gds	Gds	Gds	Gds		Gds	Gds	Gds	Gds	Gds	Gds	Gds
Engine	N15 4-6-0	H15 4-6-0	H15 4-6-0	T9 4-4-0	N15 4-6-0	N15 4-6-0	S15 4-6-0	N15 4-6-0	Hdl 4-6-0	S15 4-6-0	H15 4-6-0	Q0-6-0	43xx 2-6-0	Q0-6-0	S15 4-6-0
Shed	BM 401	NE 74	NE 64	EL 282	BM 399	BM 401	FEL 101	BM 401	SPM	FEL 108	NE 66	EL 325	WBY	EL 325	Fel 110
WEYMOUTH									01.50				03.30		
Upwey															
Wishing Well															
Dorchester Jcn									02/08				03/45		
DORCHESTER									(To B'tol)				(To W'bury)		
Moreton															
Wool															
SWANAGE															
Corfe Castle															
Worgret Jcn															
WAREHAM															
Holton Heath															
Hamworthy Jcn															
POOLE	23.30														
Parkstone															
Branksome															
BOURNEMOUTH WEST															
BOURNEMOUTH CENTRAL															
BOURNEMOUTH CENTRAL	23/56											03.20			
Central Goods															
Boscombe															
Pokesdown															
Christchurch															
Hinton Admiral															
New Milton															
Sway															
Lymington Pier															
Lymington Town															
Lymington Jcn	00/26											03/51			
BROCKENHURST	00.30						00.45					03.55		04.11	
Beaulieu Road															
Lyndhurst Road															
Fawley															
Totton															
Redbridge							01/13							04/39	
Millbrook															
SOUTHAMPTON CENTRAL							01/21							04/45	
SOUTHAMPTON CENTRAL															
Southampton T.					00.01	01.23									
Northam															
Bevois Park Yard		00.12			00.45		01.30	01.59		03.10				<u>04.53</u>	05.10
St Denys					00/50	01/27		02/03		03/14					05/14
Swaythling						01.37	01.53								
EASTLEIGH															
EASTLEIGH		00/27			01/00	01.46	02.05	02.17		03/24					05/25
Eastleigh Yard				00.45	<u>01.05</u>			<u>02.19</u>			04.40				
Shawford					(To Ports)										
WINCHESTER		00/50		01/17				02/36		03/49	05/17				05/51
Weston															
Micheldever															
Worting Jcn		01/25		01/52				03/14		04/24	05/59				06/35
Basingstoke		(To N.Elms)		(To Feltham)				(To Feltham)		(To Woking)	(To Feltham)				(To Feltham)
Waterloo															

Train	22.20	22.45	23.25	22.45		22.38	22.38	01.20	22.38	19.35	19.35	23.25	22.45	22.45
COMBINED WORKING TIME TABLE AND ENGINE WORKINGS : 1954														
From	Feltham	Woking	N. Elms	Feltham		N. Elms	N. Elms	B'tol	N. Elms	N. Elms	Kens'tn	Kens'tn	N. Elms	Feltham
Class	Gds	Gds	Gds	Gds	Gds	Gds	Gds	Gds	Gds	Gds	Pcls	Pcls	Gds	Gds
Engine	S15 4-6-0	S15 4-6-0	LN 4-6-0	S15 4-6-0	H15 4-6-0	H15 4-6-0	H15 4-6-0	Hdl 4-6-0	H15 4-6-0	N15 4-6-0	N15 4-6-0	Q0-6-0	BR4 2-6-0	S15 4-6-0
Shed	FEL 109	FEL 110	EL 251	FEL 112	EL 313	NE 73	NE 73	SPM	NE 73	EL 263	BM 399	BM 415	EL 272	FEL 110
Pilot														
Shed														
WATERLOO														
Woking														
Basingstoke														
Worting Jcn		01/01	01/30	02/10										
Micheldever														
Wallers Ash														
WINCHESTER		01/38	02/02	02/41										
Shawford														
Eastleigh Yard	01.45	02.00	02.24	03.02									03.10	03.21
EASTLEIGH														
EASTLEIGH	01/49										03.05		03/15	03/26
Swaythling														
St Denys	01/58												03/25	03/36
Bevois Park Yard														
Northam										03.30				
Southampton (T)	02D30												03D54	
SOUTHAMPTON CENT											03.19			
SOUTHAMPTON CENT											03.27			03/47
Millbrook														
Redbridge														03/54
Totton														
Lyndhurst Rd														
Beaulieu Rd														
BROCKENHURST					03.10	03.42					04.12			04.20
Lymington Jcn					03/14	03/46					04/15			
Lymington Town														
Lymington Pier														
Sway														
New Milton														
Hinton Admiral														
Christchurch											04.42			
Pokesdown														
Boscombe														
Centrl Goods					03.55						04.55			
BOURNEMOUTH CENT											04.58			
BOURNEMOUTH CENT						04/24					05.20			
BOURNEMOUTH WEST														
Branksome						04.33								
Parkstone											05.45			
POOLE						04.44	05.15				05.49			
Hamworthy Jcn							05.24		05.58					
Holton Heath														
WAREHAM									06/08					
Worgret Jcn														
Corfe Castle														
SWANAGE														
Wool														
Moreton														
DORCHESTER									06.38	07.15				
Dorchester Jcn								06/22						
Wishing Well														
Upwey														
WEYMOUTH								06.47	07.44					

sented by the six expresses (and a night train) which operated at roughly two-hourly intervals from Waterloo to Weymouth; the best of them covering the 143 miles in around three and a quarter hours. For reasons already mentioned, through engine-running was not a normal feature of the route and of the seven services, only the 13.30, 15.20 and 16.35 workings from Waterloo kept their engines; the remainder changing engines at Bournemouth Central.

Given that the Wessex main line was one of the Southern's principal routes, if not the principal route, it is strange that Merchant Navy Pacifics played only a minor role in the running of the Weymouth trains.

The 08.30 ex Waterloo started off behind a Nine Elms member of the class but gave way to a Dorchester-based U class 2-6-0 - a class of engine not often associated with first division express work – at Bournemouth. The following 10.30 from Waterloo was unique in being Merchant Navy-hauled throughout although a change of engines was still made at Bournemouth, the incoming engine being replaced by the Nine Elms Pacific that had earlier worked down with the 02.40 News.

The remainder of the Weymouth trains provided quite a feast for those who cared for locomotive variety. The 05.30 ran from Waterloo as far as Bournemouth behind a Lord Nelson 4-6-0 where it was replaced by a West Country 4-6-2 whilst the 18.30 reversed the process by having a light Pacific to Bournemouth and a Lord Nelson for the remainder of the journey to Weymouth. For those who preferred to roll the calendar back a couple of decades there were few trains like the 22.30 Mail from Waterloo which ran via Southampton Terminus where its Lord Nelson 4-6-0 gave way to an N15 King Arthur 4-6-0. Those who wanted to prolong the night and go forward from Dorchester to Weymouth on the assumption that the mail's connection, the 03.52 ex Salisbury, was bound to produce something worthwhile were doomed to disappointment.

Their undoing was the presumption that because the trains from Salisbury to Bournemouth were all worked by T9 4-4-0's, then the 03.52 to Weymouth was bound to be covered by the same link. Alas, they reckoned without the guile of the diagramming office who, having to get a Pacific to Weymouth to work the 07.34 'Royal Wessex' to Waterloo, booked a West Country to run light from Bournemouth shed to Wimborne to take over the 03.52 from its Salisbury T9.

Such pilgrimages were the better for a knowledge of engine workings and many a disappointed enthusiast made his way back to Waterloo in the up 'Wessex' – complete with

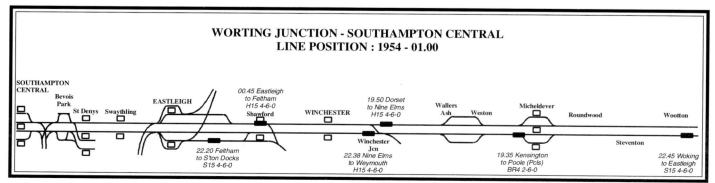

**WORTING JUNCTION - SOUTHAMPTON CENTRAL
LINE POSITION : 1954 - 01.00**

WEYMOUTH - BASINGSTOKE : WORKING TIMETABLE 1954

													05.17 Wimborne	05.17 Wimborne
Train From														
Class			Gds					ECS					ECS	
Engine	T9 4-0	M7 0-4-4T	Q1 0-6-0	M7 0-4-4T	LN 4-6-0	M7 0-4-4T	BR4 2-6-0	BR4 2-6-0	M7 0-4-4T	M7 0-4-4T	M7 0-4-4T	T9 4-4-0	R'car	LN 4-6-0
Shed	EL 283	EL 302	EL 314	EL 302	EL 251	EL 303	EL 270	EL 278	EL 298	EL 308	LYM 362	SAL 444	Wey	BM 394
WEYMOUTH													05.45	
Upwey														
Wishing Well														
Dorchester Jcn													06/01	
DORCHESTER													(To C. Cary)	
Moreton														
Wool														
SWANAGE														
Corfe Castle														
Worgret Jcn														
WAREHAM														
Holton Heath														
Hamworthy Jcn														
POOLE												05.32		
Parkstone												05.38		
Branksome												05.43		
BOURNEMOUTH WEST														
BOURNEMOUTH CENTRAL												05.48		
BOURNEMOUTH CENTRAL														05.54
Central Goods														
Boscombe														05.58
Pokesdown														06.01
Christchurch														06.06
Hinton Admiral														06.13
New Milton														06.20
Sway														06.27
Lymington Pier														
Lymington Town											06.18			
Lymington Jcn											06/26			06/31
BROCKENHURST											06.28			06.38
Beaulieu Road														
Lyndhurst Road														06.44
Fawley														
Totton							06.28							06.50
Redbridge							06.31							06.52
Millbrook							06.36							06.56
SOUTHAMPTON CENTRAL							06.39							06.59
SOUTHAMPTON CENTRAL		05.22					06.42							07.01
Southampton T.	05.15					06.04			06.43	06.57				
Northam						06.08			06.47	07.01				
Bevois Park Yard			05.35											
St Denys	05.32	05.31	05/40			06.12		06.48	06.51	07.12				07.08
Swaythling	(To	05.36				06.16		(To	06.55	(To				07.12
EASTLEIGH	Ports)	05.41				06.21		Ports)		07.00				07.17
EASTLEIGH			05/53		06.04	06.22	06.35	06.55						
Eastleigh Yard														
Shawford					06.13	06.30	06.43	07/02						
WINCHESTER			06/15		06.21	06.38	06.51	(To Chesil)						
Weston			(To Didcot)											
Micheldever						06.54	(To Alton)							
Worting Jcn						07/05								
Basingstoke						07.09								
Waterloo						08.24								

unwanted Pacific – not knowing that the following 09.20 was diagrammed to a King Arthur as far as Bournemouth and a Lord Nelson beyond.

Only slightly secondary in status to the Weymouth trains were the seven Waterloo – Bournemouth workings although, with the exception of the 12.30 Pullman, none could be described as being especially fast; a few minutes under three hours being par for the course as far as Bournemouth Central. Apart from the 11.30 ex Waterloo – a Lord Nelson working – all were booked to Bulleid Pacifics with two, the 02.40 and the Pullman, being worked by Merchant Navy Pacifics.

The provision of an express almost every hour between London and Bournemouth was not something that Waterloo had any cause to be ashamed of although, for reasons of prestige if nothing else, many railwaymen would like to have seen the level of frequency accompanied by the restoration of the two-hour timing between London and Bournemouth. Unfortunately the forces of conservatism during the first decade following nationalisation was an influence too strong to be prevailed against: the lack of water troughs and the increased scope for loss of time being the most frequent arguments brought to bear against the revivalist camp. It seemed to have escaped the memory of too many powerful influences that not only had the two-hour timing been introduced during the reign of Queen Victoria but it had been maintained by the South Western's 4-6-0's which no-one could describe as an especially flighty example of the type. Moreover the timing had been maintained – and improved on – during the 1930s by nothing larger than a Schools 4-4-0.

The objections to restoring the two-hour timing were not wholly imaginary since the pre-war working had meant running non-stop through Southampton and it was argued that given the growth of trade at Southampton since the war, the excision of the call could not be off-set by any advantage that Bournemouth would gain by an acceleration of ten minutes. The argument that Southampton should be

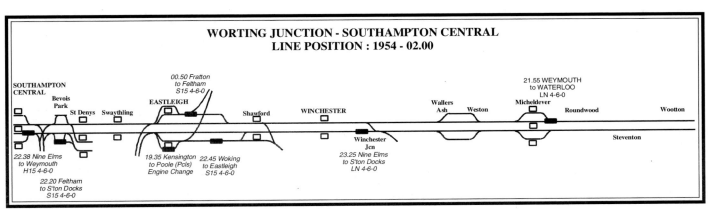

**WORTING JUNCTION - SOUTHAMPTON CENTRAL
LINE POSITION : 1954 - 02.00**

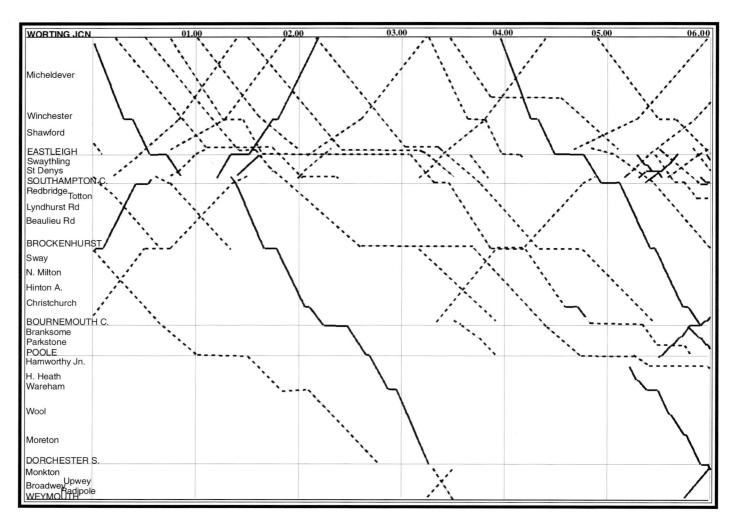

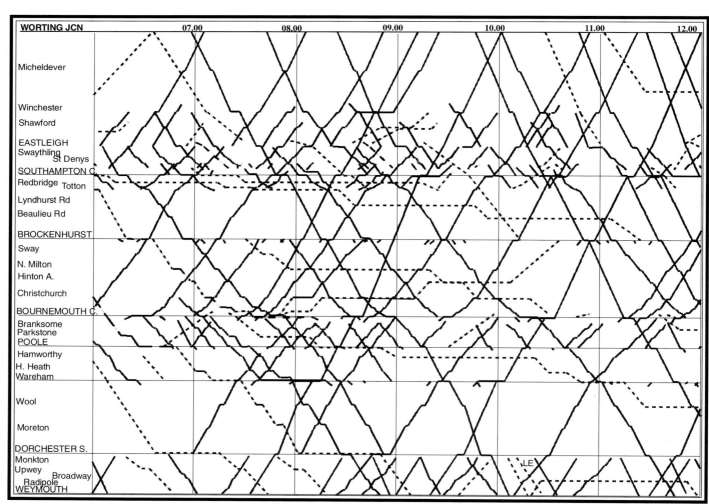

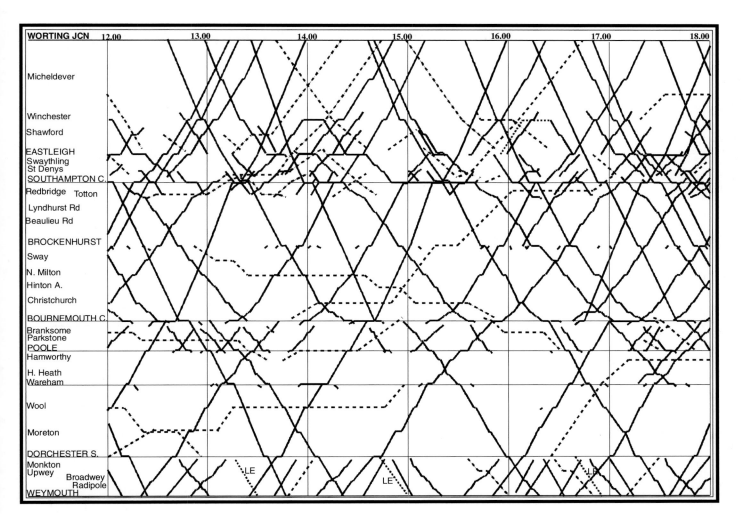

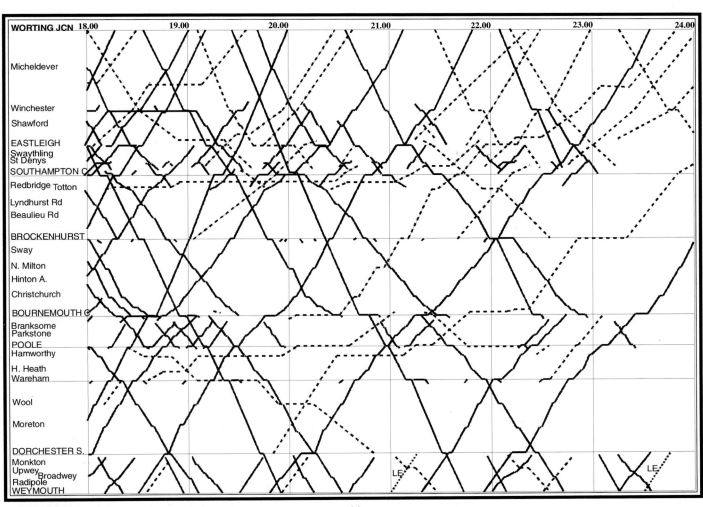

COMBINED WORKING TIME TABLE AND ENGINE WORKINGS : 1954

Station (Engine)	S15 4-6-0	Hdl 4-6-0	M7 0-4-4T	U 2-6-0	N15 4-6-0	H15 4-6-0	MN 4-6-2	MN 4-6-2	N15 4-6-0	H15 4-6-0	U 2-6-0	M7 0-4-4T	M7 0-4-4T	U 2-6-0
Train	22.45	01.20		02.10	00.45					00.45				
From	Feltham	B'tol		C.Jcn	N.Elms					N.Elms				
Class	Gds	Gds	Pcls	Fish	Gds	News	News	Gds	Gds	Gds	Fish	Gds		Fish
Shed	FEL 110	SPM	BM 407	SAL 450	EL 265	NE 79	NE 30	NE 30	EL 266	NE 79	BM 414	HAM 422	BM 408	BM 414
WATERLOO							02.40							
Woking														
Basingstoke					03.11		03.51							
Worting Jcn					03/16	03/28	03/57			03/28				
Micheldever						03.52								
Wallers Ash														
WINCHESTER					03.48		04.19			04/44				
Shawford														
Eastleigh Yard									04.50	05.06				
EASTLEIGH					03.58		04.29							
EASTLEIGH			03.40		04.08		04.36	04.46	05/05					
Swaythling					(To		(To							
St Denys					Fratton)		P'mouth)		05/16					
Bevois Park Yard														
Northam														
Southampton (T)				03.53										
SOUTHAMPTON CENT							04.56							
SOUTHAMPTON CENT							05.07	05/27						
Millbrook														
Redbridge							05/12	05/34						
Totton														
Lyndhurst Rd														
Beaulieu Rd														
BROCKENHURST	04.45						05.28	06.00						
Lymington Jcn	04/49						05/31							
Lymington Town														
Lymington Pier														
Sway														
New Milton														
Hinton Admiral														
Christchurch							05/47				06.00			
Pokesdown														
Boscombe														
Central Goods	05.27													
BOURNEMOUTH CENT							05.54							
BOURNEMOUTH CENT			05.47								06/08		06.30	
BOURNEMOUTH WEST														
Branksome			05.54										06.37	
Parkstone			05.58								06.23		06.41	
POOLE			06.04								06.27		06.46	06.50
Hamworthy Jcn			06.10									06.30		06/56
Holton Heath			06.16									(To Brock)		07.06
WAREHAM			06.26										06.42	07.28
Worgret Jcn			06.29											07/31
Corfe Castle			06.39											
SWANAGE			06.48											
Wool													(To	07.41
Moreton													Swanage)	07.52
DORCHESTER														08.02
Dorchester Jcn		06/22												
Wishing Well														
Upwey														
WEYMOUTH		06.47												

given a separate service to compensate for the proposed non-stop Bournemouth train died almost before being uttered. The cost of the extra train could not, by any stretch of the imagination, be paid for by an exercise in prestige.

Towards the end of the decade the possibility of restoring the two-hour timing with a Southampton stop was considered. The Merchant Navy engines were vastly more powerful – and faster – than anything the Southern had owned prior to 1939 and should certainly have been capable of improving on the feats put up by the Schools 4-4-0's in years gone by. Accordingly the rather meagre allocation of Merchant Navy's to Bournemouth was increased from 2 (1954) to 6 (1958), initially to take over the Royal Wessex from the Light Pacifics but latterly to play a part in the run-

ning of the two-hour trains which were reinstated – with a Southampton stop – from 1957.

Although non-stop running to Bournemouth was not resurrected in the pre-war sense, on summer Saturdays six services were given non-stop bookings from Bournemouth Central to prevent overcrowding from Southampton. Apart from the non-stop element there was little else of note about the trains and other than a note in the diagram – non-stop – no special motive power provision was made, two of the trains being worked by light Pacifics, three by Lord Nelson's and the sixth by a King Arthur. The timings were all around the two and a half-hour band in deference to the loads worked.

Whilst services to and from Waterloo were about as fast and frequent as they could be, the

same could not be said of the cross-country programme which seemed designed to deter travellers from going anywhere but London.

Services – one each daily – of a reasonably substantial nature ran from Bournemouth to Birkenhead and York (there was also an S&D service to Manchester) whilst the remainder were overdue an overhaul by at least half a century.

The route from Portsmouth (with connections from the LBSCR) to Salisbury (and on to Bristol) was as fat with opportunity as it was lean on fast trains and it is surprising that the enterprising spirit responsible for revolutionising the timetables of Sussex – admittedly after electrification – did not suggest to the Great Western that the two companies might profit from a series of limited-stop lightweight ex-

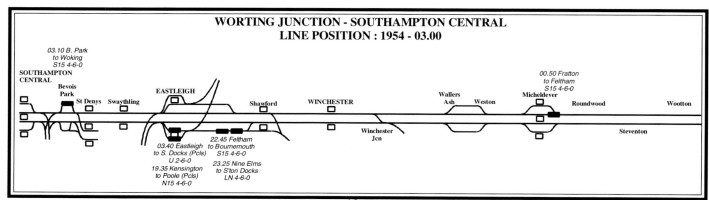

WORTING JUNCTION - SOUTHAMPTON CENTRAL
LINE POSITION : 1954 - 03.00

03.10 B. Park to Woking S15 4-6-0

SOUTHAMPTON CENTRAL — Bevois Park — St Denys — Swaythling — EASTLEIGH — Shawford — WINCHESTER — Winchester Jcn — Wallers Ash — Weston — Micheldever — Roundwood — Steventon — Wootton

00.50 Fratton to Feltham S15 4-6-0

03.40 Eastleigh to S. Docks (Pcls) U 2-6-0
19.35 Kensington to Poole (Pcls) N15 4-6-0

22.45 Feltham to Bournemouth S15 4-6-0
23.25 Nine Elms to S'ton Docks LN 4-6-0

	57xx 0-6-0T	M7&BR4 272/371	LN 4-6-0 EL 252	Q 0-6-0 EL 322	M7 0-4-4T EL 299	BR4 2-6-0 EL 277	BR4 2-6-0 EL 272	M7 0-4-4T BM 405	WC 4-6-2 BM 380	WC 4-6-2 BM 380	700 0-6-0 BM 417	M7 0-4-4T BM 409	T9 4-4-0 AJN 248	M7 0-4-4T FRA 371
WEYMOUTH – BASINGSTOKE : WORKING TIMETABLE 1954														
Train									06.25				06.45	06.45
From									Wimborne				An Jcn	An Jcn
Class				Gds					ECS		Gds			
WEYMOUTH	06.00													
Upwey														
Wishing Well														
Dorchester Jcn	06.14													
DORCHESTER	(To													
Moreton	D'ter													
Wool	W.)													
SWANAGE														
Corfe Castle														
Worgret Jcn														
WAREHAM														
Holton Heath														
Hamworthy Jcn														
POOLE														
Parkstone														
Branksome														
BOURNEMOUTH WEST									06.15					
BOURNEMOUTH CENTRAL									*06.23*					
BOURNEMOUTH CENTRAL										06.35		07.00		
Central Goods											*06.40*			
Boscombe										06.39	*06.44*	07.04		
Pokesdown										06.42		07.07		
Christchurch										06.47		07.12		
Hinton Admiral										06.54		07.19		
New Milton										07.01		07.26		
Sway										07.08		07.33		
Lymington Pier														
Lymington Town														
Lymington Jcn								07/02		07/12		07/37		
BROCKENHURST								07.03		07.15		07.38		
Beaulieu Road										07.23				
Lyndhurst Road										07.29				
Fawley														
Totton										07.35				
Redbridge										07.37			07.44	
Millbrook										07.42			07.49	
SOUTHAMPTON CENTRAL										07.45			07.52	
SOUTHAMPTON CENTRAL		07.09								07.48			07.53	
Southampton T.		07.15				07.32	07.39						07.58	08.02
Northam						07.36	07.43							08.06
Bevois Park Yard					07.25									
St Denys					07/28	07.40	07.47			07.55				08.10
Swaythling					(to	07.44	(To			07.59				(To
EASTLEIGH					F'ham)	07.49	Ports)			08.04				Ports)
EASTLEIGH			07.22		07.40	07.52				08.09				
Eastleigh Yard														
Shawford			07.30		07.48	08.01				08.17				
WINCHESTER			07.38		07.57					08.25				
Weston					(To	(To								
Micheldever			07.54		Alton)	D'cot)				08.41				
Worting Jcn			08/05							08/51				
Basingstoke			08.09							08.55				
Waterloo			09.01							10.20				

presses operating between Portsmouth and Bristol calling at Southampton and Salisbury. Instead the Southern bequeathed to later times a museum piece of workings that might have struck Roland Emmet as being absurd.

The focus of complaint was not the frequency of the service between Portsmouth and Salisbury but the fact that the South Western seemed incapable of distinguishing between a stopping train and an express. The latter was simply an example of the former but extended over a longer distance and nowhere was this better demonstrated than in 10.34 Portsmouth to Bristol which was described as an express but called at Fratton, Cosham, Fareham and so many other stations that by the time the train was pulling away from Southampton, a full hour had elapsed since leaving Portsmouth, a mere twenty-five miles away.

An interesting aspect of the Portsmouth – Salisbury axis

A further point concerning the stations on the Portsmouth – Salisbury axis was the lack of through services to London: a feature rare on the Southern. The 02.40 from Waterloo conveyed a 3-L corridor set which became the 04.36 Eastleigh to Portsmouth after which any passengers for the Fareham line had to change at Eastleigh.

Except for the three through expresses from Brighton to Bournemouth, Cardiff and Plymouth, large engines were almost as uncommon as through coaches to Waterloo. A solitary King Arthur paid a daily visit to the line whilst an early morning Portsmouth – Bristol stopping train produced an H15 4-6-0 leaving the remainder of the service to an array of motive power that had no equal even on the Southern. The thirty-four passenger departures from Fareham to Eastleigh and Southampton brought forward a medly of T9 4-4-0's (10), BR 4MT 2-6-0's (8), U 2-6-0's (4), M7 0-4-4T's and light Pacifics (3), LMS 2MT 2-6-2T's and GWR Hall 4-6-0's (2) plus an example each of an H15 4-6-0 and a BR 3MT 2-6-2T. In addition there was a fair sprinkling of goods and parcels traffic plus the Alton and Gosport branch workings.

The motive power picture was only slightly more up to date on the secondary services that operated on the main line between Southampton and Bournemouth. Bulleid Pacifics may have worked four car local trains west of Exeter but no such aggrandisement was

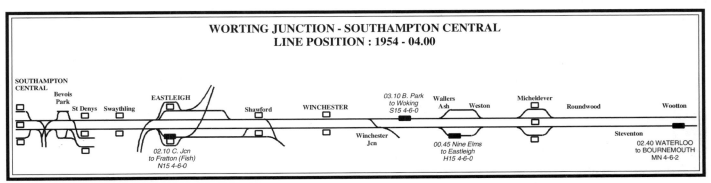

WORTING JUNCTION - SOUTHAMPTON CENTRAL
LINE POSITION : 1954 - 04.00

COMBINED WORKING TIME TABLE AND ENGINE WORKINGS : 1954

Location	(1)	(2)	(3)	(4)	(5)	(6)	(7)	(8)	(9)	(10)	(11)	(12)	(13)	(14)
From / Class	Gds	Fish		Gds		03.45 Sarum Gds	00.55 Feltham Gds	22.40 Padd	S&D	Vans				Gds
Engine	M7 0-4-4T	Q 0-6-0	T9 4-4-0	BR4 2-6-0	T9 4-4-0	Q1 0-6-0	S15 4-6-0	H dl 4-6-0	BK 71	MN 4-6-2	M7 0-4-4T	M7 0-4-4T	LN 4-6-0	Q1 0-6-0
Shed	EL 298	EL 317	EL 280	EL 270	EL 286	EL 319	FEL 102	OOC	4F 0-6-0	NE 30	BM 407	BM 410	BM 393	EL 319
WATERLOO														
Woking														
Basingstoke														
Worting Jcn							04/54							
Micheldever														
Wallers Ash														
WINCHESTER							05.40							
Shawford														
Eastleigh Yard				05.30			06.00							
EASTLEIGH														
EASTLEIGH	05.17		05.28	05/34	05/50									
Swaythling	05.23													
St Denys	05.29			05/44	05/55									
Bevois Park Yard														
Northam	05.34					06.02								
Southampton (T)	05.37		05.53											
SOUTHAMPTON CENT		05.39	05.58	05.54										
SOUTHAMPTON CENT		05.44	06.02	06.08										
Millbrook			06.08	06.13										
Redbridge		05/50												
Totton			06.03	06.12										
Lyndhurst Rd														
Beaulieu Rd														
BROCKENHURST		06.15	06.20											
Lymington Jcn		06/19												
Lymington Town		(To												
Lymington Pier		Dorset												
Sway		via B'ch)												
New Milton														
Hinton Admiral														
Christchurch												06.51		
Pokesdown												06.57		
Boscombe												07.00		
Central Goods														07.20
BOURNEMOUTH CENT												07.03		
BOURNEMOUTH CENT										06.40	06.55	07.05	07.10	07/24
BOURNEMOUTH WEST								06.48		06.48				
Branksome											07.02	07.12	07.17	07.34
Parkstone											07.06	07.16	07.24	07.39
POOLE									07.00		07.11	07.22	07.34	
Hamworthy Jcn									(To		(To	07.29	07.40	
Holton Heath									Bath)		B'stone)	07.35	07.46	
WAREHAM												07.40	07.57	
Worgret Jcn													08/00	
Corfe Castle														
SWANAGE														
Wool													08.08	
Moreton													08.18	
DORCHESTER													08.33	
Dorchester Jcn							07/44						08/35	
Wishing Well													08.43	
Upwey													08.47	
WEYMOUTH							08.05						08.50	

enjoyed in Wessex where a profusion of T9 4-4-0's, M7 tanks and a variety of 4-6-0's held sway to such an extent it was possible to travel from Southampton to Bournemouth and back – jumping off one train en route to catch the next– and hardly be hauled by the same class twice. The midday train from Eastleigh even managed to produce an H15 4-6-0 whilst a good knowledge of engine workings could result in being hauled over both the West Moors and the main line by a Q class 0-6-0 goods engine.

East of Poole the timetable was rich with trains but to the west, with the Bournemouth terminators and S&D interlopers filtered out, much of what remained consisted of the through expresses between Waterloo and Weymouth and the stopping services which ran in support.

Remoteness from the capital did not deter the Southern from giving Wareham and Dorchester the fastest service possible and where another undertaking might have made its London trains call at all stations west of Poole, the Southern ran them as expresses right through to Weymouth, the smaller intermediate stations being served by connecting trains from Bournemouth. Motive power on the latter was not quite as varied or venerable as that on the Southampton – Bournemouth trains but T9 4-4-0's and a U 2-6-0 could be found along with Lord Nelson 4-6-0's and light Pacifics. (Double-heading was not a common feature of Southern operations but blessings in pairs could be found on an evening stopper from Weymouth which ran between Wareham and Bournemouth behind an M7 0-4-4T and a U 2-6-0).

The main and secondary lines were only a part of the story and the branch lines of Wessex were a subject in themselves, ranging from the short (Lymington or Swanage) to the lengthy;

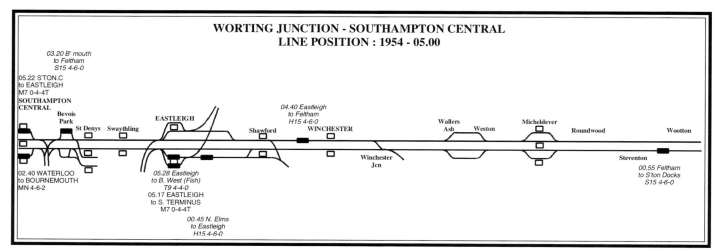

WORTING JUNCTION - SOUTHAMPTON CENTRAL
LINE POSITION : 1954 - 05.00

WEYMOUTH - BASINGSTOKE : WORKING TIMETABLE 1954

Station	06.30 B'mouth / M7 0-4-4T / BM 408	04.05 Sarum Gds / 700 0-6-0 / SAL 452	06.45 Wimborne / M7 0-4-4T / BM 406	M7 0-4-4T / LYM 362	BR4 2-6-0 / EL 273	43XX / WEY	WC 4-6-2 / NE 38	Gds / 700 0-6-0 / BM 417	Gds / 700 0-6-0 / EL 327	Gds / 700 0-6-0 / EL 326	T9&BR4 / 248/316	08.20 E'leigh / BR4 2-6-0 / EL 273	08.33 Elgh Gds / 700 0-6-0 / EL 327	08.44 Elgh Gds / 700 0-6-0 / EL 326
WEYMOUTH						07.17								
Radipole						07.21								
Upwey						07.24								
Wishing Well														
Dorchester Jcn						07/33								
DORCHESTER						(To C'ham)								
Moreton														
Wool														
SWANAGE														
Corfe Castle														
Worgret Jcn														
WAREHAM														
Holton Heath														
Hamworthy Jcn														
POOLE		06.42	06.59											
Parkstone			07.04											
Branksome			07.08											
BOURNEMOUTH WEST			07.12				07.20							
BOURNEMOUTH CENTRAL							07.28							
BOURNEMOUTH CENTRAL		07/06					07.30							
Central Goods		07.11						07.53						
Boscombe							07.35	07.57						
Pokesdown							07.38							
Christchurch							07.43							
Hinton Admiral														
New Milton							07.54							
Sway														
Lymington Pier				07.39										
Lymington Town				07.43										
Lymington Jcn	07/41			07/51			08/02							
BROCKENHURST	07.42			07.53			08.04							
Beaulieu Road														
Lyndhurst Road														
Fawley														
Totton														
Redbridge							08/18							
Millbrook														
SOUTHAMPTON CENTRAL							08.23							
SOUTHAMPTON CENTRAL							08.26							
Southampton T.											08.33			
Northam											08.37			
Bevois Park Yard														
St Denys											08.41			
Swaythling											08.45			
EASTLEIGH											08.50			
EASTLEIGH					08.20		08/36				08.55			
Eastleigh Yard													08.33	08.44
Shawford					08.28								08.50	09.01
WINCHESTER					08.35		08.48				(To A. Jcn)	08.57	09.08	09.20
Weston													(To Alton)	(to Chesil)
Micheldever												09.13		
Worting Jcn							09/11					09/24		
Basingstoke												09.28		
Waterloo							10.00					Reading		

the latter including the line from Winchester Junction to Alton – which for those with time on their hands, was an alternative route to Waterloo – and the old main line from Brockenhurst to Broadstone. The one feature they all had in common was the M7 0-4-4T which almost up to the last hour of steam working remained the standard branch engine in the district. (Diesel multiple units made an appearance – principally on the Portsmouth – Salisbury axis – from 1957 but their use was by no means universal and two years after their introduction, the Southern displayed sufficient faith in steam to replace many of the older push and pull units in the Bournemouth area with specially adapted main line Maunsell stock). At the beginning of 1960 there were as many as twenty-seven M7's operating from Eastleigh and Bournemouth and the pleasure derived from boarding an 0-4-4T-hauled train from Brockenhurst at a time when every other service of its kind elsewhere seemed to be in the hands of the ubiquitous LMS 2-6-2T needs no enlargement by the author.

Foreign engines were not a marked feature of the Wessex main line and although a few of the LMS 2-6-2T's had found their way to Eastleigh, there was no parallel to the state of affairs on other parts of the Southern where imported LMS 2-6-4T's had pretty well taken over complete routes.

On the other hand Great Western running powers were quite extensive and several times a day Hall 4-6-0's worked down the main line on services between Reading and Portsmouth or Southampton Terminus. South of Shawford the number of GW incursions increased further with arrivals from the Didcot, Newbury and Southampton. At the far end of the line the position was reversed with the Southern running over Great Western metals west of Dorchester Junction.

Most people who worked on or who knew the line had their favourite services and for many it was either the Royal Wessex, the Belle or the Birkenhead: trains remarkable for their weight, speed or both. It cannot be denied that there was something very impressive about charging up Weston bank with a Pacific and a heavy load

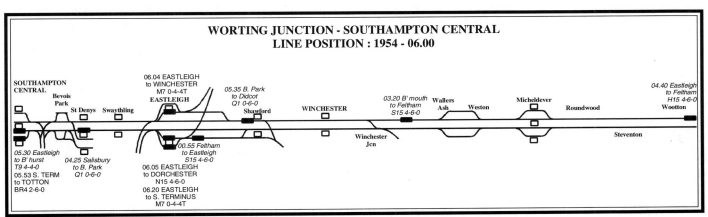

WORTING JUNCTION - SOUTHAMPTON CENTRAL
LINE POSITION : 1954 - 06.00

COMBINED WORKING TIME TABLE AND ENGINE WORKINGS : 1954

Train					*07.27*		*07.20*				06.03	*04.50*		
From					C. Cary		B'mth				Fareham	Elgh		
Class		*Fish*			*Fish*		*Gds*					*Gds*		
Engine	M7 0-4-4T	T9 4-4-0	T9 4-4-0	Railcar	U 2-6-0	N15 4-6-0	Q1 0-6-0	M7 0-4-4T	M7&T9	M7 0-4-4T	M7 0-4-4T	N15 4-6-0	M7 0-4-4T	M7 0-4-4T
Shed	BM 410	EL 280	SAL 444	Wey	BM 414	BM 401	EL 319	EL 297	362/403	BM 405	FRA 371	EL 266	SWA 421	EL 305
WATERLOO														
Woking														
Basingstoke														
Worting Jcn														
Micheldever														
Wallers Ash														
WINCHESTER														
Shawford														
Eastleigh Yard														
EASTLEIGH														
EASTLEIGH						06.05		06.20						06.35
Swaythling								06.26						06.41
St Denys								06.30		06.40				06.46
Bevois Park Yard														
Northam								06.34						
Southampton (T)								06.37						
SOUTHAMPTON CENT						06.14				06.45				06.52
SOUTHAMPTON CENT						06.20								06.59
Millbrook						06.24								07.03
Redbridge						06.29								07.08
Totton						06.33								07.11
Lyndhurst Rd						06.40								(To
Beaulieu Rd						06.46								Fawley)
BROCKENHURST		*06.35*				06.58			07.04	07.17		07.30		
Lymington Jcn		*06/38*				07/01			07/08	07/20		07/34		
Lymington Town									07.15					
Lymington Pier									07.20	(Via				
Sway						07.06				B'stone)		07.45		
New Milton		*06.52*				07.14						07.56		
Hinton Admiral		*07.00*				07.22								
Christchurch		*07.11*				07.30								
Pokesdown		*07.22*				07.38								
Boscombe		*07.32*				07.42								
Central Goods														
BOURNEMOUTH CENT		*07.35*				07.45								
BOURNEMOUTH CENT			07.42			07.53								
BOURNEMOUTH WEST														
Branksome			07.49			08.00								
Parkstone			07.53			08.04	08.08							
POOLE			07.58			08.10	*08.14*							
Hamworthy Jcn			(To			08.15								
Holton Heath			S'bury)			08.20								
WAREHAM	08.18					08.26							08.31	
Worgret Jcn	08/21					08/29							08/34	
Corfe Castle	08.31												08.44	
SWANAGE	08.40												08.53	
Wool						08/35								
Moreton						08.43								
DORCHESTER				08.50		08.52								
Dorchester Jcn				08/50		08/53								
Wishing Well														
Upwey														
WEYMOUTH				09.01		09.08								

but, provided that time did not enter the equation, for the writer's money the most interesting service was the little-known 11.36 from Bournemouth West; nominally a stopping train to Southampton but in fact one of the few LSW workings to run through to the Great Central.

By all the rules of timetable compilation the 11.36 should have run as a parcels train since most of it consisted of fish vans that had come down with the 02.05 ex Clapham Junction and were being returned to Neasden. It was a busy time of day however and by adding some passenger accommodation – a 3-car L set – it was possible to run it as an all-stations

working, connecting at Southampton into the 11.30 Weymouth – Waterloo. The engine was the Nine Elms H15 4-6-0 that had come down with the 00.45 goods to Eastleigh and the 07.46 stopping passenger to Bournemouth.

Whilst running the 11.36 as a passenger train suited the commercial people, it did not suit the operators in the London District and in order to get it past Worting at an acceptable time, it had to go through the procedure of decanting its passengers at Southampton Central and then recessing in Bevois Park sidings for half an hour before continuing forward as a passenger train from St Denys. Through passen-

gers were supposed to bridge the gap with the 13.15 Romsey – Portsmouth (no great punishment with its LSWR coaches and M7 0-4-4T) although the author – probably the only through passenger the service ever had – generally flashed his pass at the guard and enjoyed a solitary lunch at Bevois Park with one eye on the shunting and the other on the main line.

Although stopping trains were no strangers to the South Western, the 11.36 was in a class of its own. It did not, on the one hand, run as part of a regular pattern whilst, on the other, neither its stock or engine was of the type generally seen on passenger trains and the

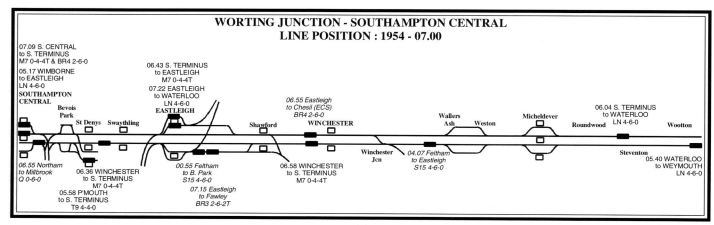

WORTING JUNCTION - SOUTHAMPTON CENTRAL
LINE POSITION : 1954 - 07.00

WEYMOUTH - BASINGSTOKE : WORKING TIMETABLE 1954													
Train	07.47		08.06	02.40					07.36		*07.53*		07.00
From	Sarum		Fawley	Bath					B'stone		*BM*		Dorset
Class				Mail	ECS						*Gas*		
Engine	T9 4-4-0	M7 0-4-4T	M7 0-4-4T	H15 4-6-0	LM5 4-6-0	M7 0-4-4T	M7 0-4-4T	Q0-6-0	M7 0-4-4T	M7 0-4-4T	700 0-6-0	T9 4-4-0	Q0-6-0
Shed	SAL 445	BM 408	EL 305	EL 313	BATH1	BM 413	BM 413	EL 324	BM 404	BM 407	*BM 417*	BM 403	EL 324
WEYMOUTH													
Upwey													
Wishing Well													
Dorchester Jcn													
DORCHESTER							07.00						
Moreton							07.09						
Wool							07.18						
SWANAGE								07.15					
Corfe Castle								07.26					
Worgret Jcn							07/23	07/35					
WAREHAM							07.26	07.38					
Holton Heath							07.32	07.44					
Hamworthy Jcn							07.38	07.50					
POOLE				07.27			(Via		07.46	07.57			
Parkstone							Ringwood)		07.52	08.03			
Branksome									07.57	08.09			
BOURNEMOUTH WEST			07.30	07.44						08.12			
BOURNEMOUTH CENTRAL			07.38						08.02				
BOURNEMOUTH CENTRAL			07.41			07.50							
Central Goods													
Boscombe			07.45								*08.09*		
Pokesdown			07.48										
Christchurch			07.53		<u>07.58</u>	08.10					*08.16*		
Hinton Admiral			08.00			08.17							
New Milton			08.08			08.24							
Sway			08.15			08.31							
Lymington Pier												08.10	
Lymington Town		08.05										08.30	
Lymington Jcn		08/13	08/19			08/35						08.39	08/49
BROCKENHURST		**08.15**	08.22			08.37						**08.41**	**08.50**
Beaulieu Road			08.30										
Lyndhurst Road			08.36										
Fawley													
Totton		08.37	08.42										
Redbridge	08.30	08.40	08.45										
Millbrook	08.34	08.44	08.50										
SOUTHAMPTON CENTRAL	08.37	08.47	08.53										
SOUTHAMPTON CENTRAL	08.41	08.50	08.55										
Southampton T.		<u>08.55</u>											
Northam													
Bevois Park Yard													
St Denys	08.48		09.02										
Swaythling	(To		09.06										
EASTLEIGH	Ports)		09.11										
EASTLEIGH													
Eastleigh Yard													
Shawford													
WINCHESTER													
Weston													
Micheldever													
Worting Jcn													
Basingstoke													
Waterloo													

sloth that generated impatience on other slow services was overshadowed by a sense of novelty: any fool could travel from Bournemouth to London in just over two hours but to do it in five took talent.

Unfortunately it was not possible to complete the journey and Woking, where the L-set was shunted off to form the following day's 07.32 to Southampton Terminus - was as far as the paying passenger could travel. The H15 gathered up its fish vans and prepared to take them over to the Great Central Yard at Neasden whilst the London-bound passenger had to conclude his journey in a multiple-unit.

The fact the 11.36 had no corresponding down working did not mean that the returning passenger had to fall in with the straitjacket of the interval timetable. Prior to electrification there had been seven through trains from Waterloo to Southampton via Alton and Alresford and although through running had become a matter of history by 1954, the timetable continued to display the note 'Via Alton' next to the Alton – Eastleigh locals as though they shared the same status as the expresses via Basingstoke. Several times a day, therefore, it was possible to get a electric from Waterloo to Alton and complete the journey behind an M7

0-4-4T, arriving in Southampton behind the same engine and in the same coaches that one's Grandfather would have used.

Variations on the theme included travelling from Alton to Southampton via Fareham but the really determined diehard could actually do the complete trip behind steam – provided he had a brake pass and didn't mind a two-hour break at Alton – although it meant leaving Waterloo in the 03.45 News, riding behind an N15X 4-6-0 as far as Woking and going forward to Alton in the 700 0-6-0-hauled 05.35 connection.

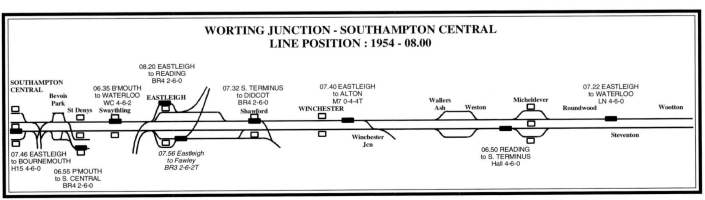

EASTLEIGH

On the other regions of BR there was a tendency for Pullman services to be times a little more slowly than ordinary expresses; it being believed that the motion sometimes imparted by high speed travel was inappropriate at mealtimes. On the Southern however the Bournemouth Belle was given the fastest timing of the day - 88 minutes non-stop to Southampton and 130 minutes to Bournemouth Central - with class 8P haulage: a Nine Elms Merchant Navy 4-6-2 being allocated to the outward and return trips. In order to provide connections for stations beyond Bournemouth, the rear portion of the 11.30 ex Waterloo recessed at Bournemouth Central for half an hour; running forward to Weymouth after being overtaken by the Pullman.

Well known as one of the busiest junctions on the South Western and a household name because of its locomotive works, one of Eastleigh's lesser-known claims to fame stems from the fact it was the location of the best practical joke ever played in railway circles.

The chief clerk at Eastleigh station had once been a booking clerk at neighbouring Fareham where over a long period of time he had upset his chief clerk – a patrician, rather sensitive man – through his appalling table manners. The Fareham chief was too much of a gentleman to go as far as telling his junior that there were sound reasons for eating with one's mouth closed and instead waited for an opportunity to settle the score for several years of having his stomach turned.

As it happened the junior man succeeded to the chief's position at Eastleigh before anything could be done and the question of revenge fell into abeyance.

Some time later the District Senior Relief clerk happened to be relieving at Fareham.

"Do you know, Mr Legg," he announced to the chief clerk. "I was at Eastleigh yesterday and had to watch their new chief eat his lunch. I've never seen anything like it."

"You don't have to tell me…"

"He needs teaching a lesson. He's put me off food for a month."

The two discussed methods of retribution but no ideas took root until the Station Master arrived for his afternoon tea. The proprieties of rank were normally very carefully observed in the Fareham district but an incident some time earlier had narrowed the difference a little

sufficiently to admit the SM into the discussion.

The District Relief Clerk was normally far too senior in rank to issue tickets but on one occasion he had been asked to do a turn of duty of Fareham so that one of the regular junior clerks could take a day off.

"Do you know, Mr Beckett," the Station Master had observed. "Having you issuing tickets at Fareham is like using the Flying Scotsman on the Gosport branch…"

After that the SM could do no wrong and the trio continued to discuss the matter at hand. The Senior Relief Clerk declared vehemently that he was not prepared to have his stomach turned every time he paid a visit to Eastleigh yet the three keenest brains known to the Fareham district were unable to come up with anything. Their reverie was interrupted by one of the platform staff who stuck his head through the ticket hatch.

"They're building one of them new supermarkets at Eastleigh…"

At intervals during the following few months a number of people, mainly from the Fareham area, seemed to go out of their way to engage the Eastleigh chief in conversation and marvel at the unbelievable concession the railway had apparently wrung out of the supermarket owners.

"If you're railway staff and show your ID card, you get everything at quarter price. Just like privilege tickets…."

At first the Eastleigh chief had told his informants that they were talking through the tops of their heads but as the rumours persisted he began to wonder until, a day or two after the supermarket's opening ceremony, the Fareham stationmaster happened to drop into Eastleigh booking office.

"Guv'nor, " asked the chief, fishing. "Have you heard about this new supermarket that's just been opened?"

"Supermarket?"

"Isn't there a rumour that railway staff can get their groceries for less than anyone else…"

"I am told that if you show your railway ID card, you only have to pay a quarter of the price."

"And I thought people were pulling my leg."

"The supermarket has been built on former railway property therefore the Transport Commission were able to wring the concession out of the owners."

"Are there any conditions?"

"Only that you've got to purchase at least thirty items and must do your shopping between five and six on a Friday evening."

"Yes, I suppose that makes sense…."

The SM disappeared from the Southern on promotion a few days later and was unable to witness the event himself but it came to his ears that half the railway establishment of Eastleigh had crowded into the new supermarket to witness the sight of the chief clerk doing his shopping. Apparently his family, relatives and overfilled trolleys were so numerous that they clogged every aisle; all movement coming to a halt as the head of the family waved a small green card under the nose of the uncomprehending shop assistant and demanded

TRAFFIC MOVEMENTS : EASTLEIGH STATION (1954)

Train	Arr	Engine	Shed	Dep	Destination
16.55 Worcester (Pds)	*00.05*	*N15 4-6-0*	*EL 266*		
23.16 Portsmouth & S.	00.07	U 2-6-0	SAL 482		
22.30 WATERLOO	00.33	LN 4-6-0	EL 253	00.43	POOLE
		U 2-6-0	*FRA 363*	*01.00*	*Portsmouth Pds*
21.55 WEYMOUTH	01.20	LN 4-6-0	NE 31	01.31	WATERLOO
01.23 S. Terminus	01.37	N15 4-6-0	BM 399		(Fwd at 01.46)
19.35 Kensington Pds	*01.40*	*BR4 2-6-0*	*EL 273*		*(Fwd at 03.05)*
(01.23 ST)		T9 4-4-0	EL 286	01.46	Portsmouth &S
		U 2-6-0	SAL 482	01.55	Yeovil
(19.35 Kensington)		*N15 4-6-0*	*BM 399*	*03.05*	*Poole Pds*
		U 2-6-0	*SAL 450*	*03.40*	*Southampton Docks Pds*
02.10 Clapham Jcn (Fish)	*03.58*	*N15 4-6-0*	*EL 265*	*04.08*	*Fratton (Fish)*
02.40 WATERLOO	04.29	MN 4-6-2	NE 30		
(02.40 WATERLOO)		BR3 2-6-2T	EL 331	04.36	PORTSMOUTH HBR
(02.40 WATERLOO)		MN 4-6-2	NE 30	04.46	BOURNEMOUTH C
		M7 0-4-4T	EL 298	05.17	S. Terminus
		T9 4-4-0	*EL 280*	*05.28*	*Bournemouth W. (Fish)*
		LM2 2-6-2T	EL 306	05.39	Fareham
05.22 Southampton C	05.41	M7 0-4-4T	EL 302	06.04	Winchester
		N15 4-6-0	BM 401	06.05	Dorchester
		M7 0-4-4T	EL 300	06.07	Romsey
		M7 0-4-4T	EL 297	06.20	S. Terminus
06.04 SOUTHAMPTON T.	06.21	LN 4-6-0	EL 251	06.22	WATERLOO
		M7 0-4-4T	EL 305	06.35	Fawley
		M7 0-4-4T	EL 303	06.36	Alton
06.30 Fareham	06.50	LM2 2-6-2T	EL 306		
06.36 Winchester	06.49	BR4 2-6-0	EL 277	06.52	S. Terminus
		BR4 2-6-0	EL 278	06.55	Winchester (Chesil) ECS
06.43 Romsey	06.57	M7 0-4-4T	EL 304		
		M7 0-4-4T	EL 304	07.00	Fareham
06.43 S. Terminus	07.00	M7 0-4-4T	EL 298		
06.58 Winchester	07.11	M7 0-4-4T	EL 302	07.13	S. Terminus
05.17 Wimborne	07.17	LN 4-6-0	BM 394		
		LN 4-6-0	EL 252	07.22	WATERLOO
		LM2 2-6-2T	EL 306	07.30	Romsey
05.40 WATERLOO	07.36	LN 4-6-0	NE 31	07.39	WEYMOUTH
		M7 0-4-4T	EL 299	07.40	Alton
		H15 4-6-0	NE 79	07.46	Bournemouth C.
		M7 0-4-4T	EL 300	07.52	Fareham
07.32 S. Terminus	07.49	BR4 2-6-0	EL 277	07.52	Didcot
		M7 0-4-4T	EL 298	07.56	Andover Jcn
06.35 BOURNEMOUTH C	08.04	WC 4-6-2	BM 380	08.09	WATERLOO
08.00 Romsey	08.15	LM2 2-6-2T	EL 306		(Fwd at 08.35)
		BR4 2-6-0	EL 273	08.20	Reading
07.29 Salisbury	08.17	H15 4-6-0	NE 72		(Fwd at 08.26)
06.50 Reading	08.21	HALL 4-6-0	RDG 66		(Fwd at 08.28)
(07.29 Salisbury)		H15 4-6-0	NE 72	08.26	Bristol
(06.50 Reading)		HALL 4-6-0	RDG 66	08.28	S. Terminus
(08.00 Romsey)		LN 4-6-0	BM 394	08.35	Weymouth
07.20 BOURNEMOUTH W.		WC 4-6-2	NE 38	08/36	WATERLOO
08.32 Romsey	08.47	AJN 249	43xx 2-6-0		(Engine Change)
07.53 Alton	08.50	M7 0-4-4T	EL 303		
08.33 S. Terminus	08.50	T9 & BR4	248/316		(Fwd at 08.55)
(08.33 S. Terminus)		T9 4-4-0	AJN 248	08.55	Andover Jcn
07.31 Woking	08.59	U 2-6-0	GUI 181		(Fwd at 09.05)
(08.32 Romsey)		LM2 2-6-2T	EL 306	09.03	Portsmouth &S.
(07.31 Woking)		43xx 2-6-0	AJN 249	09.05	S. Terminus
07.30 Bournemouth W.	09.11	H15 4-6-0	EL 313		
07.45 Newbury	09.12	22xx 0-6-0	DID 13	09.14	S. Terminus
07.34 WEYMOUTH		WC 4-6-2	BM 381	09/27	WATERLOO
09.19 S. Terminus	09.36	S15 4-6-0	FEL 102		
08.50 Portsmouth &S	09.38	BR3 2-6-2T	EL 331		(Fwd at 09.44)
(08.50 Ports)		U 2-6-0	GUI 181	09.44	Winchester
08.55 Alton	09.49	M7 0-4-4T	EL 299		
09.36 S. Terminus	09.53	M7 0-4-4T	EL 305		
		N15 4-6-0	EL 259	09.55	Southampton C
08.35 BOURNEMOUTH W.	10.10	WC 4-6-2	BM 382	10.12	WATERLOO
		M7 0-4-4T	EL 299	10.16	Alton
08.30 WATERLOO		MN 4-6-2	NE 32	10/17	WEYMOUTH
		T9 4-4-0	EL 279	10.22	Newbury
09.30 Andover Jcn	10.24	M7 0-4-4T	EL 298		
07.42 Didcot	10.28	43xx 2-6-0	DID 30	10.30	S. Terminus
10.13 S. Terminus	10.29	HALL 4-6-0	RDG 66	10.31	Reading
09.12 Reading	10.41	HALL 4-6-0	RDG 50		(Fwd at 10.49)
09.20 BOURNEMOUTH W.		N15 4-6-0	BM 399	10/45	BIRKENHEAD
(09.12 Rdg)		HALL 4-6-0	RDG 50	10.49	Portsmouth &S.
10.45 Winchester	10.55	U 2-6-0	GUI 181	11.02	S. Terminus
		M7 0-4-4T	EL 308	11.10	Romsey
09.30 WATERLOO	11.19	WC 4-6-2	NE 33	11.22	BOURNEMOUTH W.
07.43 Clapham Jcn (Pds)	*11.23*	*H15 4-6-0*	*NE 65*		
		H15 4-6-0	EL 313	11.28	Bournemouth C.
09.20 WEYMOUTH		LN 4-6-0	NE 31	11/30	WATERLOO
10.45 Portsmouth &S.	11.39	M7 0-4-4T	FRA 371		
10.30 WATERLOO		MN 4-6-2	NE 34	11/51	WEYMOUTH
11.40 S. Terminus	11.52	T9 4-4-0	BM 403		
11.45 S. Terminus	12.02	43xx 2-6-0	DID 30	12.08	Didcot
11.05 Portsmouth &S (Pds)	*12.06*	*N15 4-6-0*	*EL 265*		
12.00 S. Terminus	12.15	T9 4-4-0	EL 283		(Fwd at 12.20)
09.54 Waterloo	12.15	H15 4-6-0	NE 68		(Fwd at 12.23)
11.25 Andover Jcn	12.18	T9 4-4-0	AJN 248		
(12.00 S. Terminus)		M7 0-4-4T	FRA 371	12.20	Romsey
(09.54 W'loo)		LN 4-6-0	EL 253	12.23	Southampton C
11.05 BOURNEMOUTH W.	12.34	WC 4-6-2	BM 383	12.36	WATERLOO
11.16 BOURNEMOUTH W.	12.40	LN 4-6-0	BM 395	12.42	YORK
12.31 S. Terminus	12.48	U 2-6-0	GUI 181	12.50	Reading
12.05 Alton	13.00	M7 0-4-4T	EL 299		
		T9 4-4-0	AJN 248	13.08	Portsmouth &S.
11.30 WATERLOO		LN 4-6-0	EL 252	13/09	BOURNEMOUTH W.
		M7 0-4-4T	EL 308	13.15	S. Terminus
		H15 4-6-0	*EL 311*	*13.15*	*Clapham Jcn (Pds)*
11.30 WEYMOUTH		WC 4-6-2	BM 386	13/30	WATERLOO

Train	Arr	Engine	Shed	Dep	Destination
12.15 Reading	13.35	BR4 2-6-0	EL 273		(Fwd at 13.46)
10.50 Didcot	13.41	T9 4-4-0	EL 279		
13.23 S. Terminus	13.41	M7 0-4-4T	EL 297	13.43	Winchester
(12.15 Reading)		BR4 2-6-0	EL 273	13.46	Portsmouth &S.
13.03 Portsmouth &S.	13.47	T9 4-4-0	SAL 445	13.50	Salisbury
12.30 WATERLOO		MN 4-6-2	NE 35	13/50	BOURNEMOUTH W.
13.06 Salisbury	13.53	T9 4-4-0	SAL 443		(Fwd at 14.05)
11.36 Bournemouth W.	13.55	H15 4-6-0	NE 79	14.02	Woking
		T9 4-4-0	SAL 443	14.05	S. Terminus
13.56 S. Terminus	14.13	M7 0-4-4T	EL 308		(Fwd at 14.22)
		M7 0-4-4T	EL 302	14.15	Alton
14.05 Romsey	14.20	M7 0-4-4T	EL 329		
(13.56 S. Terminus)		43xx 2-6-0	DID 20	14.22	Didcot
14.10 Winchester	14.23	M7 0-4-4T	EL 297		(Fwd at 14.35)
13.05 BOURNEMOUTH W.	14.31	MN 4-6-2	NE 32	14.33	WATERLOO
(14.10 Winchester)		T9 4-4-0	EL 283	14.35	Bournemouth C.
13.30 WATERLOO		WC 4-6-2	BM 381	15/10	WEYMOUTH
14.55 S. Terminus	15.12	T9 4-4-0	FRA 366		
12.42 Didcot	15.15	22xx 0-6-0	DID 31	15.23	S. Terminus
13.48 Reading	15.17	HALL 4-6-0	OX 208	15.21	Portsmouth &S.
13.25 WEYMOUTH		WC 4-6-2	BM 385	15/30	WATERLOO
14.45 Portsmouth &S	15.36	HALL 4-6-0	RDG 50	15.38	Reading
		M7 0-4-4T	EL 300	15.51	Fawley
		M7 0-4-4T	EL 298	15.59	Alton
09.30 BIRKENHEAD		N15 4-6-0	BM 399	16/04	BOURNEMOUTH W.
		M7 0-4-4T	EL 297	16.06	S. Terminus
15.45 Portsmouth &S.	16.31	T9 4-4-0	AJN 248	16.34	Romsey
16.25 Winchester	16.38	M7 0-4-4T	EL 329		
15.05 BOURNEMOUTH W.	16.36	MN 4-6-2	NE 34	16.39	WATERLOO
16.25 S. Terminus	16.42	M7 0-4-4T	EL 305		
15.20 WATERLOO		WC 4-6-2	BM 382	16/45	WEYMOUTH
		M7 0-4-4T	EL 299	16.55	Alton
16.10 Alton	17.03	M7 0-4-4T	EL 302	17.08	S. Terminus
16.56 S. Terminus	17.12	22xx 0-6-0	DID 31	17.14	Didcot
15.30 WATERLOO	17.14	WC 4-6-2	NE 37	17.16	BOURNEMOUTH W.
		M7 0-4-4T	EL 305	17.16	Romsey
17.03 Romsey	17.18	2MT 2-6-2T	AJN 247	17.20	S. Terminus
14.20 Weymouth	17.22	T9 4-4-0	EL 282		(Fwd at 17.42)
10.23 YORK	17.22	LN 4-6-0	BM 395	17.25	BOURNEMOUTH W.
		H15 4-6-0	EL 310	17.27	Portsmouth &S.
16.34 BOURNEMOUTH W.		MN 4-6-2	NE 35	17/33	WATERLOO
		M7 0-4-4T	EL 329	17.34	Brockenhurst
(14.20 Weymouth)		T9 4-4-0	EL 282	17.42	Romsey
17.23 S. Terminus	17.40	M7 0-4-4T	EL 297	17.44	Alton
17.30 SOUTHAMPTON DOCKS		LN 4-6-0	EL 254	17/46	WATERLOO
14.56 Oxford	17.54	BR4 2-6-0	EL 277		(Fwd at 18.02)
17.38 S. Terminus	17.58	BR4 2-6-0	EL 321	17.59	Winchester
16.35 WATERLOO		WC 4-6-2	BM 383	18/00	WEYMOUTH
(14.56 Oxford)		BR4 2-6-0	EL 277	18.02	S. Terminus
17.50 Romsey	18.06	M7 & T9	305/248		
16.38 Bournemouth C	18.08	H15 4-6-0	EL 313		
17.05 Alton	18.09	M7 0-4-4T	EL 298		
17.17 Portsmouth &S	18.11	BR4 2-6-0	EL 273	18.15	Reading
17.16 Fawley	18.19	M7 0-4-4T	EL 300		(Fwd at 18.30)
17.10 Reading	18.26	U 2-6-0	GUI 181	18.28	S. Terminus
(17.16 Fawley)		T9 4-4-0	EL 280	18.30	Andover Jcn
17.05 BOURNEMOUTH W.	18.31	LN 4-6-0	EL 253	18.33	WATERLOO
17.38 Bournemouth C	19.04	LN 4-6-0	EL 252		
17.30 WATERLOO	19.11	WC 4-6-2	BM 380	19.14	BOURNEMOUTH W.
18.58 S. Terminus	19.14	BR3 2-6-2T	AJN 247	19.15	Alton
18.02 Alton	19.18	M7 0-4-4T	EL 299	19.20	S. Terminus
		M7 0-4-4T	EL 298	19.20	Romsey
		T9 4-4-0	AJN 248	19.22	Portsmouth &S.
17.35 WEYMOUTH		MN 4-6-2	NE 30	19/31	WATERLOO
19.00 Basingstoke (Pds)	*19.39*	*S15 4-6-0*	*FEL 106*		
		H15 4-6-0	*NE 72*	*19.45*	*Waterloo (Pds)*
18.40 Andover Jcn	19.50	L1 4-4-0	EL 309		
18.30 WATERLOO		WC 4-6-2	BM 386	19/53	WEYMOUTH
19.40 S. Terminus	19.57	M7 0-4-4T	EL 302	19.59	Winchester
		S15 4-6-0	*FEL 106*	*20.03*	*S'ton Docks (Pds)*
19.50 Romsey ECS	20.05	M7 0-4-4T	EL 298		
19.00 Reading	20.16	N15X 4-6-0	BSK 236		(Fwd at 20.22)
(19.00 Reading)		BR4 2-6-0	EL 270	20.22	S. Terminus
18.35 Bournemouth W.	20.19	N15 4-6-0	EL 266	20.24	Reading
19.25 Alton	20.25	M7 0-4-4T	EL 297		
		T9 4-4-0	SAL 443	20.28	Salisbury
20.25 Eastleigh Yard		S15 4-6-0	FEL 105	20/30	Ashford (SECR)
19.35 Andover Jcn	20.36	T9 4-4-0	EL 282		
20.04 S'ton Docks (Pds)	*20.38*	*43xx 2-6-0*	*BTL 365*		
17.55 Didcot	20.39	BR4 2-6-0	EL 278	20.42	S. Terminus
20.03 Portsmouth &S.	20.52	T9 4-4-0	EL 284		
16.35 Exeter	21.03	S15 4-6-0	FEL 101		(Fwd at 21.28)
18.30 WEYMOUTH	21.04	WC 4-6-2	NE 33	21.09	WATERLOO
19.30 WATERLOO	21.09	WC 4-6-2	BM 385	21.13	BOURNEMOUTH W.
21.05 S. Terminus	21.22	M7 0-4-4T	EL 299		
(16.35 Exeter)		S15 4-6-0	FEL 101	21.28	Portsmouth &S.
20.38 Alton	21.34	2MT 2-6-2T	AJN 247		
		43xx 2-6-0	BTL 365	21.44	Salisbury
		HALL 4-6-0	*OX 208*	*21.45*	*Crewe (Pds)*
		T9 4-4-0	EL 282	21.51	S. Terminus
19.55 Bournemouth W (Pds)	*21.56*	*N15 4-6-0*	*EL 265*		*(Fwd at 22.15)*
18.42 Waterloo (Pds)	*22.05*	*H15 4-6-0*	*NE 64*		
(B'mouth Pds)		*N15X 4-6-0*	*BSK 236*	*22.15*	*Waterloo (Pds)*
		N15 4-6-0	*EL 263*	*22.20*	*S'ton Docks (Pds)*
22.04 Southampton C	22.20	BR3 2-6-2T	EL 330		
21.48 Portsmouth &S.	22.33	T9 4-4-0	AJN 248		
21.00 WATERLOO		LN 4-6-0	NE 31	22/36	SOUTHAMPTON DOCKS
22.34 Winchester	22.47	M7 0-4-4T	EL 302	22.49	Southampton Central
19.48 Weymouth	22.49	N15 4-6-0	EL 263	22.52	Reading
19.10 Bristol	23.06	S15 4-6-0	SAL 467		
16.40 Plymouth	23.46	BR4 2-6-0	EL 272		

that the bill be reduced by three-quarters...

At other times - table manners permitting - the staff at Eastleigh could close ranks in an extraordinary display of self-help as happened one morning at Waterloo when Bernie 'Silver Fox' Briggs, the Eastleigh SM, was on his way back to Hampshire. Walking through the ticket barrier he bumped into one of his guards, also on his way back to Eastleigh, and the two men strolled past the coaches, looking for an empty compartment.

The guard was an interesting character with quite a history. An English-speaking Pole, he had effected his escape from Poland in 1939, travelling from Warsaw to Calais on foot and –

Although most shunting in Southampton Docks was performed by USA tanks, a number of turns were covered by LBSCR E1 0-6-0T's dating from 1874. In 1956 some of the latter were replaced by E2 0-6-0T's which had previously been based on the Chatham section. 32106, still sporting a Stewarts Lane shedplate, shunts at Southampton in September 1961.

COMBINED WORKING TIME TABLE AND ENGINE WORKINGS : 1954

Train	05.30					05.45	06.55			05.58	05.58	00.55		
From	Elgh					BRISTOL	H. JN			Ports	Ports	Feltham		
Class	Gds	Gds		Fish			Gds					Gds		
Engine	T9 4-4-0	Q0-6-0	M7 0-4-4T	T9 4-4-0	M7 0-4-4T	BR4 2-6-0	Hdl 4-6-0	M7 0-4-4T	M7 0-4-4T	T9 4-4-0	U 2-6-0	S15 4-6-0	M7 0-4-4T	M7 0-4-4T
Shed	EL 286	EL 323	BM 406	EL 280	BM 408	EL 277	BRD	HAM 422	BM 407	FRA 365	SAL 450	FEL 102	EL 302	BM 409
WATERLOO														
Woking														
Basingstoke														
Worting Jcn														
Micheldever														
Wallers Ash														
WINCHESTER						06.36							06.58	
Shawford						06.42							07.04	
Eastleigh Yard											07.00			
EASTLEIGH						06.49							07.11	
EASTLEIGH						06.52					07/05		07.13	
Swaything						06.58							07.19	
St Denys						07.02				07.05	07/16		07.23	
Bevois Park Yard											07/20			
Northam		06.55				07.06				07.09			07.27	
Southampton (T)						07.09				07.12	07.27		07.30	
SOUTHAMPTON CENT		07.03									07.32			
SOUTHAMPTON CENT		07.25									07.36			
Millbrook	07.13	07.30									07.40			
Redbridge	07.33										07.45			
Totton	07.37										(To S'bury)			
Lyndhurst Rd														
Beaulieu Rd														
BROCKENHURST			07.52											07.56
Lymington Jcn			07/55											07/59
Lymington Town			08.02											
Lymington Pier														
Sway														08.03
New Milton														08.10
Hinton Admiral														08.15
Christchurch														08.21
Pokesdown														08.27
Boscombe														08.29
Central Goods														
BOURNEMOUTH CENT														08.32
BOURNEMOUTH CENT														08.36
BOURNEMOUTH WEST		08.10		08.18					08.30					08.44
Branksome		08.15							08.33					
Parkstone		08.19							08.37					
POOLE		08.28							08.42					
Hamworthy Jcn		(To							08.47					
Holton Heath		Brock)							08.52					
WAREHAM							08.40		08.57					
Worgret Jcn							08/45							
Corfe Castle								09.38						
SWANAGE								09.52						
Wool														
Moreton														
DORCHESTER														
Dorchester Jcn							09/13							
Wishing Well														
Upwey							09.20							
Radipole														
WEYMOUTH							09.25							

20

Weston bank - the 22 miles from Waltham to St. Denys fell at 1 in 303 and was one of the best racing stretches on the South Western. Light Pacific 34006 'Bude' runs through Winchester with a down express.

WEYMOUTH - BASINGSTOKE : WORKING TIMETABLE 1954

Train	07.17	08.52					08.10	08.52	08.50				04.10
From	Brock	Romsey					B'mouth	Romsey	Ports				Chel'ham
Class												Gds	Gds
Engine	M7 0-4-4T	T9 4-4-0	M7 0-4-4T	MN 4-6-2	WC 4-6-2	S15 4-6-0	M7 0-4-4T	T9 4-4-0	M7 0-4-4T	U 2-6-0	M7 0-4-4T	M7 0-4-4T	43XX 2-6-0
Shed	BM 405	FRA 365	SWA 421	NE 30	BM 381	FEL 102	BM 406	FRA 365	LYM 362	GUI 181	EL 305	EL 304	CHEL 21
WEYMOUTH					07.34								
Radipole													
Upwey													
Wishing Well													
Dorchester Jcn													
DORCHESTER					07.50								
Moreton													
Wool					08.02								
SWANAGE		07.36											
Corfe Castle		07.48											
Worgret Jcn		07/57			08/07								
WAREHAM		(Via Wim) **07.59**			08.15								
Holton Heath													
Hamworthy Jcn													
POOLE	08.08				08.26								
Parkstone	08.13												
Branksome	08.18												
BOURNEMOUTH WEST	08.22		08.20										
BOURNEMOUTH CENTRAL			08.28		08.35								
BOURNEMOUTH CENTRAL					08.40								
Central Goods													
Boscombe													
Pokesdown													
Christchurch													
Hinton Admiral													
New Milton													
Sway													
Lymington Pier									09.08				
Lymington Town									09.12				
Lymington Jcn								09/18	09/20				
BROCKENHURST					08/49			**09.19**	**09.22**				
Beaulieu Road													
Lyndhurst Road													
Fawley													
Totton													
Redbridge		09/02			09/08								09/24
Millbrook													
SOUTHAMPTON CENTRAL	09.08				09.13								
SOUTHAMPTON CENTRAL					09.17			09.25					09/30
Southampton T.							09.19				09.36		
Northam							09.23				09.40		09.42
Bevois Park Yard												09.45	(to
St Denys							09.27	09.33			09.44	09/48	S'ton
Swaythling							09.31	(To			09.48		Docks)
EASTLEIGH							09.36	Ports)		09.38	09.53		
EASTLEIGH					09.26					09.44			
Eastleigh Yard												(To Bitt'n)	
Shawford										09.52			
WINCHESTER					09.38					**09.59**			
Weston													
Micheldever													
Worting Jcn					10/01								
Basingstoke													
Waterloo					10.50								

EASTLEIGH EXPRESS PASSENGER DIAGRAMS

EL 251 LN 4-6-0

Arr	Station	Dep	Train
	Eastleigh loco	04.40	Light
	Southampton T	06.04	Pass
08.24	Waterloo	10.54	Pass
13.24	Salisbury	16.05	Pass
17.07	Waterloo		
	Nine Elms Gds	23.25	Goods
02.24	Eastleigh		
02.45	Eastleigh loco		

EL 252 LN 4-6-0

Arr	Station	Dep	Train
	Eastleigh loco	06.26	Light
	Eastleigh	07.22	Pass
09.01	Waterloo	11.30	Express
14.34	Bournemouth W.		
	Bournemouth C.	17.38	Pass
19.04	Eastleigh		
07.30	Eastleigh loco		

EL 253 LN 4-6-0

Arr	Station	Dep	Train
	Eastleigh loco	11.53	Light
	Eastleigh	12.23	09.45 ex W'loo
12.40	Southampton C	14.22	Pass
15.47	Bournemouth W	17.05	Express
20.23	Waterloo	22.30	Express
00.53	Southampton T		
01.48	Eastleigh loco		

EL 254 LN 4-6-0

Arr	Station	Dep	Train
	Eastleigh loco	16.40	Light
	Southampton Docks	17.30	Boat Train
19.19	Waterloo		
	Return as required		

EL 263 N15 4-6-0

Arr	Station	Dep	Train
	Eastleigh loco	21.55	
	Eastleigh	22.30	Pds
22.55	Southampton Docks		
	S'ton Terminus	01.18	22.30 Ex Waterloo
03.16	Dorchester	07.15	22.38 ex N. Elms
07.44	Weymouth	09.20	Waterloo
10.27	Bournemouth C		
	Bournemouth loco	18.00	
	Bournemouth C	18.32	15.30 ex Waterloo
18.40	Bournemouth W.	20.35	Light
	Bournemouth C	21.22	Pass
23.39	Basingstoke		
	Basingstoke loco		

EL 264 N15 4-6-0

Arr	Station	Dep	Train
	Basingstoke loco	07.08	
	Basingstoke	07.38	Pass
08.07	Reading		Light
	Moreton Cutting (Didcot)	09.40	Goods
11.30	Basingstoke		
	Basingstoke loco	14.15	
	Basingstoke	14.45	03.30 ex Ashford
15.53	Eastleigh		
	Eastleigh loco	18.12	Light
	S'ton Docks	18.55	Goods
22.40	Nine Elms		
	Nine Elms loco		

EL 265 N15 4-6-0

Arr	Station	Dep	Train
	Nine Elms loco	01.10	Light
	Clapham Jcn	02.10	Pds
03.58	Eastleigh	04.08	Pds
05.50	Fratton	05.59	Pds
06.02	Portsmouth & S		
	Fratton loco	09.17	
	Portsmouth & S	09.35	Pds
09.40	Fratton	11.05	Pds
12.06	Eastleigh		
	Eastleigh loco	16.15	Light
	S. Terminus	17.05	Pass
18.17	Bournemouth C	18.20	
18.49	Wimborne		Light
	Bournemouth W	19.55	Pds
21.56	Eastleigh		
22.15	Eastleigh loco		

EL 266 N15 4-6-0

Arr	Station	Dep	Train
	Eastleigh loco	04.15	
	Eastleigh	04.50	Goods
10.30	Bournemouth C Goods		
	Bournemouth loco	17.30	
	Bournemouth C	17.50	Pass
17.58	Bournemouth W.	18.35	Pass
21.12	Basingstoke		
	Basingstoke loco	22.40	
	Basingstoke	23.10	Pds
00.05	Eastleigh		
	Eastleigh loco		

---

so it was said – taking a German life for every mile he covered.

"The only good-a German…" he would he would often start, finishing the sentence by drawing a finger across his throat.

He was, as the saying goes, a hard case and as the two men walked down the platform their attention was attracted by a commotion coming from inside one of the coaches.

"What's the problem?" Briggs asked the platform Inspector.

"Bleedin' Teddy-boys. There's three of them messing up the first class and threatening to do in anyone who interferes. I'm off to get the Transport Police."

Briggs and the Pole looked at each other.

"Never mind-a da Police," said the Pole. "Dey just-a fill in forms. Mr Briggs, hold-a my bag."

Before anyone could say anything he climbed into the coach and disappeared into one of the compartments. Half a minute later there was a series of dull thuds and through the door window flew the three trouble-makers, one after the other and each in a beautiful trajectory that led to a pile of unconscious bodies.

The Pole emerged from the coach calmly rubbing his hands.

"There-a you are, Mr Inspector. They give-a you no more trouble."

In the end they had to be taken away, still unconscious, by ambulance and later a police inspector gave Briggs a pretty lengthy list of injuries. He added that there had been some talk of a lawsuit but it got no further than the Waterloo police: "Now then, my lad, you've got one arm in plaster. You don't want anything happening to the other….."

"Coo," breathed the Silver Fox as he received the details. "Good job for them they weren't German."

A small town of little more than thirty thousands heads, Eastleigh's position as the operating capital of the South Western owed almost everything to history and very little to origi-nating traffic and had Southampton been fully developed at the time the railway established itself in the area, it is quite likely that Eastleigh would never have grown beyond a sleepy rural halt. It was Eastleigh's good fortune that the terrain around Southampton restricted the railway to a strip of land too narrow to allow the construction of marshalling yards and the other facilities usually associated with traffic centres. Had the route from Portsmouth to Salisbury via Southampton been fully established in the early days of the LSWR, matters might have taken a different turn but as it was Eastleigh and Southampton tended to expand independently with the result that a gap of six miles came to separate the commercial and operating functions: a feature of life exacerbated in the early years of the century when the LSWR closed the running shed at Southampton and transferred its duties to Eastleigh. The third element in the equation – Eastleigh locomotive works – was in many ways incidental to the main function of the railway and had very little bearing on operational matters.

Apart from a share in the Southampton boat traffic, which was operated by unscheduled special arrangements, Eastleigh played a very limited role in express passenger work and was the largest South Western shed not to have a significant allocation of Bulleid Pacifics; having none at all until receiving a trio of rejects from the Somerset & Dorset in late 1954.

The few main line passenger turns that the shed had were worked by its eight Lord Nelson 4-6-0's for which there were three regular diagrams and one – the Channel Islands boat train – which operated only on Mondays Wednesdays and Fridays.

Two of the regular turns worked early morning business trains from Southampton Terminus (06.04) and Eastleigh (07.22) to Waterloo, the latter being of note in that it started back from Southampton Docks on Wednesdays and Fridays and on these occasions was formed of a 10 coach Pullman set instead of its normal

---

EASTLEIGH MIXED TRAFFIC DIAGRAMS

EL 270 BR 4MT 2-6-0

Arr	Station	Dep	Train
	Eastleigh loco	04.30	light
04.49	Southampton C		
	Shunt		
	Southampton C	05.26	light
	S. Terminus	05.53	Pass
06.12	Totton	06.28	Pass
07.51	Portsmouth & S	09.33	Pass
11.05	Salisbury	12.44	10.27 ex Bristol
14.06	Portsmouth & S	14.15	ECS
14.20	Fratton		
	Fratton loco	17.30	
	Portsmouth & S	18.03	
19.13	Southampton C		Light
	Eastleigh	20.22	Pass
20.39	S. Terminus	22.00	Pass
23.08	Portsmouth & S		
23.35	Fratton loco		

EL 271 BR 4MT 2-6-0

Arr	Station	Dep	Train
	Fratton Loco	04.30	Light
	Fareham	05.06	News
05.35	Portsmouth & S	06.55	Pass
08.05	Southampton C		
	Northam Shunt		
	S. Terminus	12.23	Pass
13.42	Portsmouth & S		
	Fratton Loco	15.18	Light
	Fareham	16.11	Pass
16.36	Portsmouth & S	18.45	Pass
19.52	Southampton C	20.03	Pilot
20.19	Eastleigh		
20.30	Eastleigh loco		

EL 272 BR 4MT 2-6-0

Arr	Station	Dep	Train
	Eastleigh loco	02.30	Light
	Eastleigh	03.10	23.25 ex N. Elms
03.54	S. Docks		
	Southampton C	07.09	Pass
07.15	S. Terminus	07.39	Pass
08.51	Portsmouth & S	10.34	Pass
12.18	Salisbury	13.28	10.30 ex Cardiff
15.01	Portsmouth & S	17.45	Pass
19.25	Salisbury	23.10	16.40 ex Plymouth
23.46	Eastleigh		
00.05	Eastleigh loco		

EL 273 BR 4MT 2-6-0

Arr	Station	Dep	Train
	Eastleigh loco	07.50	
	Eastleigh	08.20	Pass
10.10	Reading	12.15	Pass
14.36	Portsmouth & S	14.50	Pds
14.55	Fratton		
	Fratton Loco	16.40	
	Portsmouth & S	17.17	Pass
20.09	Reading	23.00	Pass
23.29	Basingstoke	00.37	19.35 ex Ken' Pds
01.40	Eastleigh		
02.00	Eastleigh loco		

EL 277 BR 4MT 2-6-0

Arr	Station	Dep	Train
	Eastleigh loco	05.35	Light
	Winchester	06.36	Pass
07.09	S. Terminus	07.32	Pass
10.20	Didcot	15.35	Pass
18.19	S. Terminus		
	Shunt		
	S. Terminus	20.58	Light
21.16	Eastleigh loco		

EL 278 BR 4MT 2-6-0

Arr	Station	Dep	Train
	Eastleigh loco	06.25	Light
	Eastleigh	06.55	ECS
07.08	Winchester Chesil	07.15	Pass
08.25	Newbury	08.27	Pass
09.08	Reading		
	Reading GW loco	15.20	
	Reading	15.45	Pass
16.10	Didcot	17.55	Pass
18.41	Newbury	19.22	Pass
20.59	S. Terminus		Light
	S. Docks	22.00	Goods
22.18	Northam Yard		Light
23.05	Eastleigh loco		

EL 315 BR 4MT 2-6-0

Arr	Station	Dep	Train
	Eastleigh loco	07.30	
	Eastleigh	08.05	Goods
11.42	Gosport	14.00	Goods
16.53	Eastleigh		
17.10	Eastleigh loco		

EL 316 BR 4MT 2-6-0

Arr	Station	Dep	Train
	Eastleigh loco	03.10	Light
	Eastleigh	03.45	Goods
04.49	Fratton		
	Portsmouth & S	06.23	Pass
07.40	S. Terminus	08.33	Pilot
08.50	Eastleigh	09.30	Goods
12.05	Milford	17.00	Goods
17.35	Salisbury	19.30	Light
	Andover Jcn	20.50	15.20 ex C'ham
22.25	Eastleigh		
22.45	Eastleigh loco		

EL 321 BR 4MT 2-6-0

Arr	Station	Dep	Train
	Eastleigh loco	04.30	Light
	Eastleigh	05.06	Goods
07.10	Gosport	09.30	Light
	Fratton loco	11.49	Light
	Portsmouth & S.	12.15	Pass
12.35	Fareham	13.10	Goods
15.45	Bevois Park	16.12	Goods
16.22	S. Terminus	17.38	Pass
18.13	Winchester Chesil		Light
18.40	Eastleigh loco		

Merchant Navy 35029 ' Ellerman Lines' approaches Eastleigh with the up Bournemouth Belle and prepares for the long climb to Worting Junction..

EASTLEIGH GOODS DIAGRAMS

EL 310 H15 4-6-0

Arr	Station	Dep	Train
	Eastleigh Loco	16.55	
	Eastleigh	17.27	Pass
18.14	Portsmouth & S		
	Fratton loco	20.30	
	Fratton	20.45	Goods
22.11	Eastleigh	23.15	Goods
00.42	Basingstoke		
	Basingstoke loco	10.15	
	Basingstoke	10.48	01.30 Hoo Jcn
12.36	Eastleigh		
13.00	Eastleigh loco		

EL 311 H15 4-6-0

Arr	Station	Dep	Train
	Eastleigh loco	02.50	
	Bevois Park	03.58	Goods
05.37	Salisbury		
	Salisbury loco	08.20	
	Salisbury East	08.45	Goods
10.08	Eastleigh		
	Eastleigh loco	12.45	
	Eastleigh	13.15	Pds
16.24	Clapham Jcn		
	Nine Elms	19.25	Goods
23.32	S'ton Docks		Light
00.17	Eastleigh loco		

EL 312 H15 4-6-0

Arr	Station	Dep	Train
	Eastleigh loco	09.15	
	Bevois Park	10.23	Goods
12.52	Salisbury		
	Salisbury loco	15.45	
	Salisbury	16.18	Goods
17.52	Eastleigh		
18.20	Eastleigh loco		

EL 313 H15 4-6-0

Arr	Station	Dep	Train
	Eastleigh loco	23.05	
	Eastleigh	23.35	Goods
03.55	Bournemouth C Goods		
	Bournemouth loco	06.30	
	Bournemouth W	07.30	Pass
09.11	Eastleigh	11.28	Pass
13.24	Bournemouth C	16.38	Pass
18.08	Eastleigh		
18.20	Eastleigh loco		

EL 314 Q1 0-6-0

Arr	Station	Dep	Train
	Eastleigh Loco	04.54	
	Bevois Park	05.35	Goods
08.47	Didcot	09.50	Goods
12.47	Eastleigh		
	Eastleigh loco	14.15	
	Eastleigh	14.48	Goods
18.10	Didcot	19.00	Goods
21.43	Eastleigh		
22.15	Eastleigh Loco		

EL 318 Q1 0-6-0

Arr	Station	Dep	Train
	Eastleigh Loco	08.30	
	Eastleigh	09.15	Goods
10.13	Brockenhurst	11.20	Goods
	Via Wimborne		
12.58	Poole		Light
	Hamworthy Jcn	19.56	Goods
21.02	Bournemouth C Goods		
21.30	Bournemouth loco		

EL 319 Q1 0-6-0

Arr	Station	Dep	Train
	Bournemouth loco	06.50	
	Bournemouth C Goods	07.20	Goods
07.39	Parkstone	08.08	Goods
08.18	Poole	08.50	Goods
09.00	Hamworthy Jcn	11.20	Goods
11.29	Hamworthy Goods	12.10	Goods
12.19	Hamworthy Jcn	14.35	Goods
	Via Wimborne		
20.29	Eastleigh		
	Eastleigh loco	23.50	
	Eastleigh	00.20	Goods
01.45	Salisbury		
	Salisbury loco	03.45	
	Salisbury East	04.25	Goods
06.02	Northam Yard		Light
06.36	Eastleigh loco		

EL 320 Q1 0-6-0

Arr	Station	Dep	Train
	Eastleigh loco	09.10	
	Eastleigh	09.40	Goods
11.04	Fratton	13.35	Goods
14.45	Eastleigh		
15.10	Eastleigh loco		

EL 317 Q 0-6-0

Arr	Station	Dep	Train
	Eastleigh Loco	23.50	
	Northam Yard	00.36	Goods
01.20	Brockenhurst		
	Yard pilot		
	Brockenhurst	06.15	Goods
13.05	Dorchester		
	Dorchester loco	16.10	
	Dorchester	16.25	Goods
22.22	Bevois Park		
23.35	Eastleigh loco		

EL 322 Q 0-6-0

Arr	Station	Dep	Train
	Eastleigh loco	05.15	
	Eastleigh	06.00	Goods
06.58	Bevois Park	07.25	Goods
12.43	Fareham	13.30	Pds
13.42	Gosport	17.35	Goods
18.00	Fareham	19.55	Goods
20.35	Eastleigh		
21.00	Eastleigh loco		

EL 323 Q 0-6-0

Arr	Station	Dep	Train
	Eastleigh loco	06.10	
	Northam Yard	06.52	Goods
09.40	Totton	12.24	Goods
14.08	Eastleigh		
14.30	Eastleigh loco		

EL 324 Q 0-6-0

Arr	Station	Dep	Train
	Eastleigh loco	20.45	
	Eastleigh	21.20	Goods
22.35	Brockenhurst	23.59	Goods
02.47	Dorchester		
	Dorchester loco	06.15	
	Dorchester	07.00	Pass
	Via Wimborne		
08.50	Brockenhurst	09.32	Pass
10.12	Wimborne	11.12	Pass
11.41	Bournemouth W.		
12.08	Bournemouth loco		

EL 325 Q 0-6-0

Arr	Station	Dep	Train
	Bournemouth loco	02.50	
	Bournemouth C Goods	03.20	Goods
04.53	Bevois Park	05.51	Goods
06.02	Woolston		light
	Bevois Park pilot	21.04	light
21.19	Eastleigh loco		

EL 326 '700' 0-6-0

Arr	Station	Dep	Train
	Eastleigh Loco	05.45	
	Shunt East Yard		
	Eastleigh	08.44	Goods
09.34	Winchester Chesil	10.53	Goods
10.58	Winnd Sdgs	11.25	Goods
11.30	Winchester Chesil	13.00	
13.20	Eastleigh	13.28	Light
	Totton	18.12	Goods
19.00	Millbrook		Light
19.41	Eastleigh loco		

EL 327 '700' 0-6-0

Arr	Station	Dep	Train
	Eastleigh Loco	08.00	
	Eastleigh	08.33	Goods
11.53	Alton	13.34	Goods
16.24	Eastleigh		
17.00	Eastleigh loco		

EL 332 Z 0-8-0T

Arr	Station	Dep	Train
	Eastleigh Loco	08.15	
	Allbrook Pilot		
06.00	Eastleigh Loco		

EL 338 E4 0-6-2T

Arr	Station	Dep	Train
	Eastleigh Loco	09.15	Light
	Bevois Park pilot		
07.47	Eastleigh Loco		

EL 339 E4 0-6-2T

Arr	Station	Dep	Train
	Northam Yard pilot		

EL 340 '0395' 0-6-0

Arr	Station	Dep	Train
	Eastleigh Loco	07.16	
	Loco pilot		
18.15	Eastleigh Loco		

EL 341 E4 0-6-2T

Arr	Station	Dep	Train
	Eastleigh Loco	07.15	
	Loco pilot		
17.15	Eastleigh Loco		

For many years the most powerful class allocated to the Wessex region was the West Country Pacific, eleven of the class being based at Bournemouth to cover six daily diagrams.   Expectations that the class would quickly take over all express workings from Bournemouth did not materialise and Lord Nelson and King Arthur 4-6-0's remained in the front line until rebuilding eliminated many of the faults of the unrebuilt engines.   Local class 8 workings were covered by Nine Elms Merchant Navy 4-6-2's until 1954 when a pair of the latter were transferred to Bournemouth.   This total grew to seven engines following the reintroduction of the 120-minute timing in 1957.   34007 of Nine Elms stands on Eastleigh loco.

COMBINED WORKING TIME TABLE AND ENGINE WORKINGS : 1954

Train		04.07	06.23			05.40	09.25					05.30	04.50	
From		Feltham	Ports			W'loo	M. Ntn					ELGH	Elgh	
Class	Gds	Gds	Gds		S&D							GDS	Gds	
Engine	BR3 2-6-2T	S15 4-6-0	BR4 2-6-0	Q1 0-6-0	4F 0-6-0	LN 4-6-0	14xx	WC 4-6-2	M7 0-4-4T	M7 0-4-4T	M7 0-4-4T	H15 4-6-0	T9 4-4-0	N15 4-6-0
Shed	EL 328	FEL 103	EL 316	EL 319	BK72	NE 31	WEY	BM 386	BM 407	BM 408	LYM 362	NE 79	EL 286	EL 266
WATERLOO						05.40								
Woking														
Basingstoke						06.53								
Worting Jcn		06/35				06/58								
Micheldever						07.10								
Wallers Ash														
WINCHESTER		07/07				07.26								
Shawford														
Eastleigh Yard	07.15	07.30												
EASTLEIGH						07.36								
EASTLEIGH	07/20					07.39						07.46		
Swaythling												07.52		
St Denys	07/30		07.32									07.56		
Bevois Park Yard														
Northam			07.36											
Southampton (T)			07.40											
SOUTHAMPTON CENT						07.49						08.01		
SOUTHAMPTON CENT	07/42					07.55						08.07		
Millbrook												08.11		
Redbridge	07/50											08.16		
Totton	07.55											08.19	08.25	
Lyndhurst Rd	(To											08.26	08.34	
Beaulieu Rd	Fawley)											08.32		
BROCKENHURST						08.18				08.34	08.41	08.44		
Lymington Jcn						08/21				08/37	08/44	08/47		
Lymington Town											08.52			
Lymington Pier											08.54			
Sway										(to		08.51		
New Milton						08.30				Poole		08.58		09.19
Hinton Admiral										via		09.05		09.38
Christchurch						08.40				B'stone)		09.12		09.47
Pokesdown						08.46						09.18		
Boscombe						08.50						09.21		
Central Goods														
BOURNEMOUTH CENT						08.53						09.24		
BOURNEMOUTH CENT								08.59						
BOURNEMOUTH WEST				08.48										
Branksome				08.53				09.06						
Parkstone				08.57				09.10						
POOLE			08.50		09.03			09.17						
Hamworthy Jcn			09.00		(To			09.22						
Holton Heath					Bath)									
WAREHAM								09.31	09.47					
Worgret Jcn								09/34	09/50					
Corfe Castle									10.00					
SWANAGE									10.09					
Wool								09.40						
Moreton														
DORCHESTER								09.56						
Dorchester Jcn							09/42	09/58						
Monkton							09.45							
Wishing Well							09.50							
Upwey							09.53							
Radipole							09.57							
WEYMOUTH							10.00	10.07						

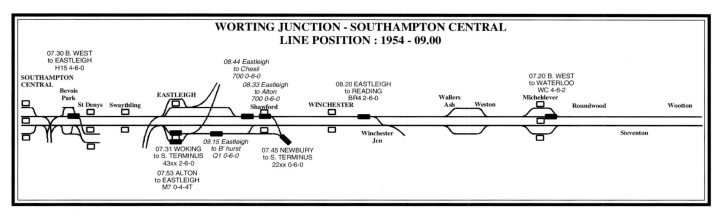

WORTING JUNCTION - SOUTHAMPTON CENTRAL
LINE POSITION : 1954 - 09.00

nine-car rake of SR stock. The provision of a Lord Nelson for the 06.04 was generous since the train was essentially a Basingstoke – Waterloo stopping train extended back for the sake of any early morning business south of Worting but the 07.22 was an express within the meaning of the act and after calling at Shawford, Winchester and Micheldever, ran non-stop from Basingstoke to Waterloo in the very respectable time of fifty-one minutes, splendidly overhauling the Pacific-hauled 06.45 Salisbury – Waterloo at Winchester in the process.

Neither of these services had a recognisable return working – the stock of the 06.04 dissolved into a number of local services whilst the 07.22 formed the 21.00 Channel Islands boat train – which caused some head-scratching amongst the diagrammers as to how to get the two Lord Nelsons back to Eastleigh since all the down Bournemouth expresses had been shared between Nine Elms and Bournemouth. The easy way out was to let the engines sit all day on Nine Elms before working back with a pair of night goods trains but this would have been at the expense of engine mileage. The problem was not helped by the fact that, apart from the 09.54 ex Waterloo, there were no slow trains on the Wessex line as there were between Waterloo and Salisbury.

In the end the decision was taken to break the reign of Pacifics on the Bournemouth services by returning the engine off the 07.22 on the 11.30 express to Bournemouth, Eastleigh being regained on an up stopping train. The other Eastleigh Lord Nelson – the engine that worked into London with the 06.04 – was returned on the 23.25 express goods from Nine Elms but spent the intervening fifteen hours by filling in on a semi-fast working to Salisbury and back; achieving thereby the distinction of being the only engine in the Eastleigh motive power district to be booked over the Basingstoke – Salisbury section.

It is a sad fact that the history of South Western locomotive development was one of

WEYMOUTH - BASINGSTOKE : WORKING TIMETABLE 1954

Train		07.53	07.15										07.28	08.34
From		BM	Sarum										T'combe	Brock
Class		Gds						Gds	Gds					
Engine	WC 4-6-2	700 0-6-0	T9 4-4-0	LN 4-6-0	Hall 4-6-0	M7 0-4-4T	T9 4-4-0	HALL 4-6-0	M7 0-4-4T	H15 4-6-0	T9 4-4-0	N15 4-6-0	2P 4-4-0	M7 0-4-4T
Shed	BM 382	BM 417	SAL 443	BM 395	WEY	EL 299	EL 279	RDG 66	SWA 421	EL 312	BM 404	BM 399	TCB 57	BM 408
WEYMOUTH					08.15									
Radipole														
Upwey														
Wishing Well														
Monkton														
Dorchester Jcn					08/28									
DORCHESTER				08.04	(To									
Moreton				08.13	Bristol)									
Wool				08.20										
SWANAGE									08.58					
Corfe Castle									09.09					
Worgret Jcn				08/24					09/18					
WAREHAM				08.28					09.20					
Holton Heath				08.34										
Hamworthy Jcn				08.39										
POOLE			08.37	08.45								09.14	09.38	
Parkstone			08.42	08.50								09.20		
Branksome			08.46	08.54								09.26		
BOURNEMOUTH WEST	08.35		08.50										09.20	09.30
BOURNEMOUTH CENTRAL	08.43			08.59								09.28		
BOURNEMOUTH CENTRAL	08.46											09.32		
Central Goods														
Boscombe	08.50											09.37		
Pokesdown	08.53											09.41		
Christchurch	08.59			09.10								09.47		
Hinton Admiral	09.06			09.19										
New Milton	09.14											09.59		
Sway	09.21													
Lymington Pier														
Lymington Town														
Lymington Jcn	09/25											10/07		
BROCKENHURST	09.28											10.11		
Beaulieu Road														
Lyndhurst Road	09.40													
Fawley														
Totton	09.46													
Redbridge	09/48										10.08	10/26		
Millbrook											10.15			
SOUTHAMPTON CENTRAL	09.52											10.31		
SOUTHAMPTON CENTRAL	09.56										10/21	10.35		
Southampton T.								10.13						
Northam								10.17						
Bevois Park Yard											10.23	10.30		
St Denys	10.03							10.21			10/27			
Swaything								10.24						
EASTLEIGH	10.10							10.29						
EASTLEIGH	10.12					10.16	10.22	10.31		10/37		10/45		
Eastleigh Yard														
Shawford						10.24	10.30	10.39		(To S'bury)				
WINCHESTER	10.26					10.33		10.47				10.59		
Weston						(to	(To							
Micheldever	10.42					Alton)	Newbury)	11.03						
Worting Jcn	10/53							11.13				11/22		
Basingstoke	10.57							11.17				11.26		
Waterloo	11.55							(Reading)				(B'head)		

Eastleigh-based U 2-6-0 31791 works a goods train across Canute Road and into Southampton Docks. From 1952 Eastleigh acquired a sizeable allocation of these engines although its diagrammed work tended to favour the BR Standard 4MT locomotives. 31791 was a rebuild of a 'River' 2-6-4T and spent many years working from Yeovil.

EASTLEIGH T9 4-4-0 DIAGRAMS

Note: 284. 366 and 403 were worked jointly with Fratton and Bournemouth sheds.

EL 227 T9 4-4-0

Arr	Station	Dep	Train
	Eastleigh loco	09.50	
	Eastleigh	10.22	Pass
11.40	Newbury	12.25	Pass
13.41	Eastleigh		
14.00	Eastleigh loco		

EL/ 280 T9 4-4-0

Arr	Station	Dep	Train
	Eastleigh loco	05.05	
	Eastleigh	05.28	Pds
08.18	Bournemouth W.	09.28	Light
10.40	Millbrook	11.00	Trip
11.20	S. Docks		Light
	Eastleigh loco	16.45	
	Eastleigh	18.30	17.15 ex Fawley
19.45	Andover Jcn		
20.00	Andover Jcn loco		

EL 246 T9 4-4-0

Arr	Station	Dep	Train
	Andover Jcn loco	05.00	
	Station pilot		
	Andover Jcn	09.37	Goods
10.14	Fullerton	10.40	Goods
11.05	Long Parish	11.45	Goods
12.00	Fullerton	12.14	Goods
14.16	Romsey	14.35	Goods
14.55	Nursling	15.14	Goods
15.24	Romsey	18.05	F
18.39	Eastleigh		
19.00	Eastleigh loco		

EL 282 T9 4-4-0

Arr	Station	Dep	Train
	Eastleigh loco	09.34	Light
	Eastleigh	09.55	Pass
10.10	S. Terminus	10.29	Pass
11.51	Bournemouth C	12.14	09.40 ex Brighton
12.22	Bournemouth W.		
	Bournemouth loco	15.40	
	Bournemouth C	15.56	14.20 ex Weymouth
17.22	Eastleigh	17.42	Pass
18.45	Andover Jcn	19.35	Pass
20.36	Eastleigh	21.51	Pass
22.08	S. Terminus		
	S. Docks	00.01	Goods
01.05	Eastleigh		
01.20	Eastleigh loco		

EL 283 T9 4-4-0

Arr	Station	Dep	Train
	Eastleigh loco	04.05	Light
	S. Terminus	05.15	Pass
06.36	Portsmouth & S		
	Fratton loco	07.22	
	Portsmouth & S	08.07	Pass
09.15	S. Terminus		
	Dock trips		
	S. Terminus	12.00	Pass
12.15	Eastleigh		
	Eastleigh loco	14.05	
	Eastleigh	14.35	Pass
17.02	Bournemouth C		
17.25	Bournemouth loco		

EL 284 T9 4-4-0

Arr	Station	Dep	Train
	Bournemouth loco	05.10	Light
05.38	Broadstone	07.14	Pass
08.19	S.disbury	09.25	Pass
10.57	Bournemouth W	12.20	Pass
12.28	Bournemouth C		Light
	Bournemouth W	13.20	Pass
14.59	S.disbury	15.05	ECS
	East sidings		
	S.disbury loco	16.35	
	S.disbury	17.07	Pass
19.09	Portsmouth & S.	20.03	Pass
20.52	Eastleigh		
21.05	Eastleigh loco		

EL 286 T9 4-4-0

Arr	Station	Dep	Train
	Eastleigh loco	04.45	
	Eastleigh	05.35	Goods
11.24	Brockenhurst	12.36	Goods
12.56	Lymington Pier		
	Shunt		
	Lymington Pier	15.00	Light
	Brockenhurst	16.08	Pass
17.50	Portsmouth & S.		
	Fratton loco	21.08	
	Portsmouth & S.	21.38	Pass
22.41	S. Terminus	23.15	Light
	Eastleigh	01.46	Pass
02.28	Portsmouth & S.		
03.20	Fratton loco		

EL 366 T9 4-4-0

Arr	Station	Dep	Train
	Fratton loco	10.40	
	Fratton	11.00	ECS
11.10	Portsmouth Harbour	11.37	Pass
12.05	Fareham	13.29	Pass
14.07	Southampton C.		Light
	S. Terminus	14.55	Pass
15.12	Eastleigh	15.30	Light
	S. Terminus	18.07	Pass
19.30	Bournemouth C		
19.50	Bournemouth loco		

EL 403 T9 4-4-0

Arr	Station	Dep	Train
	Bournemouth loco	05.45	Light
	Bournemouth C	06.40	ECS
06.52	Lymington	08.10	Pass
08.41	Brockenhurst	09.10	Light
	Redbridge	10.08	Goods
10.30	Bevois Park		Light
	S. Terminus	11.40	Pds
11.52	Eastleigh		
	Eastleigh loco	16.49	Light
	S. Terminus	17.28	Pass
18.37	Portsmouth & S	19.45	Pass
22.04	Andover Jcn		
22.15	Andover Jcn loco		

EL 248 T9 4-4-0

Arr	Station	Dep	Train
	Andover Jcn loco	06.20	
	Andover Jcn	06.45	Pass
07.58	S. Terminus	08.33	Pass
09.52	Andover Jcn	11.25	Pass
12.18	Eastleigh	13.05	Pass
13.49	Portsmouth & S		
	Fratton Loco	15.07	
	Portsmouth & S	15.45	Pass
16.49	Romsey	17.50	Pass
18.06	Eastleigh	19.22	Pass
20.07	Portsmouth & S	21.48	Pass
22.33	Eastleigh		
22.50	Eastleigh loco		

too many questionable designs interspersed with a few that deserved superlatives. The latter included the seventy-four 'King Arthur' N15 4-6-0's which, after a shaky start, had, together with the T9 4-4-0's, become the backbone of the system for twenty years until the advent of the Bulleid Pacifics.

For the most part Eastleigh's allocation of twelve N15's was held in reserve for Boat trains and summer holiday requirements although one of the class – 30788 'Sir Urre of the Mount' always seemed to be the Eastleigh contribution - worked in a three-day cyclic working balanced by engines from Nine Elms and Basingstoke.

The duties involved were a mixed bag but included some express work, the engine taking over the 22.30 Waterloo – Dorchester from a Lord Nelson at Southampton Terminus and working the train to its destination before proceeding forward to Weymouth with the 22.38 night goods from Nine Elms. Further express work followed as the engine made its way eastward with the 09.20 Weymouth – Waterloo as far as Bournemouth; the day ending with a trip to Basingstoke with the 19.48 Weymouth – Reading stopping train.

The second day of the diagram was mainly concerned with goods work although the engine had the distinction of appearing on Great Western metals as it ran light from Reading to Moreton Cutting Yard, Didcot, to work a goods service back to Basingstoke. In the afternoon it took over the 03.30 Ashford (SECR) to Eastleigh but instead of returning to the fold, was turned and sent out for an evening express goods from Southampton Docks to Nine Elms.

Very little time was spent in London and within a couple of hours of arriving on Nine Elms shed, the engine was on its way again, heading west with the 02.10 Clapham Junction to Portsmouth parcels via Eastleigh. It left the Portsmouth area for Eastleigh with the empty vans but before being allowed to finish the cycle of work was sent out for the evening stopping train from Southampton Terminus to Wimborne, returning from Bournemouth West with a late night parcels train.

A second (or fourth) Eastleigh N15 working covered an early morning stopping goods train to Bournemouth, the 18.35 slow train from Bournemouth West to Reading as far as Basingstoke and the 16.55 Worcester – Eastleigh Parcels.

The very small proportion of Eastleigh's King Arthurs in regular daily use suggested that the class was living on borrowed time although those who subscribed to that idea where guilty of being mislead. The Southern had for many years geared its resources to maximum usage rather than an optimum and at the height of the season, the N15's (and other engines) which had been taken out of traffic and placed in store, were returned to traffic with a vengeance and for the peak summer Saturdays of July, August and September, Eastleigh was required to place its entire allocation of N15's in traffic; the shed having to cover twelve diagrams, many of them for London workings, with twelve locomotives: a requirement that called for 100% availability!

If the King Arthurs were under utilised during the winter the same cannot have been said of the T9 4-4-0's; seven of the shed's ten examples being at work each day. The work shouldered by the class was purely local, ranging from Bournemouth in the south to Andover Junction in the north and Salisbury in the west. As was appropriate for a 6' 7" engine, most of the work was on passenger services although one duty consisted entirely of local goods work

EASTLEIGH M7 0-4-4T DIAGRAMS

EL 298 M7 0-4-4T (P & P fitted)

Arr	Station	Dep	Train
	Eastleigh loco	04.45	
	Eastleigh	05.17	Pass
05.37	S. Terminus	06.43	Pass
07.00	Eastleigh	07.56	Pass
08.54	Andover Jcn	09.30	Pass
10.24	Eastleigh		
	Eastleigh loco	12.45	
	Eastleigh	13.15	Pass
14.13	Alton	14.30	Pass
15.30	Eastleigh	15.59	Pass
17.00	Alton	17.05	Pass
18.09	Eastleigh	19.20	Pass
19.35	Romsey	19.50	ECS
20.05	Eastleigh		
20.20	Eastleigh loco		

EL 299 M7 0-4-4T (P & P fitted)

Arr	Station	Dep	Train
	Eastleigh loco	07.10	
	Eastleigh	07.40	Pass
08.41	Alton	08.55	Pass
09.49	Eastleigh	10.16	Pass
11.16	Alton	12.05	Pass
13.00	Eastleigh		
	Eastleigh loco	16.25	
	Eastleigh	16.55	Pass
17.52	Alton	18.02	Pass
19.37	S. Terminus	21.05	Pass
21.22	Eastleigh		
21.40	Eastleigh loco		

EL 297 M7 0-4-4T

Arr	Station	Dep	Train
	Eastleigh loco	05.45	Pass
	Eastleigh	06.20	Pass
06.37	S. Terminus		
	Dock trips		
	S. Terminus	13.23	Pass
13.57	Winchester	14.10	Pass
14.23	Eastleigh		
	Eastleigh	15.35	
	Eastleigh	16.06	Pass
16.23	S. Terminus	17.23	Pass
18.49	Alton	19.25	Pass
20.25	Eastleigh		
20.45	Eastleigh loco		

EL 297 M7 0-4-4T

Arr	Station	Dep	Train
	Eastleigh loco	05.30	
	Eastleigh	06.07	Pass
06.23	Romsey	06.43	Pass
06.57	Eastleigh	07.52	Pass
08.12	Fareham	08.36	Light
	Botley	09.40	Goods
09.55	B. Waltham	10.55	Goods
11.10	Botley	12.25	Goods
12.45	Eastleigh		
	Eastleigh loco	15.20	
	Eastleigh	15.51	Pass
16.55	Fawley	17.16	Pass
18.19	Eastleigh		
18.35	Eastleigh loco		

EL 302 M7 0-4-4T

Arr	Station	Dep	Train
	Eastleigh loco	04.30	Light
	Southampton C	05.22	Pass
05.41	Eastleigh	06.04	Pass
06.21	Winchester	06.58	Pass
07.30	S. Terminus	07.45	Light
	Eastleigh loco	13.45	
	Eastleigh	14.15	Pass
15.18	Alton	16.10	Pass
17.24	S. Terminus	19.40	Pass
20.13	Winchester	22.34	Pass
23.03	Southampton C	23.15	Light
	S. Terminus		
	S. Terminus	01.45	Light
02.03	Eastleigh loco		

EL 303 M7 0-4-4T

Arr	Station	Dep	Train
	Eastleigh loco	06.05	
	Eastleigh	06.35	Pass
07.34	Alton	07.53	Pass
08.50	Eastleigh		
	Eastleigh	16.45	
16.55	Eastleigh loco		

EL 304 M7 0-4-4T

Arr	Station	Dep	Train
	Eastleigh loco	06.10	
	Eastleigh	06.20	Pass
07.21	Fareham	08.02	Pass
08.42	Southampton C	09.06	Light
	Bevois Park	09.45	Goods
09.53	Bitterne	10.25	Goods
10.36	Bevois Park	10.50	Light
	Southampton C		Shunt
14.30	S. Terminus		Shunt
	S. Terminus	17.15	Pilot
17.21	Southampton C	17.25	Light
	Fawley	18.43	Pass
19.31	S. Terminus	20.18	Pass
20.36	Eastleigh loco		

EL 305 M7 0-4-4T

Arr	Station	Dep	Train
	Eastleigh loco	06.00	
	Eastleigh	06.35	Pass
07.43	Fawley	08.06	Pass
08.55	S. Terminus	09.36	Pass
09.53	Eastleigh		
	Eastleigh loco	13.40	
	Eastleigh	14.09	Goods
14.41	Redbridge	15.42	light
	S. Terminus	16.25	Pass
16.42	Eastleigh	17.16	Pass
17.31	Romsey	17.50	Pilot
18.06	Eastleigh		
	Station pilot		
01.40	Eastleigh loco		

EL 307 M7 0-4-4T

Arr	Station	Dep	Train
	Eastleigh loco	04.10	
	Station pilot		
03.10	Eastleigh loco		

EL 308 M7 0-4-4T

Arr	Station	Dep	Train
	Eastleigh loco	05.54	Light
07.49	Fareham	09.40	Light
	Eastleigh	11.10	Pass
11.25	Romsey	11.35	Pass
	Eastleigh	13.15	Pass
13.32	S. Terminus	13.56	Pass
14.13	Eastleigh		
14.30	Eastleigh loco		

EL 329 M7 0-4-4T

Arr	Station	Dep	Train
	Eastleigh loco	06.10	
	Eastleigh	06.40	Pilot
07.05	Romsey	10.35	Goods
11.05	Eastleigh	12.24	Goods
13.40	Romsey	14.05	Pass
14.20	Eastleigh		
	Eastleigh loco	15.55	Light
	Winchester	16.25	Pass
16.38	Eastleigh	17.34	Pass
18.18	Brockenhurst	19.45	Light
20.28	Eastleigh loco		

EASTLEIGH LOCAL DIAGRAMS

EL 306 2MT (LMS) 2-6-2T

Arr	Station	Dep	Train
	Eastleigh Loco	05.00	
	Eastleigh	05.39	Pass
05.59	Fareham	06.30	Pass
06.50	Eastleigh	07.30	Pass
07.44	Romsey	08.00	Pass
08.15	Eastleigh	09.03	08.32 ex Romsey
09.53	Portsmouth & S	11.19	Pass
13.27	Andover Jcn		
	Station Pilot		
20.10	Andover Jcn loco		

EL 247 2MT (LMS) 2-6-2T

Arr	Station	Dep	Train
	Andover Jcn loco	06.30	
	Andover Jcn	06.47	Pass
07.12	Tidworth	07.50	Pass
08.14	Andover Jcn	10.35	Pass
10.57	Tidworth	12.15	Mixed
12.22	Ludgershall	12.35	Goods
13.40	Ludgershall	14.00	Pass
14.15	Andover Jcn		
14.30	Andover Jcn loco	15.45	
	Andover Jcn	16.12	Pass
16.50	Romsey	17.03	Pass
17.35	S. Terminus	18.58	Pass
20.18	Alton	20.38	Pass
21.34	Eastleigh		
21.50	Eastleigh loco		

EL 309 L1 4-4-0

Arr	Station	Dep	Train
	Eastleigh Loco	06.10	
	Eastleigh	06.40	Goods
10.45	Andover Jcn		
	Yard Pilot		
	Andover Jcn	15.11	Light
	Andover Town	17.00	Goods
17.05	Andover Jcn	18.40	Pass
19.50	Eastleigh		
20.05	Eastleigh Loco		

EL 328 BR 3MT 2-6-2T

Arr	Station	Dep	Train
	Eastleigh Loco	06.45	
	Eastleigh	07.15	Goods
09.40	Fawley	12.30	Goods
14.00	Bevois Park	14.33	Light
	Marchwood	16.40	Goods
16.55	Totton	17.55	Light
	Fawley	19.05	Pilot
22.13	Eastleigh		
22.25	Eastleigh Loco		

EL 330 BR 3MT 2-6-2T

Arr	Station	Dep	Train
	Eastleigh Loco	07.20	
	Eastleigh	07.56	Goods
11.38	Fawley	13.15	Goods
14.35	Bevois Park	14.50	Light
15.05	Eastleigh Loco	16.20	Light
	Millbrook		
	Shunt		
	Millbrook	20.28	Goods
20.41	Bevois Park	21.12	Light
	Southampton C.	22.04	
22.20	Eastleigh		
22.35	Eastleigh loco		

EL 331 BR 3MT 2-6-2T

Arr	Station	Dep	Train
	Eastleigh Loco	03.45	
	Eastleigh	04.36	Pass
05.18	Portsmouth Hbr	05.40	ECS
05.50	Fratton		
	Fratton Loco	07.15	
	Fratton	07.30	ECS
07.35	Portsmouth & S.	07.45	Vans
07.50	Fratton	08.13	ECS
08.18	Portsmouth & S.	08.50	Pass
09.38	Eastleigh		
	Eastleigh Loco	11.25	
	Eastleigh	11.55	Goods
12.14	Northam Yard	13.20	Goods
15.50	Fawley	19.05	Goods
22.13	Eastleigh		
22.25	Eastleigh Loco		

EASTLEIGH MPD : ENGINE ARRIVALS/DEPARTURES

Train	On Shed	Engine	Shed	Off Shed	To Work
18.42 Waterloo Pds	22.20	H15 4-6-0	N. Elms 65	00.00	00.45 Felthamgoods
18.05 Netley - Fareham	19.25	U 2-6-0	Fratton 363	00.40	01.00 Portsmouth Pds
19.10 Bristol Pass	23.15	S15 4-6-0	Salisbury 467	01.15	02.10 B. Park - Salisbury Goods
14.45 Feltham - S. Docks	21.30	S15 4-6-0	Feltham 108	02.20	Light to B. Park
16.40 Plymouth Pass	00.00	BR 4MT 2-6-0	Eastleigh 272	02.30	03.10 S' Docks Goods
01.23 S. Terminus Pass	01.50	N15 4-6-0	Bournemouth 399	02.35	03.05 Bournemouth Pds
19.25 N. Elms - S. Docks	00.17	H15 4-6-0	Eastleigh 311	02.50	03.58 B. Park - Salisbury Goods
15.20 Cheltenham Goods	22.40	BR 4MT 2-6-0	Eastleigh 316	03.10	03.45 Fratton Goods
19.05 Fawley Goods	22.30	BR 3MT 2-6-2T	Eastleigh 331	03.45	04.36 Portsmouth Pass
20.03 Portsmouth Pass	21.10	T9 4-4-0	Eastleigh 283	04.05	Light to S. Terminus @ 05.15 Portsmouth
07.40 Clapham Jcn Pds	11.30	H15 4-6-0	N. Elms 66	04.10	04.40 Felthamgoods
16.55 Worcester Pds	00.20	N15 4-6-0	Eastleigh 266	04.15	04.50 Bournemouth Goods
22.20 Feltham - S. Docks	03.13	S15 4-6-0	Feltham 109	04.25	Light to B. Park
20.03 Southampton Pass	20.30	BR 4MT 2-6-0	Eastleigh 270	04.30	Light to S'ton Central
01.45 Light ex S. Terminus	02.03	M7 0-4-4T	Eastleigh 302	04.30	Light to S. Terminus @ 05.22 Eastleigh
17.38 S. Terminus - Winchester	18.40	BR 4MT 2-6-0	Eastleigh 321	04.30	05.06 Gosport Goods
21.48 Portsmouth Pass	22.45	T9 4-4-0	Eastleigh 286	04.45	05.35 Brockenhurst Goods
19.50 Romsey ECS	20.15	M7 0-4-4T	Eastleigh 298	04.45	05.17 S. Terminus Pass
23.25 Nine Elms Goods	02.35	LN 4-6-0	Eastleigh 251	04.49	06.04 S. Terminus - Waterloo
19.00 Didcot Goods	22.00	Q1 0-6-0	Eastleigh 314	04.54	05.35 Didcot Goods
20.38 Alton Pass	21.45	LM 2MT 2-6-2T	Eastleigh 306	05.00	05.39 Fareham Pass
18.05 Romsey goods	19.00	T9 4-4-0	Eastleigh 280	05.05	05.28 Bournemouth Pds
16.25 Dorchester - B. Park	23.35	Q0-6-0	Eastleigh 322	05.15	06.00 Bevois Park Goods
02.00 Fratton Goods	02.15	U 2-6-0	Fratton 363	05.20	06.00 Havant Goods
17.16 Fawley Pass	18.30	M7 0-4-4T	Eastleigh 300	05.30	06.07 Romsey Pass
22.26 Hamworthy Jcn Goods	02.35	N15 4-6-0	Bournemouth 401	05.30	06.05 Dorchester Pass
Light ex Southampton T.	21.16	BR 4MT 2-6-0	Eastleigh 277	05.35	Light to Winchester @ 06.36 S. Terminus
19.25 ex Alton Pass	22.10	M7 0-4-4T	Eastleigh 297	05.45	06.20 S. Terminus Pass
18.12 Totton - Millbrook Goods	19.41	700 0-6-0	Eastleigh 326	05.45	08.44 Chesil Goods
13.56 S. Terminus Pass	14.25	M7 0-4-4T	Eastleigh 308	05.54	Light to S. Terminus @ 06.57 Fareham
West End shunt	01.40	M7 0-4-4T	Eastleigh 305	06.00	06.35 Fawley Pass
C&W Pilot	17.10	O2 0-4-4T	Eastleigh 343	06.00	C&W Pilot
West End shunt	16.55	M7 0-4-4T	Eastleigh 303	06.05	06.35 Alton Pass
20.18 Light ex S. Terminus	20.36	M7 0-4-4T	Eastleigh 304	06.10	07.00 Fareham Pass
18.40 Andover Jcn Pass	20.00	L1 4-4-0	Eastleigh 309	06.10	06.40 Andover Jcn Goods
Light ex Brockenhurst	20.28	M7 0-4-4T	Eastleigh 329	06.10	06.40 Romsey (Pilot)
Bevois Park Pilot	21.19	Q0-6-0	Eastleigh 323	06.10	06.52 Totton Goods
Light ex Northam	23.05	BR 4MT 2-6-0	Eastleigh 278	06.25	06.55 Newbury Pass
17.38 Bournemouth Pass	19.04	LN 4-6-0	Eastleigh 252	06.26	06.22 Waterloo
19.05 Fawley Goods (Pilot)	22.30	BR 3MT 2-6-2T	Eastleigh 328	06.45	07.15 Fawley Goods
21.05 S. Terminus Pass	21.30	M7 0-4-4T	Eastleigh 299	07.10	07.40 Alton Pass
00.45 Nine Elms goods	05.25	H15 4-6-0	N. Elms 79	07.15	07.46 Bournemouth Pass
Works Pilot	17.00	E4 0-6-2T	Eastleigh 341	07.15	Works Pilot
Works Pilot	18.00	0395 0-6-0	Eastleigh 340	07.16	Works Pilot
22.05 Southampton C Pass	22.30	BR 3MT 2-6-2T	Eastleigh 330	07.20	07.56 Fawley Goods
14.00 Gosport Goods	17.05	BR 4MT 2-6-0	Eastleigh 315	07.30	08.05 Gosport Goods
Shed Pilot	17.00	E1 0-6-0T	Eastleigh 344	07.30	Shed Pilot
C&W Pilot	17.15	O2 0-4-4T	Eastleigh 342	07.35	C&W Pilot
19.35 Kensington Pds	02.00	BR 4MT 2-6-0	Eastleigh 273	07.50	08.20 Reading Pass
13.45 Alton Goods	16.35	700 0-6-0	Eastleigh 327	08.00	08.33 Alton Goods
05.54 Bournemouth Pass	07.30	LN 4-6-0	Bournemouth 394	08.10	08.35 Bournemouth Pass
Allbrook pilot	06.00	Z 0-8-0T	Eastleigh 332	08.15	Allbrook pilot
04.25 Salisbury - Northam Goods	06.36	Q1 0-6-0	Eastleigh 318	08.30	09.15 Brockenhurst Goods
13.35 Havant Goods	15.00	Q1 0-6-0	Eastleigh 320	09.10	09.40 Fratton Goods
16.18 Salisbury Goods	18.05	H15 4-6-0	Eastleigh 312	09.15	10.23 B. Park - Salisbury Goods
19.40 Bournemouth Pass	21.20	N15 4-6-0	Eastleigh 259	09.25	09.55 Bournemouth Pass
00.01 S' Docks goods	01.20	T9 4-4-0	Eastleigh 282	09.34	Light to S. Terminus @ 10.29 Bournemouth
12.25 Newbury Pass	14.00	T9 4-4-0	Eastleigh 279	09.50	10.22 Newbury Pass
07.30 Bournemouth Pass	09.25	H15 4-6-0	Eastleigh 313	10.58	11.28 Bournemouth Pass
01.45 Brighton goods	08.15	K 2-6-0	Fratton 815	11.15	11.45 Fratton goods
04.07 Feltham - S. Docks	10.15	S15 4-6-0	Feltham 103	11.20	11.50 Micheldever goods
08.50 Portsmouth Pass	09.50	BR 3MT 2-6-2T	Eastleigh 331	11.25	11.55 Northam Goods
22.30 Waterloo	01.48	LN 4-6-0	Eastleigh 253	11.53	12.23 Southampton Pass
22.45 Feltham - Eastleigh	03.20	S15 4-6-0	Feltham 104	12.00	Light to S. Docks
Light ex Southampton T.	10.23	22xx 0-6-0	Didcot 13	12.10	Light to Chesil
09.30 Andover Jcn Pass	10.30	M7 0-4-4T	Eastleigh 298	12.45	13.15 Alton Pass
08.45 Salisbury Goods	10.20	H15 4-6-0	Eastleigh 311	12.45	13.15 Clapham Jn Pds
09.36 S. Terminus Pass	10.00	M7 0-4-4T	Eastleigh 305	13.40	14.09 Redbridge Goods
07.45 Light ex S. Terminus	08.00	M7 0-4-4T	Eastleigh 302	13.45	14.15 Alton Pass
12.00 S. Terminus - Eastleigh	12.30	T9 4-4-0	Eastleigh 283	14.05	14.35 Bournemouth Pass
09.50 Didcot Goods	13.00	Q1 0-6-0	Eastleigh 314	14.15	14.48 Didcot Goods
Light ex Chesil	12.50	43xx 2-6-0	Reading 20	14.22	14.22 Newbury Goods
12.25 Botley Goods	13.00	M7 0-4-4T	Eastleigh 300	15.20	15.51 Fawley Pass
09.16 S. Terminus Pass	09.50	S15 4-6-0	Feltham 102	15.22	Light to S. Docks
14.10 ex Winchester Pass	14.30	M7 0-4-4T	Eastleigh 297	15.35	16.06 S. Terminus Pass
14.05 Romsey Pass	14.30	M7 0-4-4T	Eastleigh 329	15.55	16.25 Winchester - Eastleigh Pass
11.05 Fratton Pds	12.20	N15 4-6-0	Eastleigh 265	16.15	Light to S. Terminus
13.15 Fawley - B. Park Goods	15.05	BR 3MT 2-6-2T	Eastleigh 330	16.20	20.28 Millbrook - B. Park Goods
12.05 Alton Pass	13.10	M7 0-4-4T	Eastleigh 299	16.25	16.55 Alton Pass
11.40 S. Terminus Pds	12.00	T9 4-4-0	Bournemouth 403	16.49	17.28 S. Terminus - A. Jcn Pass
01.30 Hoo Jcn Goods	12.36	H15 4-6-0	Eastleigh 310	16.55	17.27 Portsmouth Pass
09.54 Waterloo	12.30	H15 4-6-0	N. Elms 68	16.58	17.50 B. Park - B' stoke goods
03.30 Ashford Goods	16.15	N15 4-6-0	Eastleigh 264	18.12	Light to S. Docks
17.50 Romsey Pass	18.20	T9 4-4-0	Andover Jn 248	18.55	19.22 Portsmouth Pass
11.18 Salisbury goods	16.25	H15 4-6-0	N. Elms 72	19.15	19.45 B'stoke Pds
09.45 Feltham - Eastleigh	14.50	S15 4-6-0	Feltham 105	19.45	20.25 Brighton goods
16.47 Havant Goods	19.20	Q1 0-6-0	Guildford 213	20.25	20.50 Salisbury Goods
12.24 Totton Goods	14.08	Q0-6-0	Eastleigh 324	20.45	21.20 Brockenhurst Goods
13.45 B'mouth - B. Park	17.56	S15 4-6-0	Feltham 111	20.45	21.15 Felthamgoods
19.00 Reading Pass	20.30	N15X 4-6-0	Basingstoke 236	21.45	23.26 Waterloo Parcels
19.55 Bournemouth Pds	22.15	N15 4-6-0	Eastleigh 263	21.55	22.20 S. Docks Pds
16.38 Bournemouth C Pass	18.15	H15 4-6-0	Eastleigh 313	23.05	23.35 Bournemouth Goods
14.35 Hamworthy Jcn Goods	20.40	Q1 0-6-0	Eastleigh 319	23.50	00.20 Salisbury Goods
19.55 Fareham Goods	20.50	Q0-6-0	Eastleigh 317	23.50	00.36 Brockenhurst Goods

between Andover Junction and Nursling. Most of the T9 diagrams consisted of engines that went off shed at the start of day and returned in the evening but a few of the workings were cyclical in nature and stayed overnight at a number of neighbouring sheds before getting back to Eastleigh. The star turn in this respect was the 05.35 Brockenhurst goods which alternated between Eastleigh, Fratton, Bournemouth and Andover Junction sheds; the Eastleigh engine taking up the working on a Monday, for example, not getting back to its home shed until Thursday night. Another notable duty was one that took a T9 down the DN&S to Newbury.

Turning engines off the shed for London trains may have been seen by an outsider as a mark of prestige work but with the responsibility for getting 84 engines away from the shed each weekday, the staff at Eastleigh MPD had very little time to reflect on the merits of one turn as opposed to another especially when - between four and eight each morning, for example - engines were ringing off at the rate of one every six minutes.

From the observers point of view the most interesting turns were the foreign duties that made up a quarter of Eastleigh's work, the greater number consisting of S15 and H15 4-6-0's from Nine Elms and Feltham. These were known as turn-round turns since the engines came on shed after arriving in Eastleigh Yard to be turned, coaled and watered before picking up the next leg of their duty. Generally there was very little administration involved although the engines that arrived on goods trains and went out for a passenger working after a quick turn-round needed to be watched carefully.

One such example was the Nine Elms H15 4-6-0 which came down with the 00.45 Nine Elms goods and went forward with the 07.46 Bournemouth passenger after a turn-round of less than two hours. It was always prudent to keep an engine handy for this turn since if the goods was running late from London, the Controller would be on the telephone, worrying about the 07.46. One line of attack was to keep a spare T9 or H15 up one's sleeve and, in this case, send it out for the 07.46, putting the booked H15 back into the diagram when the contingency engine returned with the 11.36 Bournemouth – Woking Parcels. Another was to send the 07.46 crew to relieve the H15 in the yard and take it direct to the passenger station, although, as sure as eggs are eggs, it would turn out to have an empty tender or something equally awkward.

Using your own engines to keep Feltham or Nine Elms turns running smoothly was – given a reasonable stock of spare locomotives – a fairly routine matter that caused little in the way of repercussion since it was usually a simple matter to put the London engine back into its working whilst the train concerned was at Eastleigh on its way North. It was less straightforward when the engines for the afternoon DN&S turns had to be replaced since the booked engines came from Didcot and Reading Great Western sheds and an Eastleigh engine that could be used elsewhere on the Great Western was not likely to be seen for some time. There was an element of injustice where the GW and

WORTING JUNCTION - SOUTHAMPTON CENTRAL
LINE POSITION : 1954 - 10.00

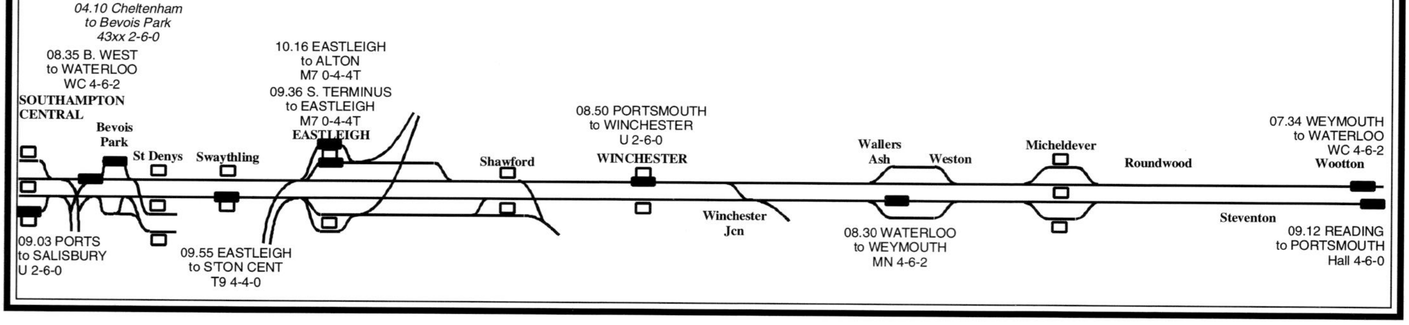

The larger BR Standard tanks did not arrive in the Eastleigh district until rather late in the day and usually as a result of being displaced by electrification. 80083, a transfer from Bricklayers Arms, works the 11.10 Southampton Terminus - Eastleigh Parcels train past Bevois Park Yard on 30th June 1964. Most of the vans in the train arrived with the 02.30 ex Waterloo and returned in the 13.15 Eastleigh to Clapham Junction.

Not having inherited a large suburban tank, the post-war Southern faced the choice of battling on with pre-grouping types or overcoming its prejudice against six-coupled tanks by embracing a number of foreign designs. Although it eventually acquired a number of the latter, the M7 0-4-4T of 1897 remained the standard suburban engine on many routes, especially those radiating from Eastleigh. 30030 arrives at Eastleigh with the 08.50 local from Portsmouth & Southsea to Winchester on 10 July 1957. This was an especially interesting service to watch since the train was brought into Eastleigh by a BR 3MT 2-6-2T but taken forward by the Guildford U 2-6-0 that had arrived in the area with the 07.31 Woking - Southampton Terminus. On this occasion 30030 worked the first part of the 3MT diagram, the booked engine taking over the duty with the 11.55 Eastleigh - Northam goods.

	06.55 Ports	06.55 Ports	Gds	08.00 Romsey	S&D	08.02 F'ham	07.30 W'bufy Milk	06.50 Reading	ECS	ECS	ECS	ECS	08.00 Romsey	04.07 Feltham Gds	06.15 Brock Gds
Engine	BR4 2-6-0	T9 4-4-0	BR3 2-6-2T	LM2 2-6-2T	LM5 4-6-0	M7 0-4-4T	43xx	HALL 4-6-0	T9 4-4-0	H15 4-6-0	M7 0-4-4T	LN 4-6-0	S15 4-6-0	Q0-6-0	
Shed	EL 271	FRA 365	EL 330	EL 306	BATH 1	EL 304	WBY	RDG 66	SAL 443	NE 79	BM 408	BM 394	FEL 103	EL 317	
Pilot															
Shed															
WATERLOO															
Woking															
Basingstoke								07.36							
Worting Jcn								07/42							
Micheldever								07.58							
Wallers Ash															
WINCHESTER								08.07							
Shawford								08.14							
Eastleigh Yard			07.56											08.39	
EASTLEIGH				08.15				08.21							
EASTLEIGH			08/10					08.28					08.35	08/43	
Swaythling								08.34					08.41		
St Denys	08.00		08/20			08.37		08.39					08.45	08/53	
Bevois Park Yard															
Northam								08.43						08.57	
Southampton (T)								08.46						09D38	
SOUTHAMPTON CENT	08.05		08.28			08.42							08.50		
SOUTHAMPTON CENT		08.13	08.32										08.54		
Millbrook													08.58		
Redbridge		08/21	08/39										09.03		
Totton		(to Romsey)	08.43										09.06		
Lyndhurst Rd													09.13		
Beaulieu Rd			(To										09.19		
BROCKENHURST			Fawley)										09.29		
Lymington Jcn													09/32		
Lymington Town															
Lymington Pier															
Sway													09.37		
New Milton													09.45		
Hinton Admiral													09.52		
Christchurch													10.00		
Pokesdown													10.06		
Boscombe													10.09		
Central Goods															
BOURNEMOUTH CENT													10.12		
BOURNEMOUTH CENT										10.10					
BOURNEMOUTH WEST				09.45							10.14	10.19			
Branksome										10.09		(To			
Parkstone										10.13		Wey-			
POOLE				09.55						10.19	10.32	mouth)			
Hamworthy Jcn				(To M'ter)						(to	(To			10.22	
Holton Heath										Scrum)	Brock)			10.40	
WAREHAM														10.49	
Worgret Jcn															
Corfe Castle															
SWANAGE															
Wool															
Moreton															
DORCHESTER															
Dorchester Jcn							10/06								
Monkton															
Wishing Well															
Upwey															
Radipole															
WEYMOUTH							10.18								

(Header: "COMBINED WORKING TIME TABLE AND ENGINE WORKINGS : 1954")

Southern was concerned: if a GW engine found its way onto an Eastleigh diagram, the wires would be red hot with demands for the engine to be sent light without delay to its home shed whereas requests to the Great Western for the return of a Southern engine seemed to stick in the bottom of their in-tray. The best recourse when a substitute was needed was to put into the diagram an engine that would scare the pants off the GW – a T9 usually did quite nicely – and ensure its early return. (One of the new BR Standard 2-6-0's arriving at Reading was quite likely to be sent to Exeter or Cardiff on a parcels train but a 4-4-0 would usually be left on the shed to be collected by a set of Southern men).

Some classes of engine were seen in very large numbers – no less than seventeen M7 0-4-4T's were booked off the shed each day – whilst others were almost rare. Numbered amongst the latter were the N15X 4-6-0's which ran most of the Basingstoke – Waterloo semi-fasts but were seldom – except at holiday times - seen outside the London outer-suburban area. Only one of the class strayed from its usual paths by working the 19.35 Reading – Southampton as far as Eastleigh and returning, after a brief visit to the shed, by relieving a King Arthur 4-6-0 on the 19.55 Bournemouth West – Waterloo parcels which the N15X worked through to London.

Southern moguls were also an uncommon element and of the several varieties at work, only two – both Fratton U 2-6-0's – were booked onto Eastleigh shed. The working of the two engines occasionally caused confusion since the first arrival came off the evening Fareham goods to work the 01.00 Portsmouth parcels whilst the other arrived with the 02.15 Fratton goods and turned round for the 06.00 goods to Havant. Both trains were covered by Fratton 363 which was an overlapping diagram

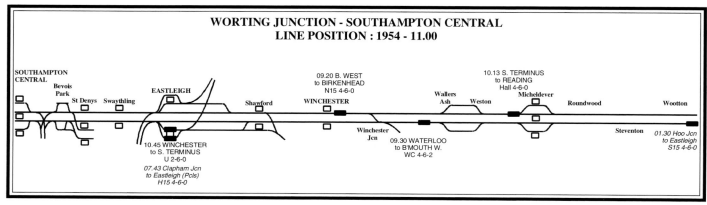

WORTING JUNCTION - SOUTHAMPTON CENTRAL
LINE POSITION : 1954 - 11.00

WEYMOUTH – BASINGSTOKE : WORKING TIMETABLE 1954

Train		07.53	08.25			09.47						09.20	08.10	10.32	
From		BM	Weymouth			Sarum						W'mouth	Bristol	Poole	
Class	Gds	Gds													
Engine	M7 0-4-4T	700 0-6-0	U 2-6-0	U 2-6-0	14xx	U 2-6-0	M7 0-4-4T	M7 0-4-4T	Castle	M7 0-4-4T	N15 4-6-0	LN 4-6-0	U 2-6-0	M7 0-4-4T	M7 0-4-4T
Shed	BM 405	BM 417	DOR 426	DOR 426	WEY	SAL 485	BM 406	BM 410	OOC	BM 409	EL 263	NE 31	SAL 450	BM 408	LYM 362
WEYMOUTH			08.25		08.40				09.00			09.20			
Radipole					08.44										
Upwey			08.30		08.48										
Wishing Well			08.34		08.53										
Monkton					08.58										
Dorchester Jcn			08/44		09/01				09/14			09/33			
DORCHESTER			08.47	08.56	(To Mdn Ntn)				(To P'ton)			09.40			
Moreton				09.07											
Wool				09.17											
SWANAGE								09.24							
Corfe Castle								09.37							
Worgret Jcn				09/22				09/46							
WAREHAM				09.26				09.48				10.04			
Holton Heath				09.32											
Hamworthy Jcn				09.38											
POOLE	09.30			09.44								10.16			
Parkstone				09.50											
Branksome	09/40			09.55											
BOURNEMOUTH WEST											10.12				
BOURNEMOUTH CENTRAL				10.00							10.20	10.27			
BOURNEMOUTH CENTRAL	09/50											10.33			
Central Goods	09.58														
Boscombe															
Pokesdown															
Christchurch															
Hinton Admiral		10.07													
New Milton															
Sway															
Lymington Pier							10.35								11.25
Lymington Town							10.39								11.29
Lymington Jcn		10/32					10/47					10/53		11/30	11/37
BROCKENHURST		10.36					10.49					10.57		11.31	11.39
Beaulieu Road															
Lyndhurst Road															
Fawley															
Totton															
Redbridge						10/29						11/11	11.21		
Millbrook													11.26		
SOUTHAMPTON CENTRAL						10.34						11.16	11.29		
SOUTHAMPTON CENTRAL						10.40						11.20	11.35		
Southampton T.															
Northam															
Bevois Park Yard															
St Denys						10.47							11.42		
Swaythling															
EASTLEIGH						(To P'mth)							(To P'mth)		
EASTLEIGH												11.30			
Eastleigh Yard															
Shawford															
WINCHESTER															
Weston															
Micheldever															
Worting Jcn												12/00			
Basingstoke															
Waterloo												12.50			

but on numerous occasions this would be overlooked and the first engine's number would be pencilled into the second turn (the 06.00 Eastleigh – Havant goods) causing the Eastleigh men who worked the train to fruitlessly scour the loco sidings looking for an engine that was twenty miles away. Trying to respond to a letter of enquiry with the excuse "10 mins delay: men looking for wrong engine" could be guaranteed to generate an entire file of correspondence.

One area in which the Southern had lagged behind the other companies was in the production of large six-coupled tank engines; a delay which meant that 0-4-4 locomotives, obsolete elsewhere, remained the standard class for a considerable proportion of local passenger work. Soon after nationalisation LMS 2-6-4T's found their way to a number of locations in Kent and it seemed to be simply a matter of time before they or their BR Standard variants appeared in large numbers at Eastleigh.

For once the pundits were wrong and the closest the area came to being bombarded with LMS locomotives came in mid-1952 when a trio of LMS 2MT 2-6-2T's arrived from the South Eastern. Initially it was suspected that they spelt the end for the M7 0-4-4T's – a suspicion fuelled by the arrival of three BR 3MT 2-6-2T's in the autumn – but in the end it was decided to retain the status quo pending the introduction of diesel multiple-units and for most of the 1950's the handful of 2-6-2T's remained the only examples of the type in the area. The LMS 2MT's, of which there were normally only two in traffic at any one time, worked a pair of cyclic diagrams that alternated between Eastleigh and Andover Junction and included the rather extraordinary 11.19 Portsmouth – Andover Junction: a train that made twenty-two calls and required no less than two hours and eight minutes – about the time it took an express to run from London to Bournemouth –for its fifty-odd mile journey.

On paper the Portsmouth – Southampton

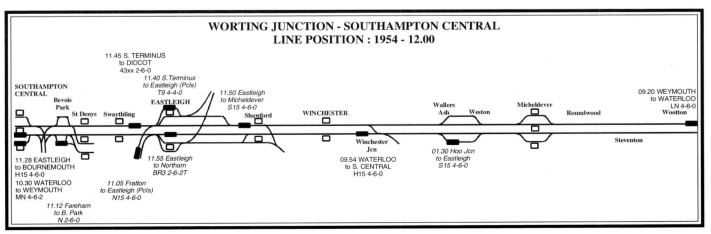

WORTING JUNCTION - SOUTHAMPTON CENTRAL LINE POSITION : 1954 - 12.00

Although Eastleigh played no part in the operations of the district, there were times when it came in very useful. Now and again the running shed would be searching fruitlessly for an engine to work a special when, out of the blue and in the knick of time, the works would announce that an ex-works engine was available for traffic. One such a happy coincidence resulted in Bournemouth-based N15 4-6-0 30743 'Lyonnesse' heading bunker towards Southampton Docks with a special goods from Eastleigh. The Portsmouth route can be seen curving away in the background.

stopping services appeared tailor-made for the M7 0-4-4T's – of which Eastleigh had an allocation of eighteen – yet in actuality with thirteen intermediate stops and at least one heavy bank, the seventy minutes of frequent starts and heavy steaming called for a performance that lay at the very threshold of M7 output whilst using a 2-6-0 – which was the next rung up the ladder – meant the trouble of either turning the engine at the end of each trip or running tender-first for half the diagram: a situation that raised understandable complaints from the footplate LDC's.

It was precisely this sort of problem that the Leader class had been conceived for although

ROMSEY : 1954

Train	Arr	Engine	Shed	Dep	Destination	Route
00.20 Eastleigh Yard		Q1 0-6-0	EL 319	00/46	S.disbury Goods	C.Ford
01.55 Eastleigh		U 2-6-0	SAL 482	02/00	Yeovil	C.Ford
01.35 S.disbury		U 2-6-0	SAL 450	02/20	Eastleigh Yard	C.Ford
02.10 Bevois Park		S15 4-6-0	SAL 467	02/52	Bristol	C.Ford
03.30 S.disbury		Q1 0-6-0	GUI 214	04/24	Chichester Gds	C.Ford
03.58 Bevois Park		H15 4-6-0	EL 311	04/38	S.disbury Goods	C.Ford
04.25 S.disbury		Q1 0-6-0	EL 319	05/24	Eastleigh Yard	C.Ford
06.07 Eastleigh	06.23	M7 0-4-4T	EL 300			
		M7 0-4-4T	EL 300	06.43	Eastleigh	C.Ford
06.40 Eastleigh Yard	07.05	L1 & M7	309/329	07.25	Andover Jcn	C.Ford
06.45 Andover Jcn	07.30	T9 4-4-0	AJN 248	07.32	S. Terminus	Nursling
07.30 Eastleigh	07.44	LM2 2-6-2T	EL 306			
05.58 Portsmouth &S.	07.57	U 2-6-0	SAL 450	07.59	S.disbury	Nursling
		LM2 2-6-2T	EL 306	08.00	Eastleigh	C.Ford
07.30 Andover Jcn	08.08	43xx 2-6-0	AJN 249			
07.56 Eastleigh	08.11	M7 0-4-4T	EL 298	08.13	Andover Jcn	C.Ford
07.47 S.disbury	08.16	T9 4-4-0	SAL 445	08.19	Portsmouth &S.	Nursling
		43xx 2-6-0	AJN 249	08.32	Eastleigh	C.Ford
06.55 Portsmouth &S.	08.32	T9 4-4-0	FRA 365			Nursling
07.29 Portsmouth &S	08.41	H15 4-6-0	NE 72	08.43	Bristol	C.Ford
		T9 4-4-0	FRA 365	08.52	Portsmouth &S.	Nursling
08.33 S. Terminus	09.10	T9 4-4-0	AJN 248	09.12	Andover Jcn	C.Ford
04.10 Cheltenham Gds		43xx 2-6-0	CH 21	09/12	Southampton Docks	Nursling
06.15 Chichester Gds		S15 4-6-0	FEL 106	09/27	Newport Gds	C.Ford
08.45 S.disbury		H15 4-6-0	EL 311	09/37	Eastleigh Yard	C.Ford
09.30 Eastleigh	09.55	BR4 2-6-0	EL 316	10.05	S.disbury Goods	C.Ford
09.30 Andover Jcn	10.08	M7 0-4-4T	EL 298	10.09	Eastleigh	C.Ford
09.03 Portsmouth &S.	10.16	U 2-6-0	FRA 368	10.18	S.disbury	Nursling
09.47 S.disbury	10.17	U 2-6-0	SAL 485	10.19	Portsmouth &S.	Nursling
10.10 S. TERMINUS	10.31	43xx 2-6-0	AJN 249	10.32	CHELTENHAM	Nursling
(Ex Yard Pilot)		M7 0-4-4T	EL 329	10.35	Eastleigh Yard	C.Ford
09.33 PORTSMOUTH &S	10.40	BR4 2-6-0	EL 270	10.42	CARDIFF	Nursling
10.23 Bevois Park		S15 4-6-0	EL 312	11/05	S.disbury Goods	C.Ford
08.10 BRISTOL	11.07		SAL 450	11.09	PORTSMOUTH &S.	Nursling
11.10 Eastleigh	11.25	M7 0-4-4T	EL 308			
		M7 0-4-4T	EL 308	11.35	Eastleigh Light	C.Ford
10.34 PORTSMOUTH &S	11.46	BR4 2-6-0	EL 272	11.48	BRISTOL	Nursling
11.30 Bevois Park		43xx 2-6-0	CH 21	12/00	Andover Jcn Gds	Nursling
11.25 Andover Jcn	12.03	T9 4-4-0	AJN 248	12.04	S.disbury	C.Ford
11.27 S.disbury Gds		S15 4-6-0	FEL 116	12/20	Chichester Gds	C.Ford
12.20 S.disbury	12.35	U 2-6-0	FRA 371			
11.19 Portsmouth &S.	12.44	LM2 2-6-2T	EL 306	12.47	Andover Jcn	C.Ford
10.27 BRISTOL		BR4 2-6-0	EL 270	13/06	PORTSMOUTH &S.	Nursling
11.00 BRIGHTON		WC 4-6-2	BTN 731	13/08	CARDIFF	Nursling
		M7 0-4-4T	FRA 371	13.15	Portsmouth &S.	C.Ford
12.24 Eastleigh Yard	13.25	M7 0-4-4T	EL 329			

Train	Arr	Engine	Shed	Dep	Destination	Route
11.00 BRIGHTON		WC 4-6-2	BTN 732	13/35	PLYMOUTH	Nursling
13.06 S.disbury	13.36	T9 4-4-0	SAL 443	13.37	S. Terminus	C.Ford
10.11 CHELTENHAM	13.42	Manor 4-6-0	CH 22	13.44	S. Terminus	Nursling
10.30 CARDIFF	13.50	BR4 2-6-0	EL 272	13.53	PORTSMOUTH &S.	Nursling
13.03 Portsmouth &S.	14.04	T9 4-4-0	SAL 445	14.05	S.disbury	C.Ford
		M7 0-4-4T	EL 329	14.05	Eastleigh	C.Ford
09.37 Andover Jcn Gds	14.16	T9 4-4-0	AJN 246	14.20	Nursling Gds	Nursling
11.18 S.disbury Gds	14.56	H15 4-6-0	NE 72		(Fwd at 15.38)	
11.00 PLYMOUTH		WC 4-6-2	BTN 731	15/17	BRIGHTON	Nursling
15.10 Nursling Goods	15.20	T9 4-4-0	AJN 246			Nursling
14.33 PORTSMOUTH &S.	15.34	U 2-6-0	SAL 485	15.36	BRISTOL	Nursling
(11.18 S.disbury Gds)		H15 4-6-0	NE 72	15.38	Eastleigh Yard	C.Ford
13.00 CARDIFF		WC 4-6-2	BTN 732	16/19	PORTSMOUTH &S.	Nursling
15.07 Nursling Gds		43xx 2-6-0	BTL 365	16/23	Southampton Docks	C.Ford
15.45 Portsmouth &S.	16.49	T9 4-4-0	AJN 248			
16.12 Andover Jcn	16.50	LM2 2-6-2T	AJN 247			
16.36 S. TERMINUS	16.58	Manor 4-6-0	CH 22	17.00	CHELTENHAM	Nursling
		LM2 2-6-2T	AJN 247	17.03	Eastleigh	C.Ford
16.18 S.disbury Gds	17.12	S15 4-6-0	EL 312	17.15	Eastleigh Yard	Nursling
13.56 Cheltenham	17.21	43xx 2-6-0	AJN 250	17.23	S. Terminus	Nursling
17.16 Eastleigh	17.31	M7 0-4-4T	EL 305		(Couple to 248 (T9))	
17.07 S.disbury	17.36	T9 4-4-0	EL 284	17.39	Portsmouth &S.	Nursling
		T9 & M7	248/305	17.50	Eastleigh	C.Ford
14.20 Weymouth	17.57	T9 4-4-0	EL 282	18.00	Andover Jcn	C.Ford
		T9 4-4-0	AJN 246	18.05	Eastleigh Yard	Nursling
16.45 Portsmouth &S.	18.14	U 2-6-0	SAL 450	18.17	S.disbury	Nursling
17.15 Fawley	18.45	T9 4-4-0	EL 280			C.Ford
17.45 PORTSMOUTH &S.	18.59	BR4 2-6-0	EL 272	19.01	CARDIFF	Nursling
		T9 4-4-0	EL 280	19.07	Andover Jcn	
18.40 Andover Jcn	19.17	L1 4-4-0	EL 309			
16.32 Bristol	19.25	U 2-6-0	FRA 368	19.28	Portsmouth &S.	Nursling
		L1 4-4-0	EL 309	19.35	Eastleigh	C.Ford
19.20 Eastleigh	19.35	M7 0-4-4T	EL 298			
		M7 0-4-4T	EL 298	19.50	Eastleigh ECS	C.Ford
19.04 S. Docks		43xx 2-6-0	AJN 250	19/59	Cheltenham Goods	Nursling
16.35 CARDIFF	20.08	U 2-6-0	SAL 482	20.09	PORTSMOUTH &S.	Nursling
19.35 Andover Jcn	20.18	T9 4-4-0	EL 282	20.21	S.disbury	
19.17 Portsmouth &S.	20.43	T9 4-4-0	SAL 443	20.45	S.disbury	
16.35 Exeter	20.46	S15 4-6-0	FEL 101	20.48	Portsmouth &S.	C.Ford
19.45 Portsmouth &S.	21.23	T9 4-4-0	BM 403	21.24	Andover Jcn	Nursling
20.50 Eastleigh Yard		Q1 0-6-0	GUI 213	21/31	S.disbury Goods	C.Ford
15.20 Cheltenham Gds		BR4 2-6-0	EL 270	21/57	Eastleigh	C.Ford
21.44 Eastleigh	21.59	43xx 2-6-0	BTL 365	22.03	S.disbury	C.Ford
19.10 Bristol	22.31	S15 4-6-0	SAL 467	22.34	Eastleigh	Nursling
16.40 Plymouth		BR4 2-6-0	EL 272	23/35	Eastleigh	C.Ford

Not everything on the Southern went in a straight line. Ignoring the direct route via Eastleigh, the 11.18 Portsmouth - Andover Junction was routed via Netley and Southampton taking more than two hours for the 52-mile trip. The Ivatt 2-6-2T that worked the service was one of three (41293 and 41304/5) allocated to Eastleigh for a two-day cyclic working which alternated between Eastleigh and Andover Junction sheds. One of the trio is seen crossing the Hamble at Bursledon on the 7th September 1957.

the Southern and its passengers would have been much better served had its Chief Mechanic Engineer grasped the job for which he was paid and produced – quickly - a conventional six-coupled tank engine.

In the event no suitable engine was produced for the line and the Southern Region had to press on with its inheritance as best it could by restricting the M7's to relatively short suburban tasks, such as Fareham – Southampton, and substituting tender engines on the longer workings. Whilst the turning requirements were irksome for the operators, the result was a delight for the enthusiast, especially those living on the Netley road where the twenty-five stopping trains had to be divided between six classes of locomotive.

Apart from the Ivatt 2-6-2T's, the only other indications of a new order at Eastleigh were a trio of BR 3MT 2-6-2T's and a growing number of 4MT 2-6-0's. The 3MT's were imported because, fortuitously, they suited conditions that had arisen on the Fawley branch whilst the 2-6-0's were drafted in as replacements for a proportion of the 46 elderly 4-4-0's that Eastleigh had on its books for local traffic.

Apart from the small number of express goods trains which ran direct to Southampton Docks or, in the case in inland traffic, to Brockenhurst and beyond, almost all freight traffic coming into the area did so via the yards at Eastleigh. The included not only London traffic but considerable amounts from the northern companies via Basingstoke and the Great Western via Salisbury, the DN&S and the MSW via Andover. The total service – twenty-four arrivals from the North and twenty-two departures for the south– was not great by some standards but it kept a Z class 0-8-0T and three 0-6-0 diesels busy for three shifts a day.

Most of the long distance traffic came in during the night – ten services arrived between midnight and six in the morning - and had to be shunted out and remarshalled for the dozen-odd services due to pull out before ten for Southampton docks and a variety of Wessex destinations. In planning the loadings to be taken by trains on other regions one usually had the luxury of being able to distinguish between priority and less urgent traffics but on the Southern, and Eastleigh in particular, because there was no core mineral traffic everything had to be regarded as being urgent with the occasional exception of traffic that Southampton Docks was unable to deal with immediately. (Much of the docks traffic was related to a particular sailing and if the ship was delayed in any way, the traffic could be held back at Eastleigh until the docks was in a position to accept).

As a general rule incoming traffic was directed into Allbrook yard where it was given a secondary classification – being shunted out by the Z 0-8-0 – and then tripped to either the Field, Tipton or Top End yards for primary classification and departure to destination.

Motive power standardisation was no more in evidence in the yards than it was anywhere else in the district and the seventy-eight main line engines booked in and out were made up from fifteen classes ranging from twenty-three workings by S15 4-6-0's to a handful of M7 0-4-4T's on the Alton and Romsey services. The Q1 0-6-0's could be found in reasonable numbers working to Salisbury, Brockenhurst, Fratton and over the DN&S to Didcot whilst the slightly older Q 0-6-0's were far less visible.

Foreign engines were less in evidence on goods than passenger work and the only booked visit by a non-Southern engine was the GW 43xx 2-6-0 of the 15.07 Salisbury – Southampton Docks which called at Allbrook for a couple of hours to attach and detach traffic. More unusual was the engine booked to the 06.40 Andover Junction goods which was an SECR L1 4-4-0, some of the class having a two-year stay at the shed as stop-gap between the withdrawal of the older LSWR 4-4-0's and the arrival of new BR 4MT locomotives.

The area between Eastleigh and Basingstoke was something of a traffic wilderness but Winchester was sufficiently busy to require a continuous shunting engine and for this purpose one Eastleigh's B4 0-4-0T's was semi-permanently allocated to the shed at Winchester City and shunted the goods sidings from 07.45 until 17.15. There was an element of comedy about the arrangement since the volume of traffic did not call for anything larger than an 0-4-0 yet, because the operators did not want a B4 running light twice a day between Eastleigh loco and Winchester for fear it would block the line either because of the time taken or, more likely, running hot (the B4's could not be coupled to any other engine, so double-heading was out of the question), the engine had to be retained at Winchester where the resident crew did not have time in their shift to work the diagram and prepare/dispose of the engine. To provide the assistance needed an Eastleigh engine cleaner was sent down to Winchester on the Weymouth mail to coal and light up the B4. At the end of the day's shunting the Winchester men were relieved by the Eastleigh crew who spent the greater part of their day working the 11.50 Eastleigh – Micheldever goods before handing over their engine to a set of Basingstoke men and disposing of the B4 : very few drivers can have started their day on an S15 4-6-0 and ended on an 0-4-0T. One sensed that there was probably a cheaper and easier way of managing matters at Winchester – especially as the far busier Chesil (DNS) yard was handled in the conventional way by traffic engines – but no-one ever came up with a workable alternative.

GOODS WORKINGS

One of the later, 1936, S15 4-6-0's 30840 of Feltham passes Sway with the 13.50 Bournemouth - Bevois Park express goods. This was the principal day service for goods traffic and brought a large quantity of goods traffic to the edge of the district where it could be remarshalled into the overnight services from Southampton and Eastleigh. The engine returned to London with the 21.15 Eastleigh - Feltham goods.

Wessex was by nature rural rather than industrial but nonetheless Southampton and Bournemouth each contained significant levels of manufacturing activity whilst a variety of plant generating sufficient traffic to be of interest to the railway, lay scattered between the New Forest area and Dorchester. In addition there was Southampton Docks, the servicing of which was to a great extent independent from other flows of traffic.

One aspect of Southern freight movements at variance with the rest of the country was the lack of distinction between goods and mineral traf-

fic. On the other regions the two were almost always segregated whilst on the Southern the amount of mineral trade was so small – chiefly small amounts to a scattered variety of destinations – that all traffic was treated as one and, in the majority of cases, moved in the same train. Another difference between the Southern and the others was in the classification of goods trains. The general policy elsewhere was to give trains a designation which reflected their running speeds and degree of continuous braking whilst on the Southern goods trains were simply goods trains whether they ran ten

BROCKENHURST YARD WORKING: 1954					
Train	Arr	Loco	Shed	Dep	Destination
22.26 Hamworthy Junction	00.30	N15 4-6-0	BM 401		(Fwd at 00.45)
23.35 Eastleigh	00.39	H15 4-6-0	EL 313		(Fwd at 03.10)
(22.26 Hamworthy Junction)		N15 4-6-0	BM 401	00.45	Eastleigh
00.36 Northam Yard	01.20	Q0-6-0	EL 317		
22.38 NINE ELMS	02.37	H15 4-6-0	NE 73		(Fwd at 03.42)
		H15 4-6-0	EL 313	03.10	BOURNEMOUTH CENTRAL
(22.38 Nine Elms)		H15 4-6-0	NE 73	03.42	WEYMOUTH
03.20 Bournemouth Central	03.55	Q0-6-0	EL 325	04.11	Bevois Park
22.45 FELTHAM	04.20	S15 4-6-0	FEL 110	04.45	BOURNEMOUTH CENTRAL
04.50 Eastleigh	06.00	N15 4-6-0	EL 266		(Fwd at 07.30)
		Q0-6-0	EL 317	06.15	Dorchester via Wimborne
(04.50 Eastleigh)		N15 4-6-0	EL 266	07.30	Bournemouth Central
09.15 Eastleigh	10.13	Q1 0-6-0	EL 318		(Fwd at 11.20)
07.53 Bournemouth Central	10.36	700 0-6-0	BM 417		
(09.15 Eastleigh)		Q1 0-6-0	EL 318	11.20	Poole via Wimborne
05.30 Eastleigh	11.24	T9 4-4-0	EL 286		(Fwd at 12.52)
		T9 4-4-0	EL 286	12.36	Lymington Town
(05.30 Eastleigh)		700 0-6-0	BM 417	12.52	Bournemouth Central
08.45 Poole via Wimborne	14.02	Q0-6-0	BM 416		
13.50 Bournemouth Central	15.16	S15 4-6-0	FEL 111	15.29	Bevois Park
16.40 Lymington	16.54	Q0-6-0	BM 416		
14.45 Hamworthy Jcn via Wimborne	17.12	Q1 0-6-0	EL 319	19.02	Eastleigh
16.25 Dorchester via Wimborne	21.20	Q0-6-0	EL 317	21.40	Bevois Park
21.20 Eastleigh	22.35	Q0-6-0	EL 324		(Fwd at 23.59)
19.50 DORCHESTER	22.51	H15 4-6-0	NE 74	23.20	NINE ELMS
(21.20 Eastleigh)		Q0-6-0	EL 324	23.59	Poole via main line

WORTING JUNCTION - SOUTHAMPTON CENTRAL
LINE POSITION : 1954 - 13.00

SOUTHAMPTON CENTRAL
Bevois Park
St Denys Swaythling
EASTLEIGH
13.15 Eastleigh to C. Jcn (Pcls) H15 4-6-0
12.31 S. TERMINUS to READING U 2-6-0
Shawford
WINCHESTER
11.16 B. WEST to YORK LN 4-6-0
Wallers Ash Weston
Micheldever
11.05 B. WEST to WATERLOO WC 4-6-2
Roundwood
Wootton

09.54 WATERLOO to S. CENTRAL LN 4-6-0
12.05 ALTON to EASTLEIGH M7 0-4-4T
13.15 EASTLEIGH to S. TERMINUS M7 0-4-4T
09.50 Didcot to Eastleigh Q1 0-6-0
11.30 WATERLOO to B. WEST LN 4-6-0
Winchester Jcn
Steventon
12.15 READING to PORTSMOUTH BR4 2-6-0

EASTLEIGH YARD WORKING: 1954

Train	Arr	Loco	Shed	Dep	Destination	Train	Arr	Loco	Shed	Dep	Destination
		Q1 0-6-0	EL 319	00.20	Sdisbury	01.30 HOO JUNCTION	12.36	S15 4-6-0	EL 310		
		H15 4-6-0	NE 64	00.45	FELTHAM	12.25 Botley	12.45	M7 0-4-4T	EL 300		
00.01 Southampton (T)	01.05	T9 4-4-0	EL 282			09.50 Didcot	12.47	Q1 0-6-0	EL 299		
22.20 FELTHAM	01.06	S15 4-6-0	FEL 109	01.45	SOUTHAMPTON DOCKS	11.27 Sdisbury	12.48	S15 4-6-0	FEL 116		(Chichester)
22.45 WOKING	02.00	S15 4-6-0	FEL 110			13.00 Winchester (Chesil)	13.20	700 0-6-0	EL 326		
22.26 Hamworthy Jcn	02.19	N15 4-6-0	BM 401			(11.27 Sdisbury)		S15 4-6-0	FEL 116	13.23	Chichester
23.25 NINE ELMS	02.24	LN 4-6-0	EL 251		(Southampton Docks)	12.24 Totton	14.08	Q0-6-0	EL 323		
01.35 BRIGHTON	02.46	U 2-6-0	SAL 450					M7 0-4-4T	EL 305	14.09	Redbridge
02.00 Fratton	03.02	U 2-6-0	FRA 363			09.45 FELTHAM	14.35	S15 4-6-0	FEL 105		
22.45 FELTHAM	03.02	S15 4-6-0	FEL 112		(Bournemouth C)	13.35 Havant	14.45	Q1 0-6-0	EL 320		
(23.25 Nine Elms)		BR4 2-6-0	EL 272	03.10	SOUTHAMPTON DOCKS			Q1 0-6-0	EL 314	14.48	Didcot
(22.45 Feltham)		S15 4-6-0	FEL 110	03.21	BOURNEMOUTH CENTRAL			H15 4-6-0	NE 65	15.09	Southampton Docks
		BR4 2-6-0	EL 316	03.45	Fratton	03.20 ASHFORD	15.53	N15 4-6-0	EL 264		
		H15 4-6-0	NE 66	04.40	FELTHAM	11.18 Sdisbury	16.11	H15 4-6-0	NE 72		
		N15 4-6-0	EL 266	04.50	Brockenhurst	12.50 Alton	16.24	700 0-6-0	EL 327		
00.45 NINE ELMS	05.06	H15 4-6-0	NE 79			15.07 Sdisbury	16.50	43xx 2-6-0	BTL 365		(Southampton Docks)
		T9 4-4-0	EL 286	05.30	Millbrook	14.40 Gosport	17.06	BR4 2-6-0	EL 315		
00.55 FELTHAM	06.00	U 2-6-0	FRA 363	06.00	Fratton	(15.07 Sdisbury)		43xx 2-6-0	BTL 365	17.42	(Southampton Docks)
		S15 4-6-0	FEL 102		(Bevois Park)	16.18 Sdisbury	17.52	S15 4-6-0	EL 312		
		L1 4-4-0	EL 309	06.40	Andover Jcn	18.05 Romsey	18.40	T9 4-4-0	AJN 246		
01.45 BRIGHTON	06.41	K 2-6-0	FRA 815			14.45 FELTHAM	18.54	S15 4-6-0	FEL 107		(Southampton Docks)
(00.55 Feltham)		S15 4-6-0	FEL 102	07.00	BEVOIS PARK	17.33 Southampton Docks	19.02	H15 4-6-0	NE 65		
		BR3 2-6-2T	EL 328	07.15	Fawley	18.25 Fareham	19.03	Q1 0-6-0	GUI 213		
04.07 FELTHAM	07.30	S15 4-6-0	FEL 103		(Southampton Docks)	(14.45 Feltham)		S15 4-6-0	FEL 107	19.21	SOUTHAMPTON DOCKS
		BR3 2-6-2T	EL 330	07.56	Fawley	14.45 Hamworthy Jcn	20.19	Q1 0-6-0	EL 319		
06.15 Chichester	08.04		FEL 106		(Newport)			S15 4-6-0	FEL 105	20.25	ASHFORD Via LBSCR
		700 0-6-0	EL 327	08.33	Alton	19.55 Fareham	20.32	Q0-6-0	EL 322		
(04.07 Feltham)		S15 4-6-0	FEL 103	08.39	SOUTHAMPTON DOCKS			Q1 0-6-0	GUI 213	20.50	Sdisbury
		700 0-6-0	EL 326	08.44	Winchester (Chesil)			S15 4-6-0	FEL 111	21.15	FELTHAM
(06.15 Chichester)		S15 4-6-0	FEL 106	09.00	Newport			Q0-6-0	EL 324	21.20	Brockenhurst
		Q1 0-6-0	EL 318	09.15	Brockenhurst	19.00 Didcot	21.43	Q1 0-6-0	EL 314		
		BR4 2-6-0	EL 316	09.30	Sdisbury	19.05 Fawley	21.56	BR3 & BR3	331/328		
		Q1 0-6-0	EL 320	09.40	Fratton	21.05 Fratton	22.11	S15 4-6-0	EL 310		
08.45 Sdisbury	10.08	H15 4-6-0	EL 311		(Fratton)	21.56 BEVOIS PARK	22.15	S15 4-6-0	FEL 106		(Feltham)
10.35 Romsey	11.03	M7 0-4-4T	EL 329			15.20 Cheltenham	22.25	BR4 2-6-0	EL 316		
(08.45 Sdisbury)		K 2-6-0	FRA 815	11.45	Fratton	(21.56 B. Park)		S15 4-6-0	FEL 106	22.30	FELTHAM
		S15 4-6-0	FEL 105	11.50	Micheldever			S15 4-6-0	EL 310	23.15	BASINGSTOKE
		BR3 2-6-2T	EL 331	11.55	Northam Yard			H15 4-6-0	EL 313	23.4	Brockenhurst
		M7 0-4-4T	EL 329	12.24	Romsey						

or one hundred and ten miles. In a handful of cases the more urgent workings were designation as semi-fitted and expected to run with a number – usually twelve – of vacuum fitted wagons next to the engine. This however was not a mandatory stipulation and in the event of the starting yard not having an adequate supply of piped wagons on hand, the train simply ran as an unfitted service, the driver running at whatever reduced speed he deemed appropriate.

A wagon of ordinary goods traffic was reckoned as one unit, a loaded mineral vehicle as one and a half whilst empties were assessed as a half, the 4-6-0's employed on the majority of long distance trains being allowed to take up to ninety units provided the number of vehicles behind the tender did not exceed about seventy between London and Moreton or sixty beyond. (The qualification 'about seventy' refers to a concession that allowed trains to take 'one or two' additional vehicles above and beyond the limit in order to minimise the need for relief services). So far as wagon numbers conveyed were concerned the system rather negated the need for large engines – most trains loaded up to 70 loads equal to 70 in length - since even a 700 0-6-0 could load up to the line maximum. A 4-6-0 was, of course, very much faster.

The focal point of express goods organisation was neither Eastleigh or Bournemouth but

Brockenhurst which, during the night shift in particular, received a number of local services from Eastleigh and Southampton and

BEVOIS PARK/NORTHAM (SOUTHAMPTON) YARD WORKING: 1954

Train	Arr	Loco	Shed	Yard	Dep	Destination
(19.50 Dorchester)		H15 4-6-0	NE 74	Bevois Park	00.12	NINE ELMS
00.01 Southampton (T)		T9 4-4-0	EL 282	Bevois Park		
		Q0-6-0	EL 317	Northam	00.36	Brockenhurst
		T9 4-4-0	EL 282	Bevois Park	00.45	Eastleigh
22.26 Hamworthy Junction	01.30	N15 4-6-0	BM 401	Bevois Park	01.59	Fawley
		S15 4-6-0	FEL 108	Bevois Park	03.10	WOKING
23.25 Nine Elms	03.30	BR4 2-6-0	EL 272	Northam	03.46	Southampton Docks
(Yard Pilot)		E4 0-6-2T	EL 339	Northam	05.00	Southampton Terminus
03.20 BOURNEMOUTH	04.53	Q0-6-0	EL 325	Bevois Park	05.10	FELTHAM
		Q1 0-6-0	EL 314	Bevois Park	05.35	Didcot
04.25 Sdisbury	06.02	C14 0-4-0T	ST 346	Northam	06.00	Town Quay
		Q1 0-6-0	EL 319	Northam		
		Q0-6-0	EL 323	Northam	06.52	Totton
00.55 FELTHAM	07.20	S15 4-6-0	FEL 102	Northam	07.36	S. Docks
		Q0-6-0	EL 322	Bevois Park	07.25	Fareham
(Stde Pilot)		E4 0-6-2T	EL 338	Bevois Park	07.30	Light to Eastleigh
04.07 FELTHAM	08.57	S15 4-6-0	FEL 103	Northam	09.30	Southampton Docks
Light ex Eastleigh	09.33	E4 0-6-2T	EL 338	Bevois Park		Yard Pilot
		M7 0-4-4T	EL 304	Bevois Park	09.45	Bitterne
		H15 4-6-0	EL 312	Bevois Park	10.23	Sdisbury
10.08 Redbridge	10.30	T9 4-4-0	BM 404	Bevois Park		
10.25 Bitterne	10.35	M7 0-4-4T	EL 304	Bevois Park		
		43xx 2-6-0	CH 21	Bevois Park	11.30	Andover Junction
11.12 Fareham	12.01	N 2-6-0	FRA 370	Bevois Park		
11.55 Eastleigh	12.14	BR3 2-6-2T	EL 331	Northam		
11.36 Bournemouth Pds	13.11	H15 4-6-0	NE 79	Bevois Park		
		BR3 2-6-2T	EL 331	Northam	13.20	Fawley
12.24 Totton	13.21	Q0-6-0	EL 323	Bevois Park		
		H15 4-6-0	NE 79	Bevois Park	13.44	Woking Pds
		Q0-6-0	EL 323	Bevois Park	13.46	Eastleigh
12.30 Fawley	14.00	BR3 2-6-2T	EL 328	Bevois Park		
13.15 Fawley	14.35	BR3 2-6-2T	EL 330	Bevois Park		
13.12 Fareham	15.41	BR4 2-6-0	EL 321	Bevois Park		
15.46 Southampton (Central)	15.55	T9 4-4-0	SAL 443	Bevois Park		
09.45 Feltham	15.34	H15 4-6-0	NE 65	Northam	16.00	Southampton Docks
		BR4 2-6-0	EL 321	Bevois Park	16.12	Southampton (T)
16.50 Millbrook	17.04	S15 4-6-0	FEL 111	Bevois Park		
15.07 Sdisbury	17.05	43xx 2-6-0	BTL 365	Northam		
		43xx 2-6-0	BTL 365	Northam	17.42	Southampton Docks
		H15 4-6-0	NE 68	Bevois Park	17.50	BASINGSTOKE
17.33 Southampton Docks		H15 4-6-0	NE 65	Bevois Park	18.38	Eastleigh
18.42 Woolston	18.52	T9 4-4-0	SAL 442	Bevois Park		
14.45 FELTHAM	19.40	S15 4-6-0	FEL 107	Northam	20.17	Southampton Docks
20.28 Millbrook	20.41	BR3 2-6-2T	EL 330	Bevois Park		
19.05 Fawley	21.22	BR3 & BR3	331/328	Bevois Park	21.36	Eastleigh
		S15 4-6-0	FEL 106	Bevois Park	22.12	FELTHAM
16.25 Dorchester	22.22	Q0-6-0	EL 317	Northam		
19.25 NINE ELMS	23.15	H15 4-6-0	EL 311	Northam	23.35	Southampton Docks
19.50 DORCHESTER	23.58	H15 4-6-0	NE 74	Bevois Park		(Nine Elms)

remarshalled their traffic into Bournemouth and Dorchester sections for attachment to the overnight services from Nine Elms and Feltham.

After the departure of the latter, the local trains then worked forward with traffic off the London train for Bournemouth via the main line and the West Moors branch. Although it was not shown as such in the timetable, in reality the trains concerned terminated at Brockenhurst to restart as fresh workings, having exchanged traffic. One particular advantage was that it allowed the Nine Elms – Weymouth service to bypass Bournemouth and thereby secure an early arrival at stations to the West of Poole. It was an ingenious way of dealing with traffic although it is a matter for regret that Brockenhurst never received much credit for the contribution that it made. Almost all the activity took place at night and there was very little evidence during daylight hours of the important role the yard played.

The star turn amongst freight trains was the 22.38 Nine Elms to Weymouth, which ran non-stop (apart from pausing for relief at Eastleigh) over the ninety-three miles to Brockenhurst where it left its Bournemouth section to be attached to the 22.45 Feltham – Bourne-

The 13.50 Bournemouth - Bevois Park passes New Milton behind 30497, one of the early LSWR S15 4-6-0's. The popularity of the 13.50 goods can be gauged from the fact that after calling at Christchurch it appears to have a full load behind the engine even though it was restricted to a maximum of thirty vehicles upon leaving Bournemouth.

COMBINED WORKING TIME TABLE AND ENGINE WORKINGS : 1954

Train	08.05	*04.50*			07.53	08.07	09.27		*05.30*	07.31	07.31		*06.15*	07.45
From	Bristol	*Elgh*			Alton	Ports	W'bury		*ELGH*	Woking	Woking		*Brock*	Newbury
Class		*Gds*						Gds	*GDS*				*Gds*	Gds
Engine	Hdl 4-6-0	*N15 4-6-0*	Q0-6-0	M7 0-4-4T	M7 0-4-4T	T9 4-4-0	43xx 2-6-0	Q0-6-0	*T9 4-4-0*	U 2-6-0	43xx 2-6-0	M7 0-4-4T	*Q0-6-0*	22XX 0-6-0
Shed	BTL	*EL 266*	EL 324	BM 406	EL 303	EL 283	WBY	E 323	*EL 286*	GUI 181	AJN 249	SWA 421	*EL 317*	DID 13
WATERLOO														
Woking														
Basingstoke										08.19				
Woking Jcn										**08/25**				
Micheldever										08.35				
Wallers Ash														
WINCHESTER					**08.36**					**08.47**				
Shawford					08.43					08.53				09.05
Eastleigh Yard														
EASTLEIGH					**08.50**					**08.59**				09.12
EASTLEIGH											09.05			09.14
Swaythling											09.11			
St Denys						09.08					09.14			09.22
Bevois Park Yard														
Northam						09.12					09.18			
Southampton (T)						09.15					09.20			09.27
SOUTHAMPTON CENT														
SOUTHAMPTON CENT														
Millbrook									*09.15*					
Redbridge									*09/20*					
Totton									*09.24*					
Lyndhurst Rd									09.46					
Beaulieu Rd									09.53					
BROCKENHURST			**09.32**	**09.39**										
Lymington Jcn			09/35	09/42										
Lymington Town				09.51										
Lymington Pier				09.53										
Sway				(To Wimbn)										
New Milton														
Hinton Admiral														
Christchurch		*10.18*												
Pokesdown														
Boscombe														
Centrl Goods		*10.30*												
BOURNEMOUTH CENT														
BOURNEMOUTH CENT														
BOURNEMOUTH WEST														
Branksome														
Parkstone														
POOLE														
Hamworthy Jcn														
Holton Heath														
WAREHAM													**11.06**	*11.14*
Worgret Jcn													11.09	*11/18*
Corfe Castle													11.19	
SWANAGE													11.28	
Wool														*11.27*
Moreton														
DORCHESTER														
Dorchester Jcn	10/22						11/13							
Monkton														
Wishing Well														
Upwey														
Radipole														
WEYMOUTH	**10.35**						11.24							

With the exception of the 22.30 Southampton to Feltham - used to return a Guildford U-class to its home shed - 2-6-0's were uncommon on scheduled main line goods services, H15 and S15 4-6-0's being used in preference. In theory the 4-6-0's had an eight-wagon advantage over the moguls although in practice both types were restricted to trains of seventy wagons north of Southampton. An N 2-6-0 passes Winchester in SR days.

mouth, picking up in exchange any traffic that had come in for stations beyond Branksome. It then ran direct to Poole to drop off S&D traffic before arriving at Hamworthy Junction where it detached sections for Swanage and the Hamworthy branch.

The volume of traffic dealt with by two branches – there were four private sidings at Corfe Castle alone – was sufficient to allow it to be centralised at Hamworthy Junction, an arrangement that allowed the 22.38 to run non-stop to Dorchester and thus save time which

would otherwise have been spent in shunting at Wareham. The last leg of the journey over the seven miles of Great Western track to Weymouth – for which an N15 4-6-0 took over from the H15 that had brought the train down from London - was the Southern's only goods

WEYMOUTH - BASINGSTOKE : WORKING TIMETABLE 1954

Train						11.05	12.00	09.25							
From						Fratton	S'ton	Sarum							
Class		Gds	Gds	Pds	Gds	Pds								Gds	
Engine	R'car	43xx	S15 4-6-0	T9 4-4-0	M7 0-4-4T	43xx 2-6-0	N15 4-6-0	T9 4-4-0	M7 0-4-4T	T9 4-4-0	WC 4-6-2	BR4 2-6-0	LN 4-6-0	Q0-6-0	U 2-6-0
Shed	WEY	WBY	Fel 105	BM 403	BM 404	DID 30	EL 265	EL 283	FRA 371	EL 284	BM 383	EL 271	BM 395	*EL 323*	GUI 181
WEYMOUTH	09.28	*09.37*													
Radipole	09.32														
Upwey	09.36														
Wishing Well	09.41														
Monkton	09.46														
Dorchester Jcn	09/49	*09/55*													
DORCHESTER	(To Yeo)	(To Yeo)													
Moreton															
Wool															
SWANAGE															
Corfe Castle															
Worgret Jcn															
WAREHAM															
Holton Heath															
Hamworthy Jcn															
POOLE										10.40					
Parkstone										10.47					
Branksome										10.53					
BOURNEMOUTH WEST										10.57	11.05		11.16		
BOURNEMOUTH CENTRAL											11.13		11.24		
BOURNEMOUTH CENTRAL					*10.46*						11.17		11.28		
Central Goods					*10.54*										
Boscombe											11.21		11.32		
Pokesdown											11.25		11.36		
Christchurch											11.30		11.42		
Hinton Admiral															
New Milton											11.42		11.59		
Sway											11.49				
Lymington Pier															
Lymington Town															
Lymington Jcn											11/53		12/00		
BROCKENHURST											11.57		12.04		
Beaulieu Road															
Lyndhurst Road															
Fawley															
Totton														*12.24*	
Redbridge											12/11		12/20	*12.29*	
Millbrook															
SOUTHAMPTON CENTRAL											12.16		12.25		
SOUTHAMPTON CENTRAL											12.20		12.29		
Southampton T.			*11.40*		11.45		12.00					12.23			12.31
Northam					11.49							12.27			12.35
Bevois Park Yard															
St Denys					11.53		12/06				12.27	12.35			12.39
Swaything					11.57		12.10								12.43
EASTLEIGH				11.52	12.02	12.06	12.15				12.34		12.40		12.48
EASTLEIGH						12.08	12.20				12.36		12.42		12.50
Eastleigh Yard			*11.50*												
Shawford						12.16		(To R'sey)	(To R'sey)			(To P'mth)			12.58
WINCHESTER			*12/20*								12.50		12.56		13.06
Weston						(To D'cot)									
Micheldever			*12.38*												13.22
Worting Jcn											13/13		13/19		13/33
Basingstoke											13.17		13.23		13.37
Waterloo											14.20		(York)		(Reading)

working west of Dorchester and operated only because a connection had to be maintained between the parent company and (no-one should imagine that nationalisation had been allowed to interfere with the habits of a century) its isolated Easton branch. Most of the traffic taken beyond Dorchester consisted of stone empties for the quarries at Portland.

Overnight express traffic excepted, the powerhouse of goods traffic was Eastleigh whose yards received forty services a day and forwarded thirty seven, the heaviest flow coming from in the London division (9 trains) and going out to Southampton (12). The volume of traffic in the opposite direction was significantly lighter (5 services in from Southampton and 5 forward to London) due to the fact that as much traffic as possible originating in Southampton docks was sent direct to either Feltham or Nine Elms and therefore avoided intermediate marshalling at Eastleigh. Most of the marshalling at Eastleigh concerned main line services which typically arrived in the East (down) or Field (up) Sidings for remarshalling at the adjacent Allbrook Yard where a Z class 0-8-0T was stationed for twenty hours a day. Trains on the Salisbury – Portsmouth route generally arrived at Eastleigh in clearly established sections, primary marshalling having been done at either Salisbury or Fratton, and these services generally called at the East Yard to remove or attach complete sections before moving on. Traffic moving between one route and the other was tripped as necessary from yard to yard.

By a considerable margin, the greater part of the traffic dealt with at Southampton was import and export trade, dealt with internally in the port. The remainder was sifted through Bevois Park (up) or Northam (down) situated between St Denys and Northam Junction. It is generally thought that Bevois Park dealt with internal (SR) traffic whilst Northam was occupied exclusively with shipping business but this is a misconception that has arisen, probably, because the preponderance of dock traffic at Northam suggested that the yard dealt with nothing else. In fact it was a traffic yard, pure and simple, for down trains although the running of the Wessex night traffic direct from Eastleigh to Brockenhurst – an operation that did not apply in the up direction – relieved it of a considerable amount of internal work.

Cross country traffic was made up very largely – but by no means exclusively - of coal traffic from South Wales, marshalled at Salisbury into eight daily services: three for the LBSCR, three for Eastleigh and two for Southampton. In addition the MSW forwarded one service each – mainly goods - for Eastleigh and Southampton Docks via Andover Junction.

Conveying a higher proportion of mineral traffic that the services on the main line, trains tended to be shorter but heavier than those on the main line and because of this, the type of locomotive used for each working had to be carefully monitored. A goods train could take up to 70 wagons from Salisbury to Eastleigh with almost any class of engine but a mineral service made up of the new loaded 16-ton vehicles was limited to a maximum load of 40. The cross-country workings were all booked for large engines including S15 and H15 4-6-0's but if a T9 4-4-0 happened to creep into one of the diagrams – and stranger things happened – then the load had to be reduced to 22 wagons amidst howls from the yard inspector as to whether the train was worth running.

In the return direction trains could take up to 70 mineral empties and the disparity resulted in an imbalance of service between Eastleigh and Salisbury with only one service running through from the LBCSR, two starting from Eastleigh and – because it was less dependant on mineral traffic – three from Southampton. Motive power remained on the heavy side with S15 and H15 4-6-0's being responsible for most of the workings.

The Salisbury – Fratton route was the closest the Southern came to having a mineral route and had traffic been heavier than it actually was, it is probable that consideration would have been given to an eight-coupled locomotive which, capable of handling a full 70-wagon load would have reduced the number of trains on the section. In the meantime full loads could only be handled by double-heading which, on the Southern, was tricky to arrange since few of the engines working the route were allowed to be coupled to another. H15, S15 and N15's could only work singly and of the engines that worked towards Eastleigh from Salisbury, only the Q1 0-6-0 could be double-headed. The chances of getting two of the class together at the same time were astronomical.

COMBINED WORKING TIME TABLE AND ENGINE WORKINGS : 1954

Train		08.55	09.03	08.00		09.33	08.30	08.30	08.30	08.30			06.15		10.25	
From		Alton	Ports	Romsey		Ports	Padd	W'loo	W-S-M	W'loo			Brock		Bitt	
Class	Gds													Gds	Gds	Gds
Engine	Q1 0-6-0	M7 0-4-4T	U 2-6-0	T9 4-4-0	WC 4-6-2	43xx 2-6-0	BR4 2-6-0	Castle	MN 4-6-2	Hal 4-6-0	U 2-6-0	M7 0-4-4T	M7 0-4-4T	Q0-6-0	M7 0-4-4T	M7 0-4-4T
Shed	EL 318	EL 299	FRA 368	EL 282	BM 385	AJN 249	EL 270	OOC	NE 32	BTL	DOR 426	BM 409	LYM 362	EL 317	BM 407	EL 304
WATERLOO									08.30							
Woking																
Basingstoke									09.40							
Worting Jcn									09/47							
Micheldever																
Wallers Ash																
WINCHESTER		09.36							10.08							
Shawford		09.42														
Eastleigh Yard	09.15															
EASTLEIGH		09.49														
EASTLEIGH	09/19			09.55					10/17							
Swaythling				10.01												
St Denys	09/34		09.53	10.05			10.18									10/33
Bevois Park Yard																10/35
Northam																
Southampton (T)			10.10													
SOUTHAMPTON CENT		09.58		10.10		10.15	10.23		10.26							
SOUTHAMPTON CENT	09/42	10.02		10.18		10.26			10.31							
Millbrook																
Redbridge	09/49	10/07		10/23		10/31	10/36									
Totton		(To S'bury)		(To Chel)		(To Cardiff)										
Lyndhurst Rd																
Beaulieu Rd																
BROCKENHURST	10.13								10.53				10.58			
Lymington Jcn									10/56				11/01			
Lymington Town													11.10			
Lymington Pier													11.12			
Sway																
New Milton																
Hinton Admiral																
Christchurch																
Pokesdown																
Boscombe																
Central Goods																
BOURNEMOUTH CENT									11.14							
BOURNEMOUTH CENT					10.23						11.20	11.25				
BOURNEMOUTH WEST												11.33				
Branksome					10.30											
Parkstone					10.34											
POOLE					10.42				(To W'mouth)		11.34					
Hamworthy Jcn					10.48						11.38					
Holton Heath					10.54											
WAREHAM					11.01						11.47				11.57	
Worgret Jcn					11/04						11/50				12/00	
Corfe Castle															12.10	
SWANAGE															12.19	
Wool					11.10						11.56			12.10		
Moreton					11.19									12.42		
DORCHESTER					11.28						12.13			12.56		
Dorchester Jcn					11/31		11/43		12/03	12/15						
Monkton																
Wishing Well																
Upwey					11.40											
Radipole																
WEYMOUTH					11.45		11.56		12.15	12.24						

Double-heading a goods train meant that double the normal load could be taken provided the limit of 70 wagon-lengths was not exceeded. This was a useful facility where bulk loads were concerned. One loaded 16 ton mineral wagon was equivalent to 2 ordinary goods vehicles and therefore the limit for a 4-6-0 was 45 wagons equal to 90 loads equal to 45 in length. Adding a second 4-6-0 increased the limit to 70 wagons = 140 loads = 70 in length. The limit that could not be exceeded was the length of seventy wagons. BR 4MT 75067 double-heads a Standard 2-6-0 through New Milton with an up goods. The leading vehicles are civil engineering vehicles for which special 'out of gauge' arrangements were needed.

WEYMOUTH - BASINGSTOKE : WORKING TIMETABLE 1954

Train				11.36			11.12	12.24			11.04	10.45			
From				B'mouth			Wimborne	Totton			Brock	Swanage			
Class				Pds	Auto			Gds	Gds				Gds			Pds	
Engine	LN 4-6-0	H15 4-6-0	Hdl 4-6-0	H15 4-6-0	14xx	Q0-6-0	Q0-6-0	M7 0-4-4T	M7 0-4-4T	M7 0-4-4T	M7 0-4-4T	M7 0-4-4T	T9 4-4-0	H15 4-6-0	WC 4-6-2
Shed	BM 393	NE 79	WEY	NE 79	WEY	EL 324	EL 323	HAM 422	BM 407	BM 406	HAM 422	SWA 421	EL 284	EL 311	BM 386
WEYMOUTH	**10.10**		**10.33**		**10.45**										**11.30**
Radipole					10.49										
Upwey	10.17				10.54										
Wishing Well					10.59										
Monkton					11.04										
Dorchester Jcn	10/26		10/47		11/07										11/43
DORCHESTER	**10.31**		**(To**		**(To**										**11.48**
Moreton	10.39		W'ton)		Dor.W)										
Wool	10.47														12.03
SWANAGE								10.45	11.06			11.36			
Corfe Castle								11.00	11.20		11.30	11.47			
Worgret Jcn	10/52								11/29		11/45	11/56			12/08
WAREHAM	**10.57**								**11.31**		**11.50**	**11.58**			**12.13**
Holton Heath	11.03							(To							
Hamworthy Jcn	11.09							W'ham)							
POOLE	**11.17**					**11.27**				**11.59**					**12.25**
Parkstone	11.23					11.32				12.04					
Branksome	11.28					11.37				12.09					
BOURNEMOUTH WEST		11.36				11.41							12.20		
BOURNEMOUTH CENTRAL	**11.34**	**11.44**								**12.14**			**12.28**		**12.35**
BOURNEMOUTH CENTRAL		**11.48**													**12.41**
Central Goods															
Boscombe		11.53													
Pokesdown		11.56													
Christchurch		12.02													
Hinton Admiral		12.09													
New Milton		12.17													
Sway		12.25													
Lymington Pier															
Lymington Town															
Lymington Jcn		12/29													12/59
BROCKENHURST		**12.32**													
Beaulieu Road		12.40													
Lyndhurst Road		12.46													
Fawley															
Totton		12.52													
Redbridge		12.55				13.00									13/10
Millbrook		13.00				13/07									
SOUTHAMPTON CENTRAL		**13.03**													**13.15**
SOUTHAMPTON CENTRAL				13.06		13/12									**13.20**
Southampton T.															
Northam															
Bevois Park Yard				13.11		13.21									
St Denys		(To													
Swaything		Woking)													
EASTLEIGH															
EASTLEIGH														13.15	**13.30**
Eastleigh Yard															
Shawford														13.26	
WINCHESTER														(To	
Weston														C. Jcn)	
Micheldever															
Worting Jcn															13/59
Basingstoke															
Waterloo															**14.49**

Eastleigh U 2-6-0 31619 works the 04.50 Eastleigh - Bournemouth at Walkford in lieu of the booked N15 4-6-0. These engines were more in evidence on workings from Eastleigh to Portsmouth and Salisbury than on the main line.

BOURNEMOUTH DISTRICT GOODS WORKINGS

BOURNEMOUTH CENTRAL GOODS

Train	Arr	Engine	Shed	Dep	Destination
		Q0-6-0	EL 325	03.20	Feltham
		Q0-6-0	BM 416	03.30	Poole
23.35 Eastleigh	03.55	H15 4-6-0	EL 313		
22.45 Feltham	05.27	S15 4-6-0	Fel 110		
		700 0-6-0	BM 417	06.37	Boscombe
04.05 Salisbury	06.08	700 0-6-0	SAL 452		
		Q1 0-6-0	EL 319	07.20	Parkstone
		700 0-6-0	BM 417	07.52	Brockenhurst
09.30 Poole	09.58	M7 0-4-4T	BM 405		
04.50 Eastleigh	10.30	N15 4-6-0	EL 266		
10.46 Bournemouth Loco	10.54	M7 0-4-4T	BM 404		
		M7 0-4-4T	BM 404	11.35	Poole
		G6 0-6-0T	BM 418	11.40	Bournemouth Loco
		S15 4-6-0	Fel 111	13.50	Bevois Park
13.45 Hamworthy Jcn	14.56	Q0-6-0	BM 415		
05.30 Eastleigh	15.30	700 0-6-0	BM 417		
		U 2-6-0	DOR 426	17.27	Dorchester
15.05 Corfe Castle	21.02	Q1 0-6-0	EL 318		
		WC 4-6-2	BATH 1	21.20	Bath
19.50 Dorchester	21.47	H15 4-6-0	NE 74	22.09	Nine Elms

POOLE GOODS YARD

Train	Arr	Engine	Shed	Dep	Destination
17.15 Bath	00.06	4F 0-6-0	BATH 18		
21.20 Eastleigh	01.01	Q0-6-0	EL 324	01.30	Dorchester
		4F 0-6-0	BATH 18	01.40	Bath
03.30 Bournemouth C.	03.55	Q0-6-0	BM 416		
22.38 Nine Elms	04.44	H15 4-6-0	NE 73	05.15	Weymouth
04.05 Salisbury	06.08	700 0-6-0	SAL 452	06.42	Bournemouth C. Goods
02.40 Bath	07.09	WC 4-6-2	BATH 1	07.27	Bournemouth West
		Q0-6-0	BM 416	08.45	Brockenhurst
		Q1 0-6-0	EL 319	08.50	Hamworthy Jcn
		M7 0-4-4T	BM 405	09.30	Bournemouth C. Goods
		B4 0-4-0T	BM 419	10.57	Broadstone
06.40 Evercreech Jcn	11.43	4F 0-6-0	TCB 55		
12.00 Broadstone	12.27	B4 0-4-0T	BM 419		
08.39 Eastleigh via W. Moors	12.58	Q1 0-6-0	EL 318		
11.35 Bournemouth C.	13.24	M7 0-4-4T	BM 404		
		Q0-6-0	BM 415	13.25	Hamworthy Jcn
13.45 Hamworthy Jcn	13.54	Q0-6-0	BM 415	14.30	Bournemouth C. Goods
		4F 0-6-0	TCB 55	15.05	Templecombe
17.27 Bournemouth C	17.51	U 2-6-0	DOR 426	18.24	Dorchester
18.08 Parkstone	18.14	M7 0-4-4T	BM 405		
15.05 Corfe Castle	20.05	Q1 0-6-0	EL 318	20.36	Bournemouth C. Goods
19.50 Dorchester	21.15	H15 4-6-0	NE 74	21.25	Nine Elms
21.20 Bournemouth C	21.45	WC 4-6-2	BATH 1	22.20	Bath
22.26 Hamworthy Jcn	22.34	N15 4-6-0	BM 401	23.30	Eastleigh

BOURNEMOUTH WEST GOODS

Train	Arr	Engine	Shed	Dep	Destination
02.40 Bath	07.44	WC 4-6-2	BATH 1		
		M7 0-4-4T	BM 405	13.34	Branksome
17.30 Branksome	17.35	M7 0-4-4T	BM 405		

HAMWORTHY GOODS

Train	Arr	Engine	Shed	Dep	Destination
07.40 Hamworthy Jcn	07.49	B4 0-4-0T	HAM 423		
11.20 Hamworthy Jcn	11.29	Q1 0-6-0	EL 319	12.10	Hamworthy Jcn
		B4 0-4-0T	HAM 423	12.10	Bank
14.00 Hamworthy Jcn	14.09	B4 0-4-0T	HAM 423	16.00	Hamworthy Jcn

HAMWORTHY JUNCTION GOODS

Train	Arr	Engine	Shed	Dep	Destination
22.28 Nine Elms	05.25	H15 4-6-0	NE 73	05.58	Weymouth
		M7 0-4-4T	HAM 422	06.30	Swanage
		B4 0-4-0T	HAM 423	07.40	Hamworthy Goods
08.50 Poole	09.00	Q1 0-6-0	EL 319		
06.15 Brockenhurst via Wimborne	10.49	Q0-6-0	EL 317	11.14	Dorchester
		Q1 0-6-0	EL 319	11.20	Hamworthy Goods
12.10 Hamworthy Goods	12.19	Q1 0-6-0	EL 319		*(Banked by B4)*
		B4 0-4-0T	HAM 423	13.00	Light to Hamworthy Goods
13.25 Poole	13.34	Q0-6-0	BM 415		
		Q0-6-0	BM 415	13.45	Bournemouth C. Goods
		B4 0-4-0T	HAM 423	14.00	Hamworthy Goods
		Q1 0-6-0	EL 319	14.35	Eastleigh via Wimborne
		B4 0-4-0T	HAM 423	15.30	Light to Hamworthy Goods
16.00 Hamworthy Goods	16.09	B4 0-4-0T	HAM 423		
12.00 Dorchester	15.47	U 2-6-0	DOR 427		
16.25 Dorchester	17.11	Q0-6-0	EL 317	18.00	Bevois Park via Wimborne
		B4 0-4-0T	HAM 423	18.24	Light to Poole Harbour
17.27 Bournemouth C. Goods	18.34	U 2-6-0	DOR 426	18.50	Dorchester
15.05 Corfe Castle	19.06	M7 0-4-4T	HAM 422		*(Fwd at 19.56)*
(15.05 Corfe Castle)		Q1 0-6-0	EL 318	19.56	Bournemouth C. Goods
19.50 Dorchester	20.28	H15 4-6-0	NE 74		*(To Nine Elms)*
20.30 Light ex Poole Harbour	20.37	B4 0-4-0T	HAM 423		
(Ex Dorchester)		H15 4-6-0	NE 74	21.08	Nine Elms via B'mouth
		N15 4-6-0	BM 401	22.26	Eastleigh via B'mouth

DORCHESTER SOUTH GOODS

Train	Arr	Engine	Shed	Dep	Destination
21.20 Eastleigh	02.47	Q0-6-0	EL 324		
22.38 Nine Elms	06.38	H15 4-6-0	NE 74		
		N15 4-6-0	EL 263	07.15	Weymouth
08.06 Dorchester West	08.11	57xx 0-6-0	WEY GW	08.14	Light to D. West
06.15 Brockenhurst	12.56	Q0-6-0	EL 317		
		U 2-6-0	DOR 427	12.00	Hamworthy Jcn
		G6 0-6-0	DOR 428	15.10	GW Exchange Sdgs
15.20 GW Exchange Sdgs	15.25	G6 0-6-0T	DOR 428		
		Q0-6-0	EL 317	16.25	Bevois Park
17.10 Weymouth	17.30	H15 4-6-0	NE 74		
17.28 Bournemouth	20.50	U 2-6-0	DOR 426	20.55	GW Exchange Sdgs
21.10 GW Exchange Sdgs	21.15	U 2-6-0	DOR 426		
		H15 4-6-0	NE 74	19.50	Nine Elms

Spoils of war.   The 14-strong USA class were powerful machines which raised the limit for trains within Southampton Docks from 37 to 45 wagons. 30068 and another member of the class sit on Southampton loco in September 1961.

For the enthusiast, Southampton was a sprawling mess where, with the possible exception of St Deny's, there was no one vantage point from which more than a small proportion of the town's activities could be viewed whilst the railwayman viewed it no more favourably since each activity was spread over such a wide area it made life difficult for anyone with a watching brief.

Main line traffic was dealt with at the Central (West) station but, consisting of four through roads and a bay on the down side at the Bourne-

mouth end, it was unsuited for terminating traffic which needed facilities for shunting and running round between trips.   One of the reasons that so many local services from the Netley direction continued forward to Romsey or Salisbury was because they could not terminate and reverse at Southampton without imperilling other traffic.

In addition to its restricting layout, another curiosity of Southampton Central was the ordering of lines through the station whereby the outside pair of running roads was reserved for

express traffic whilst the inner pair were used for local services.   The arrangement – the reverse of the normal arrangement – facilitated the working of Millbrook station whose island platform was situated between the up and down lines and at the same time allowed the more important expresses to use the outer faces of Southampton Central thus saving passengers the inconvenience of having to use the footbridge.

Although it was served by an impressive number of trains – boredom was not a feature

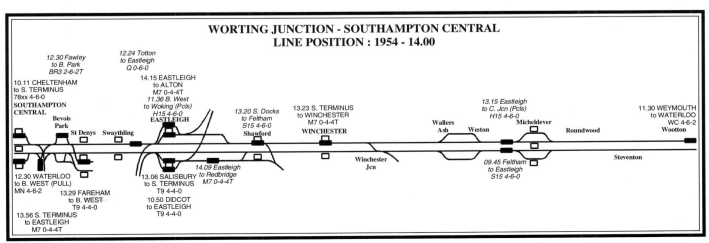

WORTING JUNCTION - SOUTHAMPTON CENTRAL
LINE POSITION : 1954 - 14.00

Although they were by no means scarce, in their unrebuilt state West Country Pacifics were never quite as commonplace as they were later to become. The class worked most of the morning expresses up to Waterloo but in the down direction none of the class were booked to make an appearance between the 19.30 Waterloo - Bournemouth and the 09.40 Brighton - Bournemouth - an interval of almost fourteen hours. 34092 'City of Wells' brings the 09.03 (Summer Saturdays) Portsmouth - Plymouth into Southampton Central.

associated with the Central – it was not an operationally satisfying location to work at since the majority of trains simply came and went with few of the attachments or engine changes that added zest to railway life. The greatest problem one had to tackle nine days out of ten was that of getting innumerable passengers and their luggage into and out of an express during the three or four minutes allowed. All long distance services were well subscribed to but the 07.20 Bournemouth – Waterloo was especially popular and passengers had to be all but manhandled into the coaches to get it away within the booked three minutes. Frequently one would ask at meetings for an additional minute to be granted but the answer was always the same – No.

Local services – and a number of longer distance stopping trains - that were based on the town tended to be run to and from Southampton terminus, a separation which on the one hand kept the lines clear for through traffic whilst, on the other, caused tremendous difficulties for the passenger who arrived from the Netley direction and wanted to go forward to Bournemouth. St Deny's was the obvious point of connection but it was felt that to advertise the fact would mean inserting an additional stop into many main line trains, costing them three of four minutes running time. There were also questions raised as to how the connections would be displayed in the timetable although, as a concession, a number of up expresses were given a stop at St. Deny's for the benefit of passengers from the Terminus or off the Netley line. No corresponding stops, however, were made by down trains.

Although taking second place in terms of status, of the two stations the author preferred

No.	74		75		76		77		79		80	
	Additional (B.98)		Additional (B.99)		Additional (B.100)		Additional (B.101)		Additional (B.102)		Additional (B.103)	
	arr	dep	arr	dep	arr	dep	arr	dep	arr	dep	arr	dep
Waterloo	TL	08 54	TL	09 29	TL	09 47	TL	10 22	TL	20 09	TL	21 00
Clapham Jn. ...	09 01		09 36		09 54		10 29		20 16		21 07	
Hampton Court Jn...	09 12		09 47		10 05		10 40		20 27		21 18	
Woking Jn.	09 23		09 58		10 16		10 52		20 38		21 29	
Brookwood	LL	..	LL	..	LL	..	LL	..	20F43	20LL48	LL	..
Farnborough	..	TL	..	TL	..	TL	..	TL	..	TL	..	TL
Worting Jn.	10 00		10 39		10 51		11 25		21 15		22 04	
Winchester Jn. ...	10 16		10 56		11 07		11 41		21 30		22 24	
Allbrook	TL	..	TL	..	TL	..	TL	..	TL	..	TL	..
Eastleigh	10 27		11 06		11 23		11 50		21 40		22 37	
St. Denys	..	..	..	..	..	..	..	..	..	..	..	..
Northam Jn.	10 40		11 12		11 31		11 56		21 49		22 49	
Northam..........	..	..	..	..	..	..	..	..	..	..	..	..
Southampton Term.	..	..	..	..	..	..	..	..	..	..	..	..
So'ton Dock Gates ..	10 47		..	11 17	..	..	..	..	..	..	..	..
So'ton Eastern Docks	10 57	..	11 27	..	..	..	..	..	..	..	..	..
Southampton Cen. ..	..	..	..	..	11 34TL		11 59TL		21 52LL		22 52LL	
Millbrook Dock Entr.	..	..	..	..	11 38		12 03		21 57		22 57	
So'ton Western Dks.	..	..	..	..	11 48	..	12 13	..	22 07	..	23 07	..

Although Boat train timings were quoted in the working book, they appeared as a reference and confirmation of those which were to operate on any particular day was done by circular and by inclusion in the weekly special traffic notice.

In the above example Trains 98 and 99 were run for the SS Chusan, 100 and 101 for the SS Canberra, 102 and 103 for the SS France. The numbers of trains required and the berth used would be notified by the shipping company. After unloading at Southampton trains 98,100 and 101 returned empty to Clapham Junction Carriage Sidings. Not all sailings required such lavish facilities and the SS Oranjefontein was dealt with by the strengthening of the 08.45 Waterloo - Lymington Pier as far as Southampton Central.

Arrivals were much lighter than departures on this particular day and the only up services were trains 104 and 105 for passengers arriving in the SS Arcadia.

No.	81		82	
	Additional (B.104)		Additional (B.105)	
	arr	dep	arr	dep
Southampton Western Docks	..	10 55	..	12 28
Millbrook Dk. Exit ..	11 05		12 38	
Millbrook	LL		TL	
Southampton Cen. ..	11 09		12 43	
Northam Jn.	11 12½		12 47	
Eastleigh	11 22LL		12 57	
Shawford U.H.S. ...	11*36			
Winchester Jn.	11 44½		13 16	
Worting Jn.	12 04½LL		13 36TL	
Farnborough	..		13 57LL	
Brookwood	..	TL	..	TL
Woking Jn.	12 38		14 12	
Hampton Court Jn. ..	12 48		14 24	
Clapham Jn.	12 59		14 34	
Waterloo	13 06	..	14 41	..

SOUTHAMPTON CENTRAL : 1954

Train	Arr	Engine	Shed	Dep	Destination
23/35 Eastleigh Yard		H15 4-6-0	EL 313	00/04	Bournemouth
21.55 WEYMOUTH	00.25	N15 4-6-0	BM 399	00.32	WATERLOO
00.36 Northam		Q0-6-0	EL 317	00/44	Brockenhurst
22.26 Hamworthy Jcn Goods		N15 4-6-0	BM 401	01/21	Bevois Park
22.30 WATERLOO		N15 4-6-0	EL 263	01/22	DORCHESTER
22.38 Nine Elms		H15 4-6-0	NE 73	02/02	Weymouth
19.35 Kensington (Pds)	03.19	N15 4-6-0	BM 399	03.27	Poole
22.45 Feltham		S15 4-6-0	Fel 110	03/47	Bournemouth
03.20 Bournemouth Goods		Q0-6-0	EL 325	04/45	Bevois Park
Light ex Eastleigh MPD	04.49	M7 0-4-4T	EL 302		(Wks 05.22 Up)
02.40 WATERLOO	04.56	MN 4-6-2	NE 30	05.07	BOURNEMOUTH C.
		M7 0-4-4T	EL 302	05.22	Eastleigh
04.50 Eastleigh Yard		N15 4-6-0	EL 266	05/27	Brockenhurst
05.28 Eastleigh (Fish)	05.39	T9 4-4-0	EL 280	05.44	Bournemouth W.
05.53 Southampton T.	05.58	BR4 2-6-0	EL 270	06.02	Totton
05.30 Eastleigh Yard	05.54	T9 4-4-0	EL 286	06.08	Millbrook Goods
06.05 Eastleigh	06.14	N15 4-6-0	BM 401	06.20	Dorchester
06.28 Totton	06.39	BR4 2-6-0	EL 270	06.42	Portsmouth
06.03 Fareham	06.45	M7 0-4-4T	FRA 371		(Wks 07.09 Up)
06.35 Eastleigh	06.52	M7 0-4-4T	EL 305	06.59	Fawley
05.17 Wimborne	06.59	LN 4-6-0	BM 394	07.01	Eastleigh
		M7 0-4-4T	FRA 371	07.09	Southampton T.
06.55 Northam Yard	07.03	Q0-6-0	EL 323	07.25	Millbrook Goods
05.58 Portsmouth	07.32	U 2-6-0	SAL 450	07.36	Salisbury
07.15 Eastleigh Yard		BR3 2-6-2T	EL 328	07/42	Fawley
06.35 BOURNEMOUTH C.	07.45	WC 4-6-2	BM 380	07.48	WATERLOO
06.45 Andover Jcn	07.52	T9 4-4-0	AJN 248	07.53	Southampton T.
05.40 WATERLOO	07.49	LN 4-6-0	NE 31	07.55	WEYMOUTH
07.46 Eastleigh	08.01	H15 4-6-0	NE 79	08.07	BOURNEMOUTH C.
06.55 Portsmouth	08.05	BR4 2-6-0	EL 271		(Engine change)
(06.55 Portsmouth)		T9 4-4-0	FRA 365	08.13	Romsey
07.20 BOURNEMOUTH W.	08.23	WC 4-6-2	NE 38	08.26	WATERLOO
07.56 Eastleigh Yard	08.28	BR3 2-6-2T	EL 330	08.32	Fawley
		BR4 2-6-0	EL 271	08.35	Light to S. Terminus
07.47 Salisbury	08.37	T9 4-4-0	SAL 445	08.41	Portsmouth
08.02 Fareham	08.42	M7 0-4-4T	EL 304		
08.06 Fawley	08.47	M7 0-4-4T	EL 305	08.50	Southampton T.
08.00 Romsey	08.50	LN 4-6-0	BM 394	08.54	Weymouth
07.30 Bournemouth W.	08.53	H15 4-6-0	EL 313	08.55	Eastleigh
		M7 0-4-4T	EL 304	09.06	Light to B. Park
08.52 Romsey	09.08	T9 4-4-0	FRA 365		(Fwd at 09.25)
07.34 WEYMOUTH	09.13	WC 4-6-2	BM 381	09.17	WATERLOO
(08.52 Romsey)		T9 4-4-0	FRA 365	09.25	Portsmouth
04.10 Cheltenham Goods		43xx 2-6-0	Chel 21	09/30	Southampton Docks
09.15 Eastleigh Yard		Q1 0-6-0	EL 318	09/42	Brockenhurst
08.35 BOURNEMOUTH W.	09.52	WC 4-6-2	BM 382	09.56	WATERLOO
09.03 Portsmouth	09.58	U 2-6-0	FRA 368	10.02	Salisbury
10.10 Southampton T.	10.15	43xx 2-6-0	AJN 249	10.18	Cheltenham
10.08 Redbridge Goods		T9 4-4-0	BM 404	10/21	Bevois Park
09.33 PORTSMOUTH	10.23	BR4 2-6-0	EL 270	10.26	CARDIFF
08.30 WATERLOO	10.26	MN 4-6-2	NE 32	10.31	WEYMOUTH
09.20 BOURNEMOUTH W.	10.31	N15 4-6-0	BM 399	10.35	BIRKENHEAD
10.29 Southampton T.	10.34	T9 4-4-0	EL 282	10.37	Bournemouth C.
09.47 Salisbury	10.34	U 2-6-0	SAL 485	10.40	Portsmouth
09.40 BRIGHTON	11.13	WC 4-6-2	BTN 730	11.16	BOURNEMOUTH W.
09.20 WEYMOUTH	11.16	LN 4-6-0	NE 31	11.20	WATERLOO
10.34 PORTSMOUTH	11.28	BR4 2-6-0	EL 272	11.32	BRISTOL
06.10 BRISTOL	11.29	U 2-6-0	SAL 450	11.35	PORTSMOUTH
09.30 WATERLOO	11.31	WC 4-6-2	NE 33	11.36	BOURNEMOUTH W.
11.30 Bevois Park		43xx 2-6-0	CH 21	11/38	Andover Jcn
11.28 Eastleigh	11.43	H15 4-6-0	EL 313		(Fwd at 12.10)
10.30 WATERLOO	11.59	MN 4-6-2	NE 34	12.04	WEYMOUTH
(11.28 Eastleigh)		H15 4-6-0	EL 313	12.10	Bournemouth C.
11.05 BOURNEMOUTH	12.16	WC 4-6-2	BM 383	12.20	WATERLOO
11.19 Portsmouth	12.26	LM2 2-6-2T	EL 306	12.28	Andover Jcn
11.16 BOURNEMOUTH W.	12.25	LN 4-6-0	BM 395	12.29	YORK
09.54 WATERLOO	12.40	LN 4-6-0	EL 253		(Wks 14.22 Down)
11.00 BRIGHTON	12.47	WC 4-6-2	BTN 731	12.54	CARDIFF
11.36 Bournemouth W	13.03	H15 4-6-0	NE 79	13.06	B. Park (Pds)
12.24 Totton Goods		Q0-6-0	EL 323	13/12	Bevois Park
11.30 WEYMOUTH	13.15	WC 4-6-2	BM 386	13.20	WATERLOO
11.30 BRIGHTON	13.16	WC 4-6-2	BTN 732	13.21	PLYMOUTH
10.27 BRISTOL	13.19	BR4 2-6-0	EL 270	13.23	Portsmouth
11.30 WATERLOO	13.18	LN 4-6-0	EL 252	13.25	BOURNEMOUTH W.
13.20 Northam		BR3 2-6-2T	EL 331	13/27	Fawley
13.15 Romsey	13.34	M7 0-4-4T	FRA 371	13.37	Portsmouth
12.30 Fawley Goods		BR3 2-6-2T	EL 328	13/50	Bevois Park

Train	Arr	Engine	Shed	Dep	Destination
10.11 Cheltenham	13.58	Manor 4-6-0	Chel 22	14.03	Southampton T.
12.30 WATERLOO (PULL)	13.58	MN 4-6-2	NE 35	14.03	BOURNEMOUTH W.
13.29 Fareham	14.07	T9 4-4-0	FRA 366		(Engine change)
10.30 CARDIFF	14.07	BR4 2-6-0	EL 272	14.12	PORTSMOUTH
13.05 BOURNEMOUTH W.	14.16	MN 4-6-2	NE 32	14.20	WATERLOO
(13.29 Fareham)		LN 4-6-0	EL 253	14.22	Bournemouth W.
13.15 Fawley Goods		BR3 2-6-2T	EL 330	14/27	Bevois Park
14.09 Eastleigh Yard		M7 0-4-4T	EL 305	14/31	Redbridge Yard
14.10 Winchester	14.50	T9 4-4-0	EL 283		(Wks 15.40 Down)
13.50 BRIGHTON	14.57	WC 4-6-2	BTN 730	15.00	BRIGHTON
14.03 Portsmouth	15.06	T9 4-4-0	FRA 365		(Shunt to bay)
13.25 WEYMOUTH	15.14	WC 4-6-2	BM 385	15.20	WATERLOO
14.33 PORTSMOUTH	15.16	U 2-6-0	SAL 485	15.20	BRISTOL
13.30 WATERLOO	15.20	WC 4-6-2	BM 381	15.24	WEYMOUTH
11.00 PLYMOUTH	15.30	WC 4-6-2	BTN 731	15.34	BRIGHTON
	15.35	N 2-6-0	FRA 370		(Wks 16.05 Up)
		T9 4-4-0	FRA 365	15.40	S. Terminus (ECS)
		T9 4-4-0	EL 283	15.40	Bournemouth C.
		N 2-6-0	FRA 370	16.05	Fareham
15.51 Eastleigh	16.02	M7 0-4-4T	EL 300	16.07	Fawley
09.30 BIRKENHEAD	16.12	N15 4-6-0	BM 399	16.17	BOURNEMOUTH W.
15.05 BOURNEMOUTH W.	16.16	MN 4-6-2	NE 34	16.22	WATERLOO
13.00 CARDIFF	16.32	WC 4-6-2	BTN 732	16.36	BRIGHTON
16.08 Brockenhurst	16.37	T9 4-4-0	EL 286	16.39	Portsmouth
16.36 Southampton T.	16.41	Manor 4-6-0	CH 22	16.44	Cheltenham
16.50 Redbridge Goods		S15 4-6-0	Fel 111	16/50	Bevois Park
15.20 WATERLOO	16.53	WC 4-6-2	BM 382	16.58	WEYMOUTH
14.20 Weymouth	17.06	T9 4-4-0	EL 282	17.08	Andover Jcn
17.05 Southampton T.	17.10	N15 4-6-0	EL 265	17.11	Wimborne
17.15 Southampton T.	17.21	M7 & T9	365/304		
16.34 BOURNEMOUTH W.	17.19	MN 4-6-2	NE 35	17.23	WATERLOO
15.30 WATERLOO	17.26	WC 4-6-2	NE 37	17.31	BOURNEMOUTH W.
		M7 0-4-4T	EL 304	17.35	Light to Fawley
10.23 YORK	17.37	LN 4-6-0	BM 395	17.40	BOURNEMOUTH W.
13.50 Cheltenham	17.42	43xx 2-6-0	AJN 250	17.44	Southampton T.
17.34 Eastleigh	17.49	M7 0-4-4T	EL 329	17.50	Brockenhurst
16.52 Pokesdown	17.49	H15 4-6-0	EL 313	17.52	Eastleigh
16.45 Portsmouth	17.52	U 2-6-0	SAL 450	17.56	Salisbury
17.07 Salisbury	17.53	T9 4-4-0	EL 284	17.58	Portsmouth
17.16 Fawley	18.00	M7 0-4-4T	EL 300	18.03	Andover Jcn
16.35 WATERLOO	18.08	WC 4-6-2	BM 383	18.12	WEYMOUTH
18.07 Southampton T.	18.12	T9 4-4-0	FRA 366	18.15	Bournemouth C.
17.05 BOURNEMOUTH W.	18.16	LN 4-6-0	EL 253	18.20	WATERLOO
		T9 4-4-0	FRA 365	18.27	Southampton T.
17.45 PORTSMOUTH	18.41	BR4 2-6-0	EL 272	18.45	CARDIFF
17.38 Bournemouth C.	18.47	LN 4-6-0	EL 252	18.49	Eastleigh
18.03 Portsmouth	19.13	BR4 2-6-0	EL 270		
17.35 WEYMOUTH	19.16	MN 4-6-2	NE 30	19.21	WATERLOO
18.43 Fawley	19.24	M7 0-4-4T	EL 304	19.26	Southampton T.
17.30 WATERLOO	19.23	WC 4-6-2	BM 380	19.28	BOURNEMOUTH W.
19.04 Southampton Docks		43xx 2-6-0	AJN 250	19/35	Cheltenham
		BR4 2-6-0	EL 270	19.40	Light to Eastleigh
16.32 BRISTOL	19.42	U 2-6-0	FRA 368	19.48	PORTSMOUTH
Light ex S. Terminus	19.45	Q0-6-0	BM 416		(Wks 20.10 Down)
14.45 Hamworthy Jcn Gds		Q1 0-6-0	EL 319	19/52	Eastleigh
18.45 Portsmouth	19.52	BR4 2-6-0	EL 271		
18.35 Bournemouth W.	19.58	N15 4-6-0	EL 266	20.03	(Attach Pilot)
(18.35 Bournemouth W.)		N15 & BR4	266/271	20.03	Reading
18.30 WATERLOO	20.01	WC 4-6-2	BM 386	20.05	WEYMOUTH
		Q0-6-0	BM 416	20.10	Bournemouth W.
16.35 CARDIFF	20.23	U 2-6-0	SAL 482	20.28	PORTSMOUTH
20.28 Redbridge Goods		BR3 2-6-2T	EL 320	20/33	Bevois Park
18.30 WEYMOUTH	20.45	WC 4-6-2	NE 33	20.50	WATERLOO
19.45 Portsmouth	20.55	T9 4-4-0	BM 403	21.01	Andover Jcn
19.05 Fawley Goods		BR3 & BR3	331/328	21/13	Eastleigh
Light ex B. Park	21.18	BR3 2-6-2T	EL 330		(Wks 22.04 Up)
19.30 WATERLOO	21.26	WC 4-6-2	BM 385	21.31	BOURNEMOUTH W.
19.55 Bournemouth W. (Pds)	21.39	N15 4-6-0	EL 265	21.42	Waterloo
21.20 Eatleigh Yard		Q0-6-0	EL 324	21/45	Brockenhurst
		BR3 2-6-2T	EL 330	22.04	Eastleigh
16.25 Dorchester Goods		Q0-6-0	EL 317	22/12	Bevois Park
19.48 Weymouth	22.31	N15 4-6-0	EL 263	22.35	Reading
19.10 BRISTOL	22.48	S15 4-6-0	SAL 467	22.50	EASTLEIGH
22.34 Winchester	23.03	M7 0-4-4T	EL 302		
		M7 0-4-4T	EL 302	23.15	Light to S. Terminus
19.50 Dorchester Goods		H15 4-6-0	NE 74	23/50	Nine Elms

– by quite a margin - the Terminus and while it dealt with less trains than the Central, each arrival involved stock movements and light engine running; all of which added to the interest.

Most of the trains were local but one or two had an international flavour and none more so than the two night mails which reversed in the Terminus – which the GPO preferred over the Central – and changed engines. The working of the mails was quite complicated since neither of the incoming engines worked the outward trains – both were blocked in by their respective coaches – and the up train was worked forward by the Lord Nelson which had arrived with the down Channel Islands Boat Train whilst the down went forward with the King Arthur off the 22.20 Parcels from Eastleigh. After the two mails had gone their respective ways the inward engine off the down service – a Lord Nelson 4-6-0 - ran light to Eastleigh loco; the King Arthur of the up train being turned and prepared for the 01.23 to Portsmouth. (This last service was peculiar in that it was formed of a three-coach GW set of vehicles, the stock working in with the 14.00 ex Cheltenham. The set returned from Portsmouth at 04.33 and travelled back to the Great Western in the 10.10 to Cheltenham).

Another large engine, a Lord Nelson 4-6-0, made an appearance in the station to work the 06.04 stopping train to Waterloo but after that most visitors were of a small to moderate size – M7 0-4-4T's and BR 4MT 2-6-0's – until the evening rush hour when an Eastleigh-based King Arthur arrived to work the 17.05 to Wimborne. 30788 'Sir Urre of the Mount' was one of the engines regularly chosen for this duty which was part of a three-day cyclic working with intervening nights spent at Basingstoke and

Nine Elms sheds.

Goods engines were understandably uncommon but a Feltham S15 4-6-0 could be seen in the mornings on the 09.19 local to Eastleigh; the engine having arrived with the 00.55 goods from Feltham to Southampton Docks.

During the formative years of the system the Great Western had attached considerable importance to being able to access Southampton and eventually succeeded via both the DN&S and the MSWJ. GW locomotives - including a Manor 4-6-0 - could be seen at various times of day but the services never amounted to very much although through coaches had once been conveyed to Glasgow via Didcot and to Sheffield via Cheltenham and the Midland Railway. The visitors provided an element of the unexpected and the best time to see the Great Western in the terminus was in the mid-afternoon when two successive GW workings pulled out: the 16.36 to Cheltenham behind a Manor 4-6-0 and the 16.50 Didcot with a 22xx 0-6-0. Other Great Western visitors included a Hall 4-6-0 on the 06.50 ex Reading and, booked to an evening Eastleigh parcels train, a Bristol-based 43xx 2-6-0.

A small number of Great Western engines also arrived on Southern workings because Andover Junction had been supplied with 43xx 2-6-0's for its workings both to Cheltenham and Southampton. One of its 43xx's worked in with the 07.31 from Woking, the 2-6-0 having taken the train over from a Guildford-based U 2-6-0 at Eastleigh. (The Guildford U later appeared in the Terminus with the 10.45 from Winchester before departing with an early afternoon train to Reading).

One of the great drawbacks was the lack of a main line motive power base at Southampton and if an engine failed at the last minute the locomotive Inspector had to use his wits and turn engines round as they arrived – a process that not only caused considerable confusion amongst the crews but guaranteed strange engines ending the day at curious places. There was no question of simply asking a running foreman to dig out a substitute engine, as would generally be the case, since Eastleigh loco was a good twenty minutes run away.

The nearest thing to an MPD was Southampton (71I) loco which existed to provide engines for the 78-mile docks' network. On paper the shed achieved the exceptional rate of 100% performance: 16 diagrams and 16 engines although in fact only twelve of the fourteen

SOUTHAMPTON MPD : ENGINE ARRIVALS/DEPARTURES					
Train	On Shed	Engine	Diagram	Off Shed	To Work
Town Yard Pilot	19.20	B4 0-4-0T	STN 346	05.30	Town Yard Pilot
Town Quay Pilot	18.00	C14 0-4-0T	STN 347	05.45	Town Quay Pilot
Front Yard Pilot	06.15*	USA 0-6-0T	STN 353	05.45*	Front Yard Pilot
Ocean Quay Pilot	22.00	USA 0-6-0T	STN 349	06.15	Ocean Quay Pilot
Ocean Quay Pilot	22.00	USA 0-6-0T	STN 350	06.15	Ocean Quay Pilot
Old Docks Pilot	22.00	USA 0-6-0T	STN 352	06.15	Old Docks Pilot
Empress Dock Pilot	22.00	USA 0-6-0T	STN 354	06.15	Empress Dock Pilot
Empress Dock Pilot	22.00	USA 0-6-0T	STN 355	06.15	Empress Dock Pilot
Old Docks Pilot	22.00	USA 0-6-0T	STN 356	06.15	Old Docks Pilot
New Docks Pilot	22.00	E1 0-6-0T	STN 359	06.15	New Docks Pilot
New Docks Pilot	22.00	USA 0-6-0T	STN 361	06.15	New Docks Pilot
New Docks Pilot	22.00	USA 0-6-0T	STN 358	06.15	New Docks Pilot
Empress Dock Pilot	05.45*	USA 0-6-0T	STN 351	06.15*	Empress Dock Pilot
New Docks Pilot	05.45*	E1 0-6-0T	STN 357	06.15*	New Docks Pilot
Old Docks Pilot	22.45	USA 0-6-0T	STN 348	07.15	Old Docks Pilot
New Docks Pilot	17.00	USA 0-6-0T	STN 360	09.15	New Docks Pilot

Off shed Mondays, returning Sundays.

Train From Class Engine Shed	07.43 Didcot M7 0-4-4T BM 406	09.12 Reading 32xx 2-6-0 DID 30	Gds HALL 4-6-0 RDG 50	S&D 700 0-6-0 SAL 452	Gds 2P 4-4-0 TCB 57	09.15 ELGH Gds G6 0-6-0T BM 418	05.30 ELGH Gds T9 4-4-0 EL 282	Q1 0-6-0 EL 318	T9 4-4-0 EL 286	09.40 Brighton M7 0-4-4T SWA 421	09.40 Brighton WC 4-6-=2 BTN 730	11.50 Yeovil T9 4-4-0 EL 282	R'car WEY	U 2-6-0 GUI 181
WATERLOO														
Woking														
Basingstoke		09.55												
Worting Jcn		10/01												
Micheldever		10.12												
Wallers Ash														
WINCHESTER		10.28												10.45
Shawford	10.22	10.34												
Eastleigh Yard														
EASTLEIGH	10.28	10.41												10.55
EASTLEIGH	10.30	10.49												11.02
Swaythling	10.36	(To												11.08
St Denys	10.40	Ports)									11/08			11.12
Bevois Park Yard														
Northam	10.44													11.16
Southampton (T)	10.47					10.29								11.19
SOUTHAMPTON CENT						10.34					11.13			
SOUTHAMPTON CENT						10.37					11.16			
Millbrook						10.41								
Redbridge						10.45					11/21			
Totton						10.48								
Lyndhurst Rd						10.55								
Beaulieu Rd						11.00								
BROCKENHURST	11.04					11.10	11.20	11.24			11.37			
Lymington Jcn	11/07					11/12	11/24				11/40			
Lymington Town														
Lymington Pier														
Sway	(To					11.17	(To							
New Milton	B'mouth)					11.25	Poole				11.48			
Hinton Admiral						11.32	via							
Christchurch						11.39	Wim)				11.57			
Pokesdown						11.45					12.03			
Boscombe						11.48					12.06			
Central Goods			11.35		11.40									
BOURNEMOUTH CENT					11.45	11.51					12.09			
BOURNEMOUTH CENT		11/38										12.14		
BOURNEMOUTH WEST				11.40								12.22		
Branksome		11.47												
Parkstone														
POOLE				11.50										
Hamworthy Jcn				(To										
Holton Heath				Bristol)										
WAREHAM											12.16			
Worgret Jcn											12/19			
Corfe Castle											12.29			
SWANAGE											12.38			
Wool														
Moreton														
DORCHESTER														
Dorchester Jcn														12/43
Monkton														12.46
Wishing Well														12.51
Upwey														12.54
Radipole														12.58
WEYMOUTH														13.01

COMBINED WORKING TIME TABLE AND ENGINE WORKINGS : 1954

West Country Pacific 34041 'Wilton' - one of the class later based at Eastleigh - is diagrammed for a 1964 special traffic duty which involves running a train of vans from Eastleigh to Southampton Central. Once most of the class had been rebuilt, Pacifics started appearing on all sorts of strange duties - ten years earlier an M7 0-4-4T would have been considered adequate for a handful of parcels vehicles.

WEYMOUTH - BASINGSTOKE : WORKING TIMETABLE 1954

Train From			13.15 Eastleigh	10.27 Bristol			11.36 B'mouth	13.15 Romsey	11.36 B'mouth	12.24 Totton			13.56 S'ton	09.05 Bristol
Class		Gds	Pcls	Pcls		Gds	Pcls		Pcls	Gds				
Engine	43xx	U 2-6-0	H5 4-6-0	BR4 2-6-0	M7 0-4-4T	S15 4-6-0	H15 4-6-0	M7 0-4-4T	H15 4-6-0	Q0-6-0	M7 0-4-4T	M7 0-4-4T	43xx 2-6-0	LM5 4-6-0
Shed	WBY	DOR 427	EL 311	EL 270	EL 297	Fel 104	NE 79	FRA 371	NE 79	EL 323	EL 308	EL 302	DID 20	BATH2
WEYMOUTH	11.40													
Radipole														
Upwey														
Wishing Well														
Monkton														
Dorchester Jcn	11/54													
DORCHESTER	(To W'bury)	12.10												
Moreton		12.28												
Wool														
SWANAGE														
Corfe Castle														
Worgret Jcn		(To												
WAREHAM		Ham												
Holton Heath		Jcn)												
Hamworthy Jcn														
POOLE														12.43
Parkstone														
Branksome														
BOURNEMOUTH WEST														12.53
BOURNEMOUTH CENTRAL														
BOURNEMOUTH CENTRAL														
Central Goods														
Boscombe														
Pokesdown														
Christchurch														
Hinton Admiral														
New Milton														
Sway														
Lymington Pier														
Lymington Town														
Lymington Jcn														
BROCKENHURST														
Beaulieu Road														
Lyndhurst Road														
Fawley														
Totton														
Redbridge				13/14						13.26					
Millbrook										13.31					
SOUTHAMPTON CENTRAL				13.19						13.34					
SOUTHAMPTON CENTRAL				13.23						13.37					
Southampton T.				13.23	13D20									13.56	
Northam				13.27										14.00	
Bevois Park Yard							13.44			13.46					
St Denys				13/28	13.32	13/37	13.48	13.50	13/50		14.04				
Swaythling				(To P'mth)	13.36			(To P'mth)			14.08				
EASTLEIGH					13.41		13.55				14.13				
EASTLEIGH					13.43	13/50			14.02	14/03				14.15	14.22
Eastleigh Yard										14.08					
Shawford			13.38					(To W'king)	14.11					14.23	14.30
WINCHESTER			13.57			14/11			14.22					14.30	
Weston														(To Alton)	(To D'cot)
Micheldever										14.39					
Worting Jcn			14.20				14/44		14/51						
Basingstoke			14.27					F'ham)	14.55						
Waterloo			C. Jcn						W'king						

With the third rail in place BR Standard 4MT 2-6-0 76005 runs into Southampton Central with the 07.37 Bournemouth Central - Eastleigh on 20th June 1966.

COMBINED WORKING TIME TABLE AND ENGINE WORKINGS : 1954

Train	07.43	10.34	11.35							01.30		10.30	10.30
From	C. Jcn	Ports	BM							Hoo Jn		W'loo	W'loo
Class	Pds		Gds			Gds				Gds	Gds		
Engine	H15 4-6-0	BR4 2-6-0	700 0-6-0	M7 0-4-4T	WC 4-6-2	43xx 2-6-0	H15 4-6-0	M7 0-4-4T	M7 0-4-4T	S15 4-6-0	M7 0-4-4T	MN 4-6-2	MN 4-6-2
Shed	NE 65	EL 272	SAL 452	BM 409	NE 33	CH 21	EL 313	LYM 362	BM 408	EL 310	HAM 422	NE 34	NE 30
WATERLOO					09.30							10.30	
Woking													
Basingstoke	10.19				10.41								
Worting Jcn	10.27				10/47					11/02		11/27	
Micheldever													
Wallers Ash										11.26			
WINCHESTER					11.08								
Shawford													
Eastleigh Yard													
EASTLEIGH	11.00				11.19								
EASTLEIGH					11.22			11.28					11/51
Swaythling								11.34					
St Denys		11.23						11.38					
Bevois Park Yard						11.30							
Northam													
Southampton (T)													
SOUTHAMPTON CENT	11.28				11.31		11.43						11.59
SOUTHAMPTON CENT	11.32				11.36	11/38							12.04
Millbrook													
Redbridge		11/37			11/41	11/45							
Totton	(To B'tol)												
Lyndhurst Rd													
Beaulieu Rd						(To A. Jn)							
BROCKENHURST					11.57			12.02	12.12				
Lymington Jcn					12/00			12/05	12/15			12/24	
Lymington Town								12.13					
Lymington Pier								12.15					
Sway									(To B'mouth)				
New Milton					12.09								
Hinton Admiral													
Christchurch					12.19								
Pokesdown					12.26								
Boscombe													
Centrl Goods													
BOURNEMOUTH CENT					12.30							12.43	
BOURNEMOUTH CENT					12.35								12.49
BOURNEMOUTH WEST				12.35	12.43								
Branksome			12.16	12.39									
Parkstone			12.21	12.43								(To Weymouth)	13.02
POOLE				12.49									
Hamworthy Jcn				(To Brock)									
Holton Heath													
WAREHAM											13.05		13.16
Worgret Jcn											13/09		13/19
Corfe Castle											13.23		
SWANAGE													
Wool													
Moreton													
DORCHESTER													13.36
Dorchester Jcn													13/38
Monkton													
Wishing Well													
Upwey													
Radipole													
WEYMOUTH													13.47

USA tank 30073 works a vanfit away from the Ocean Liner Terminal in Southampton Old Docks.   The terminal was a long-overdue facility which arrived too late in the day to have much more than a decade's full-time service.

WEYMOUTH - BASINGSTOKE : WORKING TIMETABLE 1954

	EL 318	EL 328	CHM 22	EL 272	LYM 362	BM 409	NE 32	BM 408	BM 416	BM 407	BM 401	DOR 427	BRD	WEY	EL 330
Train	03.20		10.11	10.30	12.55			12.12	08.45			12.00			
From	ELGH		Ch'ham	Cardiff	B'mouth			B'rock	Poole			Dorset			
Class	Gds	Gds							Gds			Gds			Gds
Engine	Q1 0-6-0	BR3 2-6-2T	78xx 4-6-0	BR4 2-6-0	M7 0-4-4T	M7 0-4-4T	MN 4-6-2	M7 0-4-4T	Q0-6-0	M7 0-4-4T	N15 4-6-0	U 2-6-0	Hal 4-6-0	R'car	BR3 2-6-2T
WEYMOUTH											12.20		12.35	12.50	
Radipole														12.54	
Upwey											12.26			12.58	
Wishing Well														13.03	
Monkton														13.08	
Dorchester Jcn											12/36		12/54	13/11	
DORCHESTER											12.41		(To B'tol)	(To Yeovil)	
Moreton											12.52	13.07			
Wool											13.01	13.16			
SWANAGE										12.42					
Corfe Castle										12.53					
Worgret Jcn										13/02	13/06				
WAREHAM		(Via Wim)								13.04	**13.10**				
Holton Heath											13.16				
Hamworthy Jcn											13.21				
POOLE	_12.58_							13.08			13.28				
Parkstone								13.14			13.33				
Branksome								13.18			13.38				
BOURNEMOUTH WEST							13.05	13.22							
BOURNEMOUTH CENTRAL							13.13				13.43				
BOURNEMOUTH CENTRAL							13.18								
Central Goods															
Boscombe							13.22								
Pokesdown							13.25		(Via Wim)						
Christchurch							13.30								
Hinton Admiral															
New Milton							13.42								
Sway							13.49								
Lymington Pier					13.22										
Lymington Town					13.25										
Lymington Jcn					13/35	13/41	13/53		13/58						
BROCKENHURST					**13.36**	13/42	13.57		**14.02**						
Beaulieu Road															
Lyndhurst Road															
Fawley		12.30													13.15
Totton		13.40													14.15
Redbridge		13/45	13/53	14/02			14/11								14/20
Millbrook															
SOUTHAMPTON CENTRAL			13.58	14.07			14.16								
SOUTHAMPTON CENTRAL		_13/50_	14.03	14.12			14.20								_14/27_
Southampton T.			_14.08_												
Northam															
Bevois Park Yard		_14.00_													_14.35_
St Denys				14/18											
Swaythling				(To P'mth)											
EASTLEIGH							14.31								
EASTLEIGH							14.33								
Eastleigh Yard															
Shawford															
WINCHESTER							14.47								
Weston															
Micheldever															
Worting Jcn							15/10								
Basingstoke							15.14								
Waterloo							16.14								

ST. DENYS : 1954

Train	Arr	Engine	Shed	Dep	Destination
19.50 Dorchester (Gds)		H15 4-6-0	NE 74	00/15	Nine Elms
00.01 Southampton T.		T9 4-4-0	EL 282	00/50	Eastleigh Yard
22.30 WATERLOO		LN 4-6-0	EL 253	00/50	DORCHESTER
21.55 WEYMOUTH		LN 4-6-0	NE 31	01/14	WATERLOO
01.23 Southampton T.	01.28	N15 4-6-0	BM 399	01.29	Ports via Eastleigh
19.55 Kensington (Pds)		BR4 2-6-0	EL 273	01/35	Poole
22.38 Nine Elms (Gds)		H15 4-6-0	NE 73	01/45	Weymouth
22.20 Feltham (Gds)		S15 4-6-0	FEL 109	01/58	Southampton Docks
22.26 Hamworthy Jcn		N15 4-6-0	BM 401	02/03	Eastleigh Yard
03.10 Bevois Park (Gds)		S15 4-6-0	FEL 108	03/14	Woking
23.25 Nine Elms (Gds)		BR4 2-6-0	EL 272	03/25	Southampton Docks
22.45 Feltham (Gds)		S15 4-6-0	FEL 110	03/36	Bournemouth C.
03.40 Eastleigh (Pds)		U 2-6-0	SAL 450	03/48	Southampton Docks
02.40 WATERLOO		MN 4-6-2	NE 30	04/51	BOURNEMOUTH C.
03.20 Bournemouth (Gds)		S15 4-6-0	FEL 110	05/14	Feltham
04.50 Eastleigh Yard		N15 4-6-0	EL 266	05/16	Brockenhurst
05.15 Southampton T.	05.21	T9 4-4-0	EL 283	05.32	Portsmouth
05.17 Eastleigh	05.26	M7 0-4-4T	EL 298	05.29	Southampton T.
05.22 Southampton C	05.28	M7 0-4-4T	EL 302	05.31	Eastleigh
05.28 Eastleigh (Fish)		T9 4-4-0	EL 280	05/34	Bournemouth W.
05.35 Bevois Park		Q1 0-6-0	EL 314	05/40	Didcot
05.30 Eastleigh Yard		T9 4-4-0	EL 286	05/44	Millbrook
04.25 Salisbury Goods		Q1 0-6-0	EL 319	06/00	Northam Yard
06.05 Eastleigh		N15 4-6-0	BM 401	06/10	Dorchester
06.04 Southampton T	06.11	LN 4-6-0	EL 251	06.12	Waterloo
06.20 Eastleigh	06.29	M7 0-4-4T	EL 297	06.30	Southampton T.
06.03 Fareham	06.39	M7 0-4-4T	FRA 371	06.40	Southampton C
06.35 Eastleigh	06.44	M7 0-4-4T	EL 305	06.46	Fawley
06.28 Totton	06.48	BR4 2-6-0	EL 270	06.49	Portsmouth
06.43 Southampton T	06.50	M7 0-4-4T	EL 298	06.51	Eastleigh
06.36 Winchester	07.01	BR4 2-6-0	EL 277	07.02	Southampton T.
05.58 Portsmouth	07.02	T9 4-4-0	FRA 365	07.05	Salisbury
06.57 Southampton T	07.04	M7 0-4-4T	EL 308		(Fwd at 07.12)
05.17 Wimborne	07.07	LN 4-6-0	BM 394	07.08	Eastleigh
(06.57 Southampton T)		M7 0-4-4T	EL 308	07.12	Fareham
00.55 Feltham (Gds)		S15 4-6-0	FEL 102	07/16	Southampton Docks
06.58 Winchester	07.22	M7 0-4-4T	EL 302	07.23	Southampton T.
07.25 Bevois Park		Q 0-6-0	EL 322	07/28	Fareham
07.15 Eastleigh Yard		BR3 2-6-2T	EL 328	07/30	Fawley
06.23 Portsmouth	07.29	BR4 2-6-0	EL 316	07.32	Southampton T.
07.32 Southampton T.	07.39	BR4 2-6-0	EL 277	07.40	Didcot
05.40 WATERLOO		LN 4-6-0	NE 31	07/44	WEYMOUTH
07.39 Southampton T	07.46	BR4 2-6-0	EL 272	07.47	Portsmouth
06.35 BOURNEMOUTH C	07.54	WC 4-6-2	BM 380	07.55	WATERLOO
07.46 Eastleigh	07.55	H15 4-6-0	NE 79	07.56	BOURNEMOUTH C.
06.55 Portsmouth	07.59	BR4 2-6-0	EL 271	08.00	Romsey
06.45 Andover Jcn	08.09	M7 0-4-4T	FRA 371	08.10	Portsmouth
07.56 Eastleigh Yard		BR3 2-6-2T	EL 330	08/20	Fawley
07.20 BOURNEMOUTH W.		WC 4-6-2	NE 38	08.31	WATERLOO
08.02 Fareham	08.36	M7 0-4-4T	EL 304	08.37	Southampton C
06.50 Reading	08.37	HALL 4-6-0	RDG66	08.39	Southampton T.
08.33 Southampton T.	08.40	T9 & BR4	248/316	08.41	Andover Jcn
08.00 Romsey	08.44	LN 4-6-0	BM 394	08.45	Weymouth
07.47 Salisbury	08.47	T9 4-4-0	SAL 445	08.48	Portsmouth
04.07 Feltham (Gds)		S15 4-6-0	FEL 103	08/53	Southampton Docks
07.30 Bournemouth W	09.01	H15 4-6-0	EL 313	09.02	Eastleigh
08.07 Portsmouth	09.07	T9 4-4-0	EL 283	09.08	Southampton T.
07.31 Woking	09.13	43xx 2-6-0	AJN 249	09.14	Southampton T.
07.34 WEYMOUTH		WC 4-6-2	BM 381	09/22	WATERLOO
07.45 Newbury	09.21	22xx 0-6-0	DID 13	09.22	Southampton T.
09.19 Southampton T.	09.26	S15 4-6-0	FEL 102	09.27	Eastleigh
08.52 Romsey	09.32	T9 4-4-0	FRA 365	09.33	Portsmouth
09.15 Eastleigh Yard		Q1 0-6-0	EL 318	09/34	Brockenhurst
09.36 Southampton T	09.43	M7 0-4-4T	EL 305	09.44	Eastleigh
09.45 Bevois Park		M7 0-4-4T	EL 304	09/48	Bitterne
09.03 Portsmouth	09.52	U 2-6-0	FRA 368	09.53	Salisbury
08.35 BOURNEMOUTH W.	10.02	WC 4-6-2	BM 382	10.03	WATERLOO
09.55 Eastleigh	10.04	N15 4-6-0	EL 259	10.05	Southampton C
09.33 PORTSMOUTH	10.17	BR4 2-6-0	EL 270	10.18	CARDIFF
10.13 Southampton T	10.20	HALL 4-6-0	RDG66	10.21	Reading
08.30 WATERLOO		MN 4-6-2	NE 32	10/21	WEYMOUTH
10.23 Bevois Park		H15 4-6-0	EL 312	10/27	Salisbury
10.25 Bitterne		M7 0-4-4T	EL 304	10/33	Northam Yard
07.42 Didcot	10.39	22xx 0-6-0	DID 30	10.40	Southampton T.
09.20 BOURNEMOUTH W.		N15 4-6-0	BM 399	10/40	BIRKENHEAD
09.47 Salisbury	10.46	U 2-6-0	SAL 485	10.47	Portsmouth
09.40 BRIGHTON		WC 4-6-2	BTN 730	11/08	BOURNEMOUTH W.
10.45 Winchester	11.11	U 2-6-0	GUI 181	11.12	Southampton T.
10.34 PORTSMOUTH	11.22	BR4 2-6-0	EL 272	11.23	BRISTOL
09.20 WEYMOUTH		N15 4-6-0	EL 263	11/25	WATERLOO
09.30 WATERLOO		WC 4-6-2	NE 33	11/26	BOURNEMOUTH W.
11.28 Eastleigh	11.37	H15 4-6-0	EL 313	11.38	Bournemouth C.
08.10 BRISTOL	11.41	U 2-6-0	SAL 450	11.42	PORTSMOUTH
11.40 Southampton T (Pds)		T9 4-4-0	BM 403	11/46	Eastleigh
11.45 Southampton T.	11.52	43xx 2-6-0	DID 30	11.53	Didcot
10.30 WATERLOO		MN 4-6-2	NE 34	11/54	WEYMOUTH
11.12 Fareham Goods		N 2-6-0	FRA 370	11/59	Northam Yard
12.00 Southampton T.	12.05	T9 4-4-0	EL 283	12.06	Romsey
11.55 Eastleigh Yard		BR3 2-6-2T	EL 331	12/10	Northam Yard
11.19 Portsmouth	12.19	LM2 2-6-2T	EL 306	12.21	Andover Jcn
11.05 BOURNEMOUTH W.	12.26	WC 4-6-2	BM 383	12.27	WATERLOO
12.23 Southampton T	12.30	BR4 2-6-0	EL 271	12.35	Portsmouth
11.16 BOURNEMOUTH W		LN 4-6-0	BM 395	12/34	YORK
09.54 Waterloo	12.33	LN 4-6-0	EL 253	12.35	Southampton C
12.31 Southampton T.	12.38	U 2-6-0	GUI 181	12.39	Reading
11.00 BRIGHTON		WC 4-6-2	BTN 731	12/42	CARDIFF
11.30 BRIGHTON		WC 4-6-2	BTN 732	13/11	PLYMOUTH
11.30 WATERLOO		LN 4-6-0	EL 252	13/14	BOURNEMOUTH W.
13.15 Eastleigh	13.24	M7 0-4-4T	EL 308	13.25	Southampton T.
11.30 WEYMOUTH		WC 4-6-2	BM 386	13/25	WATERLOO
10.27 BRISTOL		BR4 2-6-0	EL 270	13/28	PORTSMOUTH
13.23 Southampton T	13.31	M7 0-4-4T	EL 297	13.32	Winchester
13.20 Southampton Docks		S15 4-6-0	FEL 104	13/37	Feltham
13.15 Romsey	13.43	M7 0-4-4T	FRA 371		(Fwd at 13.50)
11.36 Bournemouth W	13.46	H15 4-6-0	NE 79	13.48	Woking
(13.15 Romsey)		M7 0-4-4T	FRA 371	13.50	Portsmouth
12.24 Totton		Q 0-6-0	EL 323	13/50	Eastleigh Yard
12.30 WATERLOO		MN 4-6-2	NE 35	13/53	BOURNEMOUTH W.
13.29 Fareham	14.01	T9 4-4-0	FRA 366	14.02	Bournemouth W.
13.56 Southampton T	14.03	M7 0-4-4T	EL 308	14.04	Didcot
13.06 Salisbury	14.14	T9 4-4-0	SAL 443	14.15	Southampton T.
10.30 CARDIFF		BR4 2-6-0	EL 272	14/18	PORTSMOUTH
14.09 Eastleigh Yard		M7 0-4-4T	EL 305	14/21	Redbridge
13.05 BOURNEMOUTH W.		MN 4-6-2	NE 32	14/25	WATERLOO
14.10 Winchester	14.44	T9 4-4-0	EL 283	14.45	Bournemouth C.
14.03 Portsmouth	15.00	T9 4-4-0	FRA 365	15.01	Southampton C
14.55 Southampton T	15.02	T9 4-4-0	FRA 366	15.03	Eastleigh
13.50 BOURNEMOUTH W.	15.06	MN 4-6-2	BTN 730	15.07	BRIGHTON
14.33 PORTSMOUTH		U 2-6-0	SAL 485	15/11	BRISTOL
13.30 WATERLOO		WC 4-6-2	BM 381	15/15	WEYMOUTH
15.09 Eastleigh Yard		H15 4-6-0	NE 65	15/20	Southampton Docks
13.25 WEYMOUTH		WC 4-6-2	BM 385	15/25	WATERLOO
13.12 Fareham Goods		BR4 2-6-0	EL 321	15/30	Northam Yard
12.42 Didcot	15.32	22xx 0-6-0	DID 31	15.33	Southampton T.
11.00 PLYMOUTH		WC 4-6-2	BTN 732	15/39	BRIGHTON
15.51 Eastleigh		M7 0-4-4T	EL 300	15/57	Fawley
09.30 BIRKENHEAD		N15 4-6-0	BM 399	16/07	BOURNEMOUTH W.
16.05 Southampton C	16.11	N 2-6-0	FRA 370		(Fwd at 16.17)
16.06 Eastleigh	16.15	M7 0-4-4T	EL 297	16.16	Southampton T.
(16.05 Southampton C)		N 2-6-0	FRA 370	16.17	Fareham
15.05 BOURNEMOUTH W.	16.28	MN 4-6-2	NE 34	16.29	WATERLOO
16.26 Southampton T	16.32	M7 0-4-4T	EL 305	16.33	Eastleigh
16.20 Southampton Docks		S15 4-6-0	FEL 102	16/38	Feltham
13.00 CARDIFF		WC 4-6-2	BTN 731	16/41	BRIGHTON
16.08 Brockenhurst	16.45	T9 4-4-0	EL 286	16.46	Portsmouth
15.20 WATERLOO		WC 4-6-2	BM 382	16/48	WEYMOUTH
15/07 Salisbury Goods		43xx 2-6-0	BTL 365	17/01	Northam Yard
16.56 Southampton T	17.03	M7 0-4-4T	EL 299	17.04	Didcot
14.20 Weymouth	17.14	T9 4-4-0	EL 282	17.15	Andover Jcn
16.10 Alton	17.17	M7 0-4-4T	EL 302	17.18	Southampton T.
15.30 WATERLOO		WC 4-6-2	NE 37	17/21	BOURNEMOUTH W.
16.34 BOURNEMOUTH		MN 4-6-2	NE 35	17/28	WATERLOO
17.03 Romsey	17.28	LM2 2-6-2T	AJN 247	17.29	Southampton T.
17.23 Southampton T	17.30	M7 0-4-4T	EL 297	17.31	Alton
10.23 YORK		LN 4-6-0	BM 395	17/32	BOURNEMOUTH W.
17.28 Southampton T	17.35	T9 4-4-0	BM 403	17.36	Portsmouth
17.34 Eastleigh	17.43	M7 0-4-4T	EL 329	17.44	Brockenhurst
16.45 Portsmouth	17.46	U 2-6-0	SAL 450	17.47	Salisbury
17.38 Southampton T	17.45	BR4 2-6-0	EL 321	17.49	Winchester
17.50 Bevois Park (Gds)		H15 4-6-0	NE 68	17/55	Basingstoke
16.38 Bournemouth C	17.58	H15 4-6-0	EL 313	17.59	Eastleigh
16.35 WATERLOO		WC 4-6-2	BM 383	18/03	WEYMOUTH
17.07 Salisbury	18.06	T9 4-4-0	EL 284		(Fwd at 18.14)
17.16 Fawley	18.09	M7 0-4-4T	EL 300	18.10	Andover Jcn
14.56 Oxford	18.11	BR4 2-6-0	EL 277	18.12	Southampton T.
(17.07 Salisbury)		T9 4-4-0	EL 284	18.14	Portsmouth
17.05 BOURNEMOUTH W		LN 4-6-0	EL 253	18/25	WATERLOO
17.45 PORTSMOUTH	18.35	BR4 2-6-0	EL 272	18.36	CARDIFF
17.10 Reading	18.37	U 2-6-0	GUI 181	18.38	Southampton T.
17.33 Southampton Docks		H15 4-6-0	NE 65	18/42	Eastleigh Yard
18.40 Southampton T	18.47	T9 4-4-0	FRA 365	18.48	Portsmouth
18.42 Woolston Goods		T9 4-4-0	SAL 443	18/50	Northam Yard
17.38 Bournemouth C	18.55	LN 4-6-0	EL 252	18.56	Eastleigh
18.58 Southampton T	19.05	LM2 2-6-2T	AJN 247	19.06	Alton
18.03 Portsmouth	19.07	BR4 2-6-0	EL 270	19.08	Southampton C
18.55 Southampton Docks		N15 4-6-0	EL 264	19/11	Nine Elms
17.30 WATERLOO		WC 4-6-2	BM 380	19/18	BOURNEMOUTH W.
17.35 WEYMOUTH		MN 4-6-2	NE 30	19/26	WATERLOO
18.02 Alton	19.29	M7 0-4-4T	EL 299	19.30	Southampton T.
14.45 Feltham (Gds)		S15 4-6-0	FEL 107	19/35	Southampton Docks
18.45 Portsmouth	19.45	BR4 2-6-0	EL 271	19.47	Bournemouth W.
19.40 Southampton T	19.47	M7 0-4-4T	EL 302	19.48	Winchester
16.32 BRISTOL	19.54	U 2-6-0	FRA 368	19.55	PORTSMOUTH
18.30 WATERLOO		WC 4-6-2	BM 386	19/56	WEYMOUTH
18.35 Bournemouth W	20.09	N15 4-6-0	EL 266	20.10	Reading
20.03 Eastleigh (Pds)		S15 4-6-0	FEL 106	20/11	Southampton Docks
20.04 Southampton Dock (Pds)		43xx 2-6-0	BTL 365	20/31	Eastleigh
19.00 Reading	20.31	BR4 2-6-0	EL 270	20.32	Southampton T.
16.35 CARDIFF	20.34	U 2-6-0	SAL 482	20.37	PORTSMOUTH
19.45 Portsmouth	20.49	T9 4-4-0	BM 403	20.50	Andover Jcn
17.55 Didcot	20.51	BR4 2-6-0	EL 278	20.52	Southampton T.
18.30 WEYMOUTH	20.56	WC 4-6-2	NE 33	20.57	WATERLOO
21.05 Southampton T	21.12	M7 0-4-4T	EL 299	21.13	Eastleigh
19.30 WATERLOO	21.20	WC 4-6-2	BM 385	21.21	BOURNEMOUTH W.
21.20 Eastleigh Yard		Q 0-6-0	EL 324	21/35	Brockenhurst
19.05 Fawley		2 x BR3	331/328	21/40	Eastleigh Yard
19.55 Bournemouth W (Pds)		N15 4-6-0	EL 265	21/48	Waterloo
21.56 Bevois Park		S15 4-6-0	FEL 106	22/00	Feltham
21.51 Eastleigh	22.00	T9 4-4-0	EL 282	22.01	Southampton T.
22.00 Southampton T	22.07	BR4 2-6-0	EL 270		(Fwd at 22.14)
22.04 Southampton C	22.10	BR3 2-6-2T	EL 330	22.10	Eastleigh
(22.00 Southampton T)		BR4 2-6-0	EL 270	22.14	Portsmouth
22.20 Eastleigh (Pds)		N15 4-6-0	EL 263	22/27	Southampton Docks
21.38 Portsmouth	22.35	T9 4-4-0	EL 286	22.36	Southampton T.
21.00 WATERLOO		LN 4-6-0	NE 31	22/40	SOUTHAMPTON TERM.
19.48 Weymouth	22.41	N15 4-6-0	EL 263	22.42	Reading
22.30 Southampton Docks		U 2-6-0	GUI 181	22/47	Feltham
22.34 Winchester	22.57	M7 0-4-4T	EL 302	22.58	Southampton C
19.10 Bristol	22.56	S15 4-6-0	SAL 467	23.01	Eastleigh
19.25 Nine Elms		H15 4-6-0	EL 311	23/10	Southampton Docks
23.35 Eastleigh		H15 4-6-0	EL 313	23.53	Brockenhurst

Without question, the best vantage point in the Southampton area was St. Denys where the London and Portsmouth lines merged and with an average of about eight trains an hour passing through, it was not only one of the busiest junctions outside the electrified area but had a variety of motive power that probably had no equal anywhere else. Amongst several especial points of interest was the imbalance of London services: almost all the Waterloo - Bournemouth expresses ran through non-stop whilst many of those in the opposite direction were booked to stop.

The introduction of the Hastings diesel multiple-units caused a motive power upheaval in parts of Kent with Schools class 4-4-0's moving from Hastings and London to Ashford and displacing numbers of N15 4-6-0's, some of which found their way onto the South Western. King Arthur N15 30803 'Sir Harry le Fise Lake' which moved from Ashford to Eastleigh in June 1959, restarts the 06.05 Eastleigh - Dorchester South from New Milton. 30803 was one of the N15's built in 1929 for the LBSCR and therefore had the distinction of working on all three constituents of the Southern.

SOUTHAMPTON TERMINUS : 1954

Train	Arr	Engine	Shed	Dep	Destination
21.55 WEYMOUTH	00.37	N15 4-6-0	BM 399		(Fwd at 01.10)
22.30 WATERLOO	00.53	LN 4-6-0	EL 253		(Fwd at 01.18)
(21.55 Weymouth)		LN 4-6-0	NE 31	01.10	WATERLOO
(22.30 W'loo)		N15 4-6-0	EL 263	01.18	POOLE
		N15 4-6-0	BM 399	01.23	Portsmouth & S
		LN 4-6-0	EL 253	01.30	Light Eastleigh loco
03.40 Eastleigh (Pds)	03.53	U 2-6-0	SAL 450		(Fwd at 04.34)
Light ex Eastleigh loco	04.25	T9 4-4-0	EL 283		(Wks 05.15 Ports)
(Eastleigh Pds)		B4 0-4-0T	STN 346	04.34	Docks
Light ex Docks	04.40	BR4 2-6-0	EL 272		
Light ex Eastleigh loco	05.09	LN 4-6-0	EL 251		(Wks 06.04 Waterloo)
		T9 4-4-0	EL 283	05.15	Portsmouth & S
05.17 Eastleigh	05.37	M7 0-4-4T	EL 298		
Light ex S. Centrd	05.40	BR4 2-6-0	EL 270		(Wks 05.53 Totton)
		BR4 2-6-0	EL 270	05.53	Totton
		LN 4-6-0	EL 251	06.04	WATERLOO
Light ex Eastleigh loco	06.13	M7 0-4-4T	EL 308		(Wks 06.57 Fareham)
06.20 Eastleigh	06.37	M7 0-4-4T	EL 297		(Stn pilot until 13.23)
		M7 0-4-4T	EL 298	06.43	Eastleigh
		M7 0-4-4T	EL 308	06.57	Fareham
06.30 Winchester	07.09	BR4 2-6-0	EL 277		
05.58 Portsmouth & S	07.12	T9 4-4-0	FRA 365		(Fwd at 07.27)
07.09 Southampton Centrd	07.15	U 2-6-0	FRA 371		
(05.58 Ports)		U 2-6-0	SAL 450	07.27	Salisbury
		B4 0-4-0T	STN 346	07.30	Docks (Pds)
06.58 Winchester	07.30	M7 0-4-4T	EL 302		
		BR4 2-6-0	EL 277	07.32	Didcot
		BR4 2-6-0	EL 272	07.39	Portsmouth & S
06.23 Portsmouth & S	07.40	BR4 2-6-0	EL 316		
		M7 0-4-4T	EL 302	07.45	Light to Eastleigh loco
		T9 4-4-0	FRA 365	07.47	Light to S. Cent
06.45 Andover Jcn	07.58	T9 4-4-0	AJN 248		(Fwd at 08.02)
(06.45 A. Jcn)		M7 0-4-4T	EL 371	08.02	Portsmouth & S
Light ex Docks	08.30	S15 4-6-0	FEL 102		(Wks 09.19 Eastleigh)
		T9 & BR4	248/316	08.33	Andover Jcn
Light ex S. Centrd	08.40	BR4 2-6-0	EL 271		(Wks Docks Trip)
06.50 Reading	08.46	HALL 4-6-0	RDG 66		
08.06 Fawley	08.55	M7 0-4-4T	EL 305		
08.07 Portsmouth & S	09.15	T9 4-4-0	EL 283		(Pilot to 12.00)
		S15 4-6-0	FEL 102	09.19	Eastleigh
07.31 Woking	09.20	43xx 2-6-0	AJN 249		
07.45 Newbury	09.27	22xx 0-6-0	DID 13		
(Off Docks trip)		BR4 2-6-0	EL 271	09.40	Light to Northam Yard
		M7 0-4-4T	EL 305	09.36	Eastleigh
Light ex Eastleigh loco	09.53	T9 4-4-0	EL 282		(Wks 10.29 B'mouth)
		22xx 0-6-0	DID 13	10.00	Light to Eastleigh loco
		43xx 2-6-0	AJN 249	10.10	Cheltenham
		HALL 4-6-0	RDG 66	10.13	Reading
		T9 4-4-0	EL 282	10.29	Bournemouth Centrd
07.42 Didcot	10.47	22xx 0-6-0	DID 30		
Light ex Bevois Park	10.55	T9 4-4-0	BM 403		
		U 2-6-0	GUI 181		
10.45 Winchester	11.19				
		T9 4-4-0	BM 403	11.40	Eastleigh (Pds)
		22xx 0-6-0	DID 30	11.45	Didcot
Light ex Northam Yard	11.55	BR4 2-6-0	EL 271		(Wks 12.23 Ports)
		T9 4-4-0	EL 283	12.00	Romsey
		BR4 2-6-0	EL 271	12.23	Portsmouth & S
		U 2-6-0	GUI 181	12.31	Reading
(Ex Station Pilot)		M7 0-4-4T	EL 297	13.23	Winchester
13.15 Eastleigh	13.32	M7 0-4-4T	EL 308		
		M7 0-4-4T	EL 308	13.56	Didcot
10.11 Cheltenham	14.08	78xx 4-6-0	CHM 22		
13.06 Salisbury	14.22	T9 4-4-0	SAL 443		(Wks Docks trips)
Light ex S. Centrd	14.30	T9 4-4-0	FRA 366		(Wks 14.55 Eastleigh)
		T9 4-4-0	FRA 366	14.55	Eastleigh
		T9 4-4-0	SAL 443	15.30	Light to S. Centrd
12.42 Didcot	15.40	22xx 0-6-0	DID 31		
15.40 Southampton Central ECS	15.45	T9 4-4-0	FRA 365		
Light ex Eastleigh	15.50	T9 4-4-0	FRA 366		(Wks 18.07 B'mouth)
Light ex Redbridge	15.53	M7 0-4-4T	EL 305		(Wks 16.25)
16.12 Goods ex Bevois Park	16.22	BR4 2-6-0	EL 321		(Wks 17.38 W'ter)
16.06 Eastleigh	16.23	M7 0-4-4T	EL 297		
		M7 0-4-4T	EL 305	16.25	Eastleigh
Light ex Eastleigh Loco	16.30	N15 4-6-0	EL 265		(Wks 17.05)
		78xx 4-6-0	CHM 22	16.36	Cheltenham
		22xx 0-6-0	DID 31	16.56	Didcot
		N15 4-6-0	EL 265	17.05	Wimborne
		T9 4-4-0	FRA 365	17.15	Southampton Centrd
		M7 0-4-4T	EL 297	17.23	Alton
16.10 Alton	17.24	M7 0-4-4T	EL 302		(Pilot to 19.40)
		T9 4-4-0	BM 403	17.28	Portsmouth & S
17.03 Romsey	17.35	2MT 2-6-2T	AJN 247		
		BR4 2-6-0	EL 321	17.38	Winchester
13.56 Cheltenham	17.49	43xx 2-6-0	AJN 250		
		T9 4-4-0	FRA 366	18.07	Bournemouth Centrd
14.56 Oxford	18.19	BR4 2-6-0	EL 277		(Pilot until 20.57)
18.27 Southampton Central	18.32	T9 4-4-0	FRA 365		(Fwd at 18.40)
		43xx 2-6-0	AJN 250	18.35	Light to Docks
(18.27 ex S'ton C)		T9 4-4-0	FRA 365	18.40	Portsmouth & S
17.10 Reading	18.45	U 2-6-0	GUI 181		(Pilot until 21.55)
		2MT 2-6-2T	AJN 247	18.58	Alton
18.43 Fawley	19.31	M7 0-4-4T	EL 304		
18.02 Alton	19.37	M7 0-4-4T	EL 299		(Pilot until 21.05)
(Ex Pilot)		M7 0-4-4T	EL 302	19.40	Winchester
20.09 Docks Pds	20.11	B4 0-4-0T	STN 346		(Fwd at 20.25)
		M7 0-4-4T	EL 304	20.18	Light to Eastleigh loco
(Eastleigh Pds)		B4 0-4-0T	STN 346	20.20	Docks
19.00 Reading	20.39	BR4 2-6-0	EL 270		
		BR4 2-6-0	EL 277	20.57	Light to Eastleigh loco
17.55 Didcot	20.59	BR4 2-6-0	EL 278		
(Ex Pilot)		M7 0-4-4T	EL 299	21.05	Eastleigh
		BR4 2-6-0	EL 278	21.30	Light to Docks
		U 2-6-0	GUI 181	21.55	Light to Docks
		BR4 2-6-0	EL 270	22.00	Portsmouth & S
21.51 Eastleigh	22.08	T9 4-4-0	EL 282		
22.20 Eastleigh (Pds)	22.32	N15 4-6-0	EL 263		(Fwd at 22.50)
		T9 4-4-0	EL 282	22.45	Light to Docks
(Eastleigh Pds)		B4 0-4-0T	STN 346	22.50	Docks
23.50 Light ex Docks	23.55	LN 4-6-0	NE 31		(Wks Up Mail)

The quadrupled track layout between Southampton Central and Redbridge was unusual in that the fast lines were the outer pair of tracks. The pattern was dictated by Millbrook station which was an island platform located between the centre pair of tracks. Light Pacific 34027 'Taw Valley' accelerates away from Southampton towards Redbridge where the Salisbury line diverged and four-line running ceased.

U.S.A. 0-6-0T's had booked daily duties whilst two 0-4-0T workings for a B4 and a C13 were nominally supplied by Eastleigh which also provided two ex-LBSCR E1 0-6-0T's for the New Docks and a pair of E4 0-6-2T's for the Bevois Park and Northam yard pilots.

The shed had no relevance to main line workings; its allocation was confined to the dock area and took no part in ordinary traffic movements except for rare occasions when the controller wanted an urgent special run to Eastleigh and there was no main line engine available.

Whilst a number of ships – principally those of the Cunard line – operated to a predictable timetable, estimated arrival times of ships was a rather fluid concept quite unlike those the railway was familiar with. Boat trains, goods and passenger, were therefore operated as special services with details only being given to the operating department at the last minute.

Long experience of the vagaries of marine timekeeping had allowed the railway to hone the business of arranging and running boat trains to a fine art although one could rarely tell more than a day or two in advance how many trains were going to operate on a particular day. In principle, the shipping companies would advise the railway of the approach or departure of a ship, quoting the number of passengers; information which the trains office would in-

COMBINED WORKING TIME TABLE AND ENGINE WORKINGS : 1954

Station / Train	11.12			10.30			11.28			05.30	10.30	11.35		
From	F'ham			W'loo			E'leigh			ELGH	Padd	BM		
Class	Gds		S&D			Gds		Gds	Gds		Gds			Gds
Engine	N 2-6-0	M7 0-4-4T	4F 0-6-0	MN 4-6-2	LN 4-6-0	T9 4-4-0	H15 4-6-0	BR3 2-6-2T	700 0-6-0	43xx	700 0-6-0	Q 0-6-0	T9 4-4-0	M7 0-4-4T
Shed	FRA 370	BM 407	BK72	NE 34	BM 393	EL 286	EL 313	EL 331	BM 417	WEY	SAL 452	BM 415	EL 284	BM 405
WATERLOO														
Woking														
Basingstoke														
Worting Jcn														
Micheldever														
Wallers Ash														
WINCHESTER														
Shawford														
Eastleigh Yard									11.55					
EASTLEIGH														
EASTLEIGH									12/00					
Swaything														
St Denys	11/59								12/10					
Bevois Park Yard	12.01													
Northam									12.14					
Southampton (T)														
SOUTHAMPTON CENT														
SOUTHAMPTON CENT							12.10							
Millbrook							12.14							
Redbridge							12.19							
Totton							12.22							
Lyndhurst Rd							12.29							
Beaulieu Rd							12.35							
BROCKENHURST						12.36	12.44	12.52						
Lymington Jcn						12/40	12/47	12/56						
Lymington Town						12.50								
Lymington Pier														
Sway							12.51	13.15						
New Milton							12.59	13.24						
Hinton Admiral							13.05							
Christchurch							13.12							
Pokesdown							13.18							
Boscombe							13.21							
Central Goods														
BOURNEMOUTH CENT							13.24							
BOURNEMOUTH CENT			12.55	13.01										
BOURNEMOUTH WEST			12.55	13.03									13.20	13.24
Branksome			13.00	13.08									13.25	13.40
Parkstone			13.04	13.12						13.18			13.29	
POOLE			13.09	13.19						13.24	13.27		13.37	
Hamworthy Jcn			(To Bath)	13.25							13.36		(To S'bury)	
Holton Heath				13.30										
WAREHAM		13.30		13.37										
Worgret Jcn		13/33		13/40										
Corfe Castle		13.45												
SWANAGE		13.54												
Wool				13.46										
Moreton				13.55										
DORCHESTER				14.06										
Dorchester Jcn				14/08						14/56				
Monkton														
Wishing Well														
Upwey				14.16										
Radipole														
WEYMOUTH				14.21						15.07				

terpret in terms of trains, issuing engine diagrams and carriage workings to the operating departments concerned. In addition a team of booking clerks from Southampton – a highly sought-after job - would be sent across the channel to Cherbourg to meet selected incoming transatlantic vessels and issue rail tickets for the boat trains from the railway booking office that the larger ships possessed.

So far as timekeeping was concerned, the worst offenders were arrivals from the Far East and it was by no means unusual for the associated boat trains to have to be put back by several days; engines and men being conjured up by the district control at about the time the ship was passing Calshot. (The comings and goings of passenger ships were listed in a variety of documents put out by the shipping lines but the most concise – invariably cut out and pinned to most office walls in the district – was the table given showing the weeks arrivals in the Sunday Express).

Some carriage stock was retained at Southampton but most was held at Clapham Junction, the ideal being that each set would form a return boat train in the same day. In reality the pattern of shipping seldom allowed such niceties of working and a high proportion of boat train stock either stayed in Southampton until a return working could be found or returned empty with its engine and crew to London.

Passenger shipping enjoyed a good postwar innings and for many it remained the standard means of foreign travel until 1960. The busiest year on record for transatlantic booking was 1953 and the subsequent decline was very gradual until jet aircraft became established during the early 1960's. A typical 1950's day would produce ships literally as far as the eye could see yet by 1970 Southampton Docks were all but defunct.

				WEYMOUTH - BASINGSTOKE : WORKING TIMETABLE 1954										
Train From				12.23 T'combe						12.58 Sarum		13.45 H. Jcn		11.00 Plymouth
Class	Gds		Gds			Gds						Gds		
Engine	S15 4-6-0	M7 0-4-4T	Q1 0-6-0	T9 4-4-0	WC 4-6-2	2P 4-4-0	Q0-6-0	M7 0-4-4T	M7 0-4-4T	T9 4-4-0	WC 4-6-2	Q0-6-0	43xx	WC 4-6-2
Shed	FEL 111	LYM 362	EL 314	FRA 366	BTN 730	TCB 59	BM 415	SWA 421	BM 408	SAL 444	BM 385	BM 415	WBY	BTN 731
WEYMOUTH											13.25		13.40	
Radipole														
Upwey														
Wishing Well														
Monkton														
Dorchester Jcn											13/38		13/53	
DORCHESTER											13.44		(To	
Moreton													W'bury)	
Wool											14.00			
SWANAGE							13.33							
Corfe Castle							13.44							
Worgret Jcn							13/54				14/06			
WAREHAM							13.56				14.12			
Holton Heath														
Hamworthy Jcn							13.45							
POOLE						13.51	13.54			14.16	14.25	14.30		
Parkstone						13.57	(To			14.22				
Branksome						14.01	B'mouth)			14.27				
BOURNEMOUTH WEST					13.50	14.05			14.20	14.31				
BOURNEMOUTH CENTRAL					13.58				14.28		14.35			
BOURNEMOUTH CENTRAL					14.01						14.40	14/50		
Central Goods	13.50											14.56		
Boscombe					14.05									
Pokesdown					14.08									
Christchurch	14.06				14.13									
Hinton Admiral	(To				14.20									
New Milton	B. Park)				14.27									
Sway														
Lymington Pier		14.18												
Lymington Town		14.21												
Lymington Jcn		14/29			14/35						14/58			
BROCKENHURST		14.31			14.38									
Beaulieu Road														
Lyndhurst Road														
Fawley														
Totton														
Redbridge					14/52						15/09		15/25	
Millbrook														
SOUTHAMPTON CENTRAL					14.57						15.14		15.30	
SOUTHAMPTON CENTRAL					15.00						15.20		15.34	
Southampton T.				14.55										
Northam				14.59										
Bevois Park Yard														
St Denys				15.03	15.07								15/39	
Swaything				15.07										
EASTLEIGH				15.12										
EASTLEIGH											15/30			
Eastleigh Yard		14.48		(To									(To	
Shawford		15/02		Brighton)									Brighton)	
WINCHESTER														
Weston				(To										
Micheldever				Didcot)										
Worting Jcn											15/59			
Basingstoke														
Waterloo											16.50			

The services from Bournemouth West to York and Birkenhead were the only express connections with foreign parts, the Southern provisding engines as far as Oxford and coaching stock on alternate days.   The two trains were amongst the last long-distance express passenger duties for pre-nationalisation engines, the Birkenhead service being booked to a King Arthur 4-6-0 whilst a Lord Nelson was diagrammed to the York.   Substituting for the normal 4-6-0, light Pacific 34103 'Calstock' wheels the 09.30 Birkenhead - Bournemouth into Southampton.

COMBINED WORKING TIME TABLE AND ENGINE WORKINGS : 1954

Train	11.19		09.54	11.00	01.30	09.50	12.05	11.30			15.00	05.30		
From	Ports		W'loo	Brighton	Hoo Jn	DID	Alton	Brighton			D. West	ELGH		
Class					Gds	Gds						Gds		Gds
Engine	LM2 2-6-2T	H15 4-6-0	LN 4-6-0	WC 4-6-2	S15 4-6-0	Q1 0-6-0	M7 0-4-4T	WC 4-6-2	LN 4-6-0	M7 0-4-4T	R'car	700 0-6-0	M7 0-4-4T	BR3 2-6-2T
Shed	EL 306	NE 68	EL 253	BTN 731	EL 310	EL 314	EL 299	BTN 732	EL 252	LYM 362	WEY	BM 417	SWA 421	EL 331
WATERLOO			09.54					11.30						
Woking														
Basingstoke			11.36					12.30						
Worting Jcn			11/42					12/36						
Micheldever														
Wallers Ash					12.00									
WINCHESTER			12.04		12/10		12.47	12.58						
Shawford					12/34		12.53							
Eastleigh Yard					12.36	12.47								
EASTLEIGH			12.15				13.00							
EASTLEIGH				12.23			13/09							
Swaythling				12.30										
St Denys	12.21			12.35		12/42	13/11							
Bevois Park Yard														
Northam														13.20
Southampton (T)														
SOUTHAMPTON CENT	12.26		12.40	12.47			13.16	13.19						
SOUTHAMPTON CENT	12.28			12.54			13.21	13.25						13/27
Millbrook														
Redbridge	12.34			12/59			13/26							13.43
Totton	(To			(To			(To							13.47
Lyndhurst Rd	A. Jcn)			Plymouth)			Cardiff)							(To
Beaulieu Rd														Fawley)
BROCKENHURST								13.47		13.52				
Lymington Jcn								13/50		13/55				
Lymington Town										14.03				
Lymington Pier										14.05				
Sway														
New Milton								13.59				14.34		
Hinton Admiral												14.42		
Christchurch								14.08						
Pokesdown								14.15						
Boscombe								14.19						
Central Goods														
BOURNEMOUTH CENT								14.22						
BOURNEMOUTH CENT								14.26						
BOURNEMOUTH WEST								14.34						
Branksome														
Parkstone														
POOLE														
Hamworthy Jcn														
Holton Heath														
WAREHAM													14.16	
Worgret Jcn													14/19	
Corfe Castle													14.29	
SWANAGE													14.38	
Wool														
Moreton														
DORCHESTER														
Dorchester Jcn											15/01			
Monkton											15.04			
Wishing Well											15.09			
Upwey											15.12			
Radipole											15.16			
WEYMOUTH											15.20			

It came as a considerable surprise to learn that both classes of Pacifics were to be extensively (and expensively) rebuilt after little more than a decade's service. The rebuilds saw not only a return to more orthodox aesthetics but a return to much improved availability with the result that Pacifics started to appear on duties that had previously been handled by Lord Nelson and King Arthur locomotives. Rebuilt West Country 34013 'Okehampton' runs into Southampton with the 09.30 Waterloo - Bournemouth West.

WEYMOUTH - BASINGSTOKE : WORKING TIMETABLE 1954

				14.45 Ports		13.50 BM		13.50 BM		14.35 B'mouth					14.05 Brock
Train															
From															
Class	ECS	Gds		Gds		Gds		Gds				Gds			
Engine	T9 4-4-0	28xx	Hdl 4-6-0	T9 4-4-0	S15 4-6-0	M7 0-4-4T	S15 4-6-0	M7 0-4-4T	M7 0-4-4T	Q1 0-6-0	N 2-6-0	MN 4-6-2	M7 0-4-4T	M7 0-4-4T	
Shed	FRA 365	E Jn	RDG 50	SAL 443	FEL 111	EL 298	FEL 111	BM 406	LYM 362	EL 319	FRA 370	NE 34	EL 305	BM 409	
WEYMOUTH		13.48													
Radipole															
Upwey															
Wishing Well															
Monkton															
Dorchester Jcn		14/03													
DORCHESTER		(To													
Moreton		R'stone)													
Wool															
SWANAGE															
Corfe Castle															
Worgret Jcn															
WAREHAM															
Holton Heath															
Hamworthy Jcn										14.45					
POOLE										(To Brock					15.04
Parkstone										via Wim)					15.11
Branksome															15.16
BOURNEMOUTH WEST												15.05			15.19
BOURNEMOUTH CENTRAL												15.13			
BOURNEMOUTH CENTRAL												15.18			
Central Goods															
Boscombe												15.22			
Pokesdown												15.26			
Christchurch					14.50							15.32			
Hinton Admiral															
New Milton												15.44			
Sway															
Lymington Pier								15.35							
Lymington Town								15.38							
Lymington Jcn					15/12			15/44	15/46			15/53			
BROCKENHURST					15.16		15.29	15.45	15.48			15.57			
Beaulieu Road					(To										
Lyndhurst Road					B. Park)										
Fawley															
Totton															
Redbridge							15/57					16/11			
Millbrook							16.02								
SOUTHAMPTON CENTRAL												16.16			
SOUTHAMPTON CENTRAL	15.40			15.46							16.05	16.22			
Southampton T.	15.45												16.25		
Northam													16.29		
Bevois Park Yard				15.55											
St Denys											16.17	16.29	16.33		
Swaything													16.37		
EASTLEIGH		15.36										16.36	16.42		
EASTLEIGH		15.38				15.59						16.39			
Eastleigh Yard											(To				
Shawford		15.46				16.07					Fareham)				
WINCHESTER		15.55				16.15						16.53			
Weston						(To									
Micheldever		16.11				Alton)						17.10			
Worting Jcn		16/22										17/21			
Basingstoke		16.28										17.25			
Waterloo		(Reading)										18.29			

NEW FOREST LINES

Bournemouth-based services were either too heavy for anything but a class 7 or 8 engine or else suited to an M7 and intermediate engines such as the 2-6-0 classes were less in evidence than on other parts of the Southern. The BR standard 2-6-0's appeared at Bournemouth in 1960 as replacements for the shed's trio of U moguls; the principal duty for which was to work the 06.00 parcels from Christchurch to Weymouth and the 17.41 Weymouth - Bournemouth (double-headed with an M7 0-4-4T from Wareham) stopping train. In between the 2-6-0 sat at Dorchester ready to assist with any failures that might occur on the main line.

Taking a rather liberal interpretation of the New Forest area as encompassing anything that lay between Southampton and Bournemouth, the first point of interest west of Southampton was the Fawley branch which diverged from the main line at Totton to serve an industrial area which included an Esso Petroleum complex.

FAWLEY : 1954					
Train	Arr	Engine	Shed	Dep	Destination
06.35 Eastleigh	07.43	M7 0-4-4T	EL 305	08.06	S. Terminus
07.15 Eastleigh	09.40	BR3 2-6-2T	EL 328		Shunt as required
07.56 Eastleigh	11.38	BR3 2-6-2T	EL 330		
		BR3 2-6-2T	EL 328	12.30	Bevois Park
		BR3 2-6-2T	EL 330	13.15	Bevois Park
13.20 Northam	15.50	BR3 2-6-2T	EL 331		Shunt as required
15.51 Eastleigh	16.55	M7 0-4-4T	EL 300	17.16	Andover Jcn
LE ex S. Central	18.18	M7 0-4-4T	EL 304	18.43	S Terminus
LE ex Totton	18.25	BR3 2-6-2T	EL 328		
		2 x BR3	331/328	19.05	Eastleigh

In 1954 oil traffic to the railway was of no more importance than any other commodity and although the volume that originated at Fawley was regarded as impressive at the time, within eight years the growth was such that seven block trains were leaving the refinery daily for destinations in the West Midlands. This, as it turned out, was a false prophet of growth and by the end

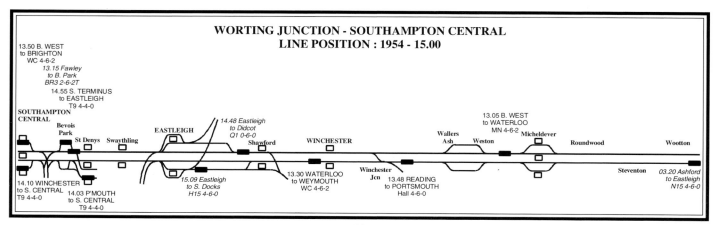

WORTING JUNCTION - SOUTHAMPTON CENTRAL
LINE POSITION : 1954 - 15.00

The largest Southern engines were familiar sights at Bournemouth Central although until 1959 the Pacifics were generally confined to the more prestigious workings. From that date, the electrification of the LCDR released sufficient Pacifics to permit the withdrawal of older locomotives with the result that the West Country engines regularly appeared on duties that would once have been the province of a much smaller engine. Rebuilt West Country 34088 '213 Squadron' shunts coaching stock at Bournemouth Central shortly after its transfer from Stewarts Lane. Other refugees from the Chatham side included a number of BR 5MT 4-6-0's which very quickly took over most of the N15 King Arthur workings. 73082 runs into Bournemouth Central on the 10th August 1963 with the 07.20 (Summer Saturdays) Waterloo - Weymouth.

Class N 2-6-0's were seldom seen in the Bournemouth area until a number made redundant by the SER electrification of 1959/62, were moved to Salisbury and Weymouth. 31405 of Salisbury - an Ashford engine for many years - waits for the right-away with a stopping train for Southampton Terminus.

of the 1960's Fawley's fate was sealed as the oil companies realised that it was far more economical to move oil by pipeline that it was by train.

Even the early growth in oil traffic presented problems since more engines were prohibited that were allowed on the branch from Totton. The ideal would have been to have the heavier freight services worked by a Q or Q1 0-

COMBINED WORKING TIME TABLE AND ENGINE WORKINGS : 1954

Train From				13.00 Chesil Gds	12.15 Reading	08.40 B'tol Gds	10.50 Didcot	Pullman			05.30 ELGH Gds	13.29 Fareham	S&D	12.30 Padd	13.29 Fareham
Class	M7 0-4-4T	M7 0-4-4T	M7 0-4-4T	700 0-6-0	BR4 2-6-0	Hal 4-6-0	T9 4-4-0	MN 4-6-2	WC 4-6-2	M7 0-4-4T	700 0-6-0	T9 4-4-0	LM5 4-6-0	Castle	LN 4-6-0
Shed	BM 406	EL 308	BM 409	EL 326	EL 273	4-6-0	EL 279	NE 35	NE 33	BM 407	BM 417	FRA 366	BATH 2	OOC	EL 253
WATERLOO								12.30							
Woking															
Basingstoke					12.53										
Worting Jcn					12/59		13/26								
Micheldever					13.09										
Wallers Ash															
WINCHESTER					13.22										
Shawford				13/10	13.28		13.35								
Eastleigh Yard				*13.20*											
EASTLEIGH					13.35		13.41								
EASTLEIGH	13.15				13.46			13/50							
Swaything	13.21				(To										
St Denys	13.25				Ports)							14.02			
Bevois Park Yard															
Northam	13.29														
Southampton (T)	13.32														
SOUTHAMPTON CENT							13.58					14.07			
SOUTHAMPTON CENT							14.03								14.22
Millbrook															14.26
Redbridge															14.31
Totton															14.34
Lyndhurst Rd															14.41
Beaulieu Rd															14.47
BROCKENHURST				14.05											14.56
Lymington Jcn				14/08				14/22							14/58
Lymington Town															
Lymington Pier															
Sway				(To											15.03
New Milton				B'mouth)											15.11
Hinton Admiral											14.55				15.17
Christchurch											15.04				15.23
Pokesdown															15.29
Boscombe															15.32
Central Goods															
BOURNEMOUTH CENT							14.40								15.35
BOURNEMOUTH CENT							14.44	14.57							15.39
BOURNEMOUTH WEST	14.35						14.52						15.35		15.47
Branksome	14.39							15.04					15.43		
Parkstone	14.42							15.08					15.49		
POOLE	14.49							15.14					(To		
Hamworthy Jcn	(To							15.19					Bristol)		
Holton Heath	Brock)							15.24							
WAREHAM								15.31	15.37						
Worgret Jcn								15/34	15/40						
Corfe Castle									15.50						
SWANAGE									15.59						
Wool								15.40							
Moreton								15.49							
DORCHESTER								16.00							
Dorchester Jcn					15/35			16.02							16/20
Monkton															
Wishing Well								16.10							
Upwey								16.14							
Radipole															
WEYMOUTH					*16.00*			16.18							16.33

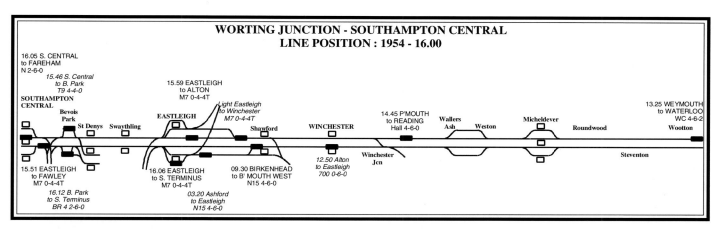

16.05 S. CENTRAL
to FAREHAM
N 2-6-0

15.46 S. Central
to B. Park
T9 4-4-0

SOUTHAMPTON
CENTRAL

Bevois Park

St Denys Swaythling

15.59 EASTLEIGH
to ALTON
M7 0-4-4T

EASTLEIGH

Light Eastleigh
to Winchester
M7 0-4-4T

Shawford

WINCHESTER

14.45 P'MOUTH
to READING
Hall 4-6-0

Wallers
Ash   Weston

Micheldever

Roundwood

13.25 WEYMOUTH
to WATERLOO
WC 4-6-2

Wootton

Winchester
Jcn

Steventon

15.51 EASTLEIGH
to FAWLEY
M7 0-4-4T

16.12 B. Park
to S. Terminus
BR 4 2-6-0

16.06 EASTLEIGH
to S. TERMINUS
M7 0-4-4T

03.20 Ashford
to Eastleigh
N15 4-6-0

09.30 BIRKENHEAD
to B' MOUTH WEST
N15 4-6-0

12.50 Alton
to Eastleigh
700 0-6-0

6-0 – the largest engines permitted - but tender engines were limited to a maximum of 15 mph, a speed that narrowed the threshold of stalling to an unacceptable degree. The only large tank engines cleared over the branch were the Z class 0-8-0T's but these were limited to shunting movements only and in any case had a 25 mph ceiling. As luck would have it, at the time the dilemma was shaping itself, the Railway Executive were trying to 'sell' their new designs to the regions and – it was probably one of the few instances of a standard engine actually solving an existing problem – it transpired that the 3MT 2-6-2T's were the ideal engine for the branch.

Three of the class arrived at Eastleigh during the autumn of 1952 – a fourth came the following spring – and very quickly became responsible for all goods traffic on the line with three of the four engines appearing daily. Traffic on the branch was such that three services from Fawley and one from Marchwood were needed daily, one of the Fawley workings requiring the efforts of two 2-6-2T's.

Demand for passenger services at Fawley was not great and in view of its proximity to Southampton, it was surprising how isolated the area was. The author once sat at a meeting

where the question of getting weekly operating notices (for some reason they could not be delivered by train) to one of the signalboxes or crossings was raised. After half an hour's fruitless discussion about the dearth of population and lack of roads, the writer decided that the only solution was transfer the problem to the Royal Mail by placing the notices in an envelope and putting the envelope in a post box.

The few passenger trains using the line were for the benefit of employees in the industrial complex and consisted of two trains in and three out, the imbalance rising from the fact that whilst everyone started work at the same time, finish-

		13.00 Cardiff								12.00 Dorset	13.50 BM		12.00 Dorset	14.20 Weymouth
WEYMOUTH - BASINGSTOKE : WORKING TIMETABLE 1954														
Train														
From														
Class	Gds					Gds				Gds	Gds		Gds	
Engine	S15 4-6-0	WC 4-6-2	T9 4-4-0	M7 0-4-4T	M7 0-4-4T	M7 0-4-4T	S15 4-6-0	R'car	22xx 0-6-0	U 2-6-0	S15 4-6-0	U 2-6-0	U 2-6-0	T9 4-4-0
Shed	FEL 102	BTN 732	EL 286	BM 407	EL 299	LYM 362	FEL 103	WEY	DID 31	DOR 427	FEL 111	DOR 426	DOR 427	EL 282
WEYMOUTH								14.05						14.20
Radipole								14.09						
Upwey								14.13					14.27	
Wishing Well								14.18						
Monkton								14.23						
Dorchester Jcn								14/26					14/36	
DORCHESTER								(To					14.44	
Moreton								Dor W)					14.56	
Wool										14.45			15.04	
SWANAGE			14.42											
Corfe Castle			14.57											
Worgret Jcn			15/06							14/53			15/09	
WAREHAM		15.08								14.58			15.14	15.21
Holton Heath													15.20	
Hamworthy Jcn													15.26	
POOLE													15.34	15.47
Parkstone													15.40	
Branksome													15.44	
BOURNEMOUTH WEST														
BOURNEMOUTH CENTRAL													15.50	
BOURNEMOUTH CENTRAL													(To A. Jn)	15.56
Central Goods														
Boscombe														16.00
Pokesdown														16.03
Christchurch														16.08
Hinton Admiral														16.15
New Milton														16.22
Sway														16.29
Lymington Pier						16.18								
Lymington Town						16.21								
Lymington Jcn						16/29								16/33
BROCKENHURST			16.08			16.31								16.36
Beaulieu Road			16.17											16.44
Lyndhurst Road			16.22											16.50
Fawley														
Totton			16.28											16.55
Redbridge		16/27	16.31											16.58
Millbrook												16.50		17.03
SOUTHAMPTON CENTRAL		16.32	16.37											17.06
SOUTHAMPTON CENTRAL		16.36	16.39											17.08
Southampton T.	16D20										16.56	16/56		
Northam											16.59			
Bevois Park Yard												17.04		
St Denys	16/38	16/41	16.46								17.03			17.15
Swaythling											17.07			
EASTLEIGH											17.12			17.22
EASTLEIGH	16/47	(To	(To	16.55			17.14							(To
Eastleigh Yard		B'ton)	P'mth)											A. Jn)
Shawford					17.03		17.22							
WINCHESTER	17/09			17.10	17.18	(To								
Weston					(To	17.33	D'cot)							
Micheldever					Alton)	(To								
Worting Jcn	17/44					B'stoke)								
Basingstoke	(To													
Waterloo	F'ham)													

Train From		13.06 Sarum	Gds	Gds	05.30 ELGH	15.45 BM	Gds	14.10 Winch	12.55 Yeo	09.45 Feltham	16.30 M. N'ton	14.03 Ports	11.10 W'ton	14.33 Ports	13.30 W'loo	
Class			Gds	Gds	Gds	Gds	Gds		Gds	Gds						
Engine	M7 0-4-4T	T9 4-4-0	M7 0-4-4T	M7 0-4-4T	700 0-6-0	M7 0-4-4T	M7 0-4-4T	T9 4-4-0	43xx	S15 4-6-0	R'car	T9 4-4-0	Castle	U 2-6-0	WC 4-6-2	M7 0-4-4T
Shed	LYM 362	SAL 443	EL 305	BM 404	BM 417	EL 297	BM 404	EL 283	WEY	FEL 105	WEY	FRA 365	SRD	SAL 485	BM 381	BM 409
WATERLOO															13.30	
Woking																
Basingstoke															14.30	
Worting Jcn										13/41					14/36	
Micheldever																
Wallers Ash																
WINCHESTER								14.10		14/13					14.58	
Shawford								14.16								
Eastleigh Yard			14.09							*14.35*						
EASTLEIGH		13.53						14.23								
EASTLEIGH		14.05	*14/11*					14.35							15/10	
Swaything		14.11						14.41								
St Denys		14.15	*14/21*					14.45				15.01		15/11		
Bevois Park Yard																
Northam		14.19														
Southampton (T)		14.22														
SOUTHAMPTON CENT								14.50				15.06		15.16	15.20	
SOUTHAMPTON CENT			*14/31*											15.20	15.24	
Millbrook																
Redbridge			*14.41*											15/25	15/29	
Totton														(To Bristol)		
Lyndhurst Rd																
Beaulieu Rd																
BROCKENHURST	15.06														15.46	
Lymington Jcn	15.09														15/49	
Lymington Town	15.17															
Lymington Pier	15.19															
Sway																
New Milton															15.58	
Hinton Admiral																
Christchurch					15.37										16.08	
Pokesdown															16.14	
Boscombe															16.17	
Central Goods					15.45		*15.52*									
BOURNEMOUTH CENT															16.20	
BOURNEMOUTH CENT					15/50										16.24	
BOURNEMOUTH WEST																16.30
Branksome					15.57											16.33
Parkstone							16.21									16.38
POOLE							*16.31*								16.36	16.43 (To Brock)
Hamworthy Jcn																
Holton Heath																
WAREHAM															16.48	
Worgret Jcn															16/51	
Corfe Castle																
SWANAGE																
Wool															16.57	
Moreton															17.06	
DORCHESTER															17.18	
Dorchester Jcn									16/27		16/48		17/00		17/20	
Monkton											16.51					
Wishing Well											16.56					
Upwey											16.59		17.28			
Radipole											17.03					
WEYMOUTH									16.50		17.07		17.14		17.33	

ing hours were staggered needing two trains in the evening. The stock requirements were covered by splitting the morning arrival at Fawley and sending an engine light from Southampton to work the third train away. All passenger services were worked by M7 0-4-4T's giving the line the distinction of having its goods workings covered by engines half a century newer than those on the passenger trains.

The rest of the New Forest section was a connoisseur's line rather than an enthusiast's, interest lying not so much in the volume of trade – a little less than that a few miles to the North – as in its antiquity. Southampton and Eastleigh sported a great many pre-grouping classes but they also had an uncomfortably high influx of postwar designs on which aesthetic opinion was divided. (Controversy on matters of appearance were not new: in 1935 a Kings Cross driver seeing an A4 Pacific for the first time described the engines as soulless monstrosities completely lacking the appearance of a steam engine. For many Southern lineside observers the Bulleid Pacifics added a new chapter on the subject).

West of Southampton the main line workings were augmented by a respectable number of stopping trains – a feature lacking between Eastleigh and Basingstoke – which kept modernisation at bay to the extent that it was possible for the observer to regress a generation or so. Although Bulleid Pacifics were visible, they tended to be bunched together; the majority travelling up to London with the Bournemouth business trains and returning at intervals during the evening.

Conscious that one may have ruffled a few air-smoothed feathers, it is salient to recall how controversial the Bulleid Pacifics were during their first decade. Their unconventional appearance appealed to some and offended others; the latter repeatedly asking the question: why, given the schedules the Southern operated to at the time, was streamlining necessary and, if it was, why was it in such an unsightly form? There was no answer to the second half of the question but what was undeniable was their mercurial performance. The author rode up on 34110 one morning and enjoyed the sort of run that made you want to stand up and cheer. Coming back with 34054 on a Boat Train, the engine steamed so badly that it had to stop at Brookwood – hardly a Beattock – to blow up and this was the very essence of the class: brilliant one day and hopeless the next. One could point at a King Arthur and claim without fear of contradiction: "There is a good engine." But one could not do the same for a Spam nor, come to that, a Lord Nelson although they did not have the appeal of appearance

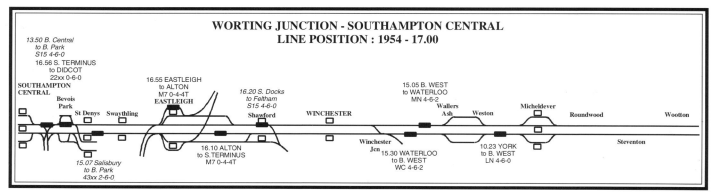

WORTING JUNCTION - SOUTHAMPTON CENTRAL
LINE POSITION : 1954 - 17.00

Train From								14.45 H.Jn		17.18 W'ter		13.50 Ch'ham		
Class		Gds	Gds	Gds	Pullman			Gds	Boat	Gds				Gds
Engine	H15 4-6-0	43xx	Q0 0-6-0	700 0-6-0	MN 4-6-2	M7 0-4-4T	T9 4-0	Q1 0-6-0	LN 4-6-0	S15 4-6-0	BR4 2-6-0	43xx 2-6-0	H15 4-6-0	H15 4-6-0
Shed	EL 313	WBY	BM 416	SAL 451	NE 35	EL 297	BM 403	EL 319	EL 254	FEL 103	EL 321	AJN 250	NE 68	EL 313
WEYMOUTH		14.30												
Radipole														
Upwey														
Wishing Well														
Monkton														
Dorchester Jcn		14/44												
DORCHESTER		(To												
Moreton		W'bury)												
Wool														
SWANAGE														
Corfe Castle														
Worgret Jcn														
WAREHAM														
Holton Heath														
Hamworthy Jcn														
POOLE				16.25										
Parkstone				(To										
Branksome				Wim)										
BOURNEMOUTH WEST					16.34									
BOURNEMOUTH CENTRAL					16.42									
BOURNEMOUTH CENTRAL	16.38				16.45									
Centrd Goods														
Boscombe	16.42													
Pokesdown	16.44													16.52
Christchurch														16.57
Hinton Admiral														17.03
New Milton														17.10
Sway														17.16
Lymington Pier														
Lymington Town			16.40											
Lymington Jcn			16/50		17/03		17/08							17/20
BROCKENHURST			16.54				17.12							17.22
Beaulieu Road														17.30
Lyndhurst Road														17.35
Fawley														
Totton														17.40
Redbridge												17.34		17/42
Millbrook												17.39		17.46
SOUTHAMPTON CENTRAL					17.19							17.42		17.49
SOUTHAMPTON CENTRAL					17.23							17.44		17.52
Southampton T.						17.23	17.28		17.30D		17.38	17.49		
Northam						17.27	17.32				17.42			
Bevois Park Yard													17.50	
St Denys						17.31	17.36				17.49		17/55	17.59
Swaythling						17.35	(To P'mth)				17.53			18.03
EASTLEIGH						17.40					17.58			18.08
EASTLEIGH					17/33	17.44			17/46		17.59		18/04	
Eastleigh Yard														
Shawford						17.53					18.06			
WINCHESTER						18.06					18.13		18/27	
Weston						(to Alton)				18.07			18.38	
Micheldever													(To B'stk)	
Worting Jcn					18/01				18/17	18/30				
Basingstoke									18.21	18.38				
Waterloo					18.50				19.19					

that the Light Pacifics had).

They also had the disconcerting habit of sticking in mid-gear – not, it is true, the only three-cylinder engines to do so – and one recalls trying to get away from Southampton with one of the class; the driver fruitlessly opening and shutting the regulator with nothing happening. After the third attempt he wound the engine into reverse, backed the train a coach-length up the platform, closing the regulator and screwing the engine into reverse whilst propelling the backwards. The engine changed direction efficiently enough but, judging from the almighty crash of crockery in the dining car clearly audible on the engine, at a tremendous cost to the catering division. The driver would have been a little more careful had it been an isolated experience but like most of his colleagues he was tired of having had to do it so often.

By the time the engine was passing the end of the platform, an Arthur would have been at Millbrook – and Arthur's didn't have the habit of catching fire, which is more than can be said for a Spam.

Thus it was that many wandered down to the New Forest to recapture the flavour of the prewar Southern and few were disappointed. Spams gasped their way northwards with the more important business trains from Bournemouth and Weymouth but for the most part traffic was handled by engines of an older and more stable generation.

The local service between Southampton and Bournemouth was surprisingly regular; the hourly stopping trains being augmented by the Waterloo semi-fast workings which called at the more important points along the coast.

With Bulleid Pacifics operating most of the

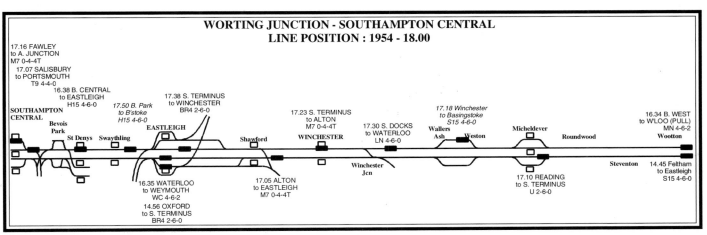

WORTING JUNCTION - SOUTHAMPTON CENTRAL
LINE POSITION : 1954 - 18.00

The only booked working through the New Forest for a GW engine occurred on Summer Sundays when a Hall 4-6-0 was diagrammed to work an unadvertised excursion from the Slough area to Bournemouth and back. Other instances were rare but not unknown and usually took place when the WR proposed a through special train and the Southern were unable to diagram a fresh engine from Basingstoke.   A filthy Taunton Hall, 4-6-0 4989 'Cherwell Hall', runs through Boscombe on the last lap of a York - Bournemouth journey.

BROCKENHURST : 1954

Train	Arr	Engine	Shed	Dep	Destination
21.55 WEYMOUTH	00.01	N15 4-6-0	BM 399	00.06	WATERLOO
22.30 WATERLOO	01.40	N15 4-6-0	EL 263	01.47	DORCHESTER
19.35 Kensington Pds	*03.52*	*N15 4-6-0*	*BM 399*	*04.12*	*Poole*
02.40 WATERLOO	05.25	MN 4-6-2	NE 30	05.28	BOURNEMOUTH CENTRAL
05.28 Eastleigh (Fish)	*06.20*	*T9 4-4-0*	*EL 280*		*(Fwd at 06.35)*
06.18 Lymington Town PP	06.28	M7 0-4-4T	LYM 362		
05.17 Wimborne	06.32	LN 4-6-0	BM 394	06.38	Eastleigh
(05.28 Eastleigh (Fish))		T9 4-4-0	EL 280	06.35	Bournemouth West
06.05 Eastleigh	06.53	N15 4-6-0	BM 401	06.58	Dorchester
06.25 Wimborne via Ringwood PP	07.03	M7 0-4-4T	BM 405		
		M7 & T9	362/403	07.04	Lymington Pier (Mixed)
06.35 BOURNEMOUTH CENTRAL	07.13	WC 4-6-2	BM 380	07.15	WATERLOO
		M7 0-4-4T	BM 405	07.17	B. West via Ringwood (PP)
07.00 Bournemouth Centrd	07.38	M7 0-4-4T	BM 409		
06.30 B. Centrd via Ringwood PP	07.42	M7 0-4-4T	BM 408	07.52	Lymington Town (PP)
07.39 Lymington Pier PP	07.53	M7 0-4-4T	LYM 362		
		M7 0-4-4T	BM 409	07.56	B. West via Ringwood.
07.20 BOURNEMOUTH WEST	08.03	WC 4-6-2	NE 38	08.04	WATERLOO
08.05 Lymington Town PP	08.15	M7 0-4-4T	BM 408		
05.40 WATERLOO	08.15	LN 4-6-0	NE 31	08.18	WEYMOUTH
07.30 Bournemouth West	08.20	H15 4-6-0	EL 313	08.22	Eastleigh
		M7 0-4-4T	BM 408	08.34	Poole via Ringwood (PP)
08.10 Christchurch	08.37	M7 0-4-4T	BM 413		
		M7 0-4-4T	LYM 362	08.41	Lymington Town (PP)
08.10 Lymington Pier.	08.41	T9 4-4-0	BM 403		
07.46 Eastleigh	08.39	H15 4-6-0	NE 79	08.44	Bournemouth Centrd
07.00 Dorchester via Ringwood	08.50	Q0-6-0	EL 324		
07.34 WEYMOUTH	09.19	WC 4-6-2	BM 381	09/00	WATERLOO
08.10 B. West via Ringwood PP	09.19	M7 0-4-4T	BM 406		
09.08 Lymington Pier PP	09.22	M7 0-4-4T	LYM 362		
08.35 BOURNEMOUTH WEST	09.26	WC 4-6-2	BM 382	09.28	WATERLOO
08.00 Romsey	09.26	LN 4-6-0	BM 394	09.29	Weymouth
		Q0-6-0	EL 324	09.32	Wimborne
		M7 0-4-4T	BM 406	09.39	Lymington Town (PP)
09.20 BOURNEMOUTH WEST	10.08	N15 4-6-0	BM 399	10.11	BIRKENHEAD
10.35 Lymington Pier PP	10.49	M7 0-4-4T	BM 406		
08.30 WATERLOO	10.50	MN 4-6-2	NE 32	10.53	WEYMOUTH
09.20 WEYMOUTH	10.54	LN 4-6-0	NE 31	10.57	WATERLOO
		M7 0-4-4T	LYM 362	10.58	Lymington Town (PP)
		M7 0-4-4T	BM 406	11.04	B. Centrd via Ringwood (PP)
10.29 Southampton Terminus	11.07	T9 4-4-0	EL 282	11.10	Bournemouth Centrd.
10.32 Poole via Ringwood PP	11.31	M7 0-4-4T	BM 408		
09.40 BRIGHTON	11.35	WC 4-6-2	BTN 730	11.37	BOURNEMOUTH WEST
11.25 Lymington	11.39	M7 0-4-4T	LYM 362		
11.05 BOURNEMOUTH WEST	11.54	WC 4-6-2	BM 383	11.57	WATERLOO
09.30 WATERLOO	11.55	WC 4-6-2	NE 33	11.57	BOURNEMOUTH WEST
		M7 0-4-4T	LYM 362	12.02	Lymington Town (PP)
11.16 BOURNEMOUTH WEST	12.02	LN 4-6-0	BM 395	12.04	YORK
		M7 0-4-4T	BM 408	12.12	B. Centrd via Ringwood (PP)
		MN 4-6-2	NE 34	12/22	WEYMOUTH
10.30 WATERLOO					
11.36 Bournemouth West	12.30	H15 4-6-0	NE 79	12.32	Woking
11.28 Eastleigh	12.42	H15 4-6-0	EL 313	12.44	Bournemouth Centrd
11.30 WEYMOUTH		WC 4-6-2	BM 386	13/01	WATERLOO
13.22 Lymington Pier PP	13.36	M7 0-4-4T	LYM 362		
12.35 Bournemouth West PP	13.42	M7 0-4-4T	BM 409		
11.30 WATERLOO	13.44	LN 4-6-0	EL 252	13.47	BOURNEMOUTH WEST.
		M7 0-4-4T	LYM 362	13.52	Lymington Town (PP)
13.05 BOURNEMOUTH WEST	13.54	MN 4-6-2	NE 32	13.57	WATERLOO
		M7 0-4-4T	BM 409	14.05	B. West via Ringwood (PP)
12.30 WATERLOO (PULLMAN)		MN 4-6-2	NE 35	14/20	BOURNEMOUTH WEST.
14.18 Lymington Pier PP	14.31	M7 0-4-4T	LYM 362		
13.50 BOURNEMOUTH WEST	14.36	WC 4-6-2	BTN 730	14.38	BRIGHTON
13.20 Fareham	14.54	LN 4-6-0	EL 253	14.56	Bournemouth West.
13.25 WEYMOUTH		WC 4-6-2	BM 385	15/00	WATERLOO
		M7 0-4-4T	LYM 362	15.06	Lymington Town (PP)
14.35 B. West via Ringwood PP	15.45	M7 0-4-4T	BM 406		
15.35 Lymington Pier PP	15.48	M7 0-4-4T	LYM 362		
13.30 WATERLOO	15.43	WC 4-6-2	BM 381	15.46	WEYMOUTH
13.05 BOURNEMOUTH WEST	15.54	MN 4-6-2	NE 34	15.57	WATERLOO
		Q0-6-0	BM 416	16.01	Lymington Pier
		T9 4-4-0	EL 286	16.08	Portsmouth & S
		M7 0-4-4T	BM 406	16.10	B. West via Ringwood (PP)
14.10 Winchester City	16.13	T9 4-4-0	EL 283	16.19	Bournemouth Centrd
16.18 Lymington Pier PP	16.31	M7 0-4-4T	LYM 362		
14.20 Weymouth	16.34	T9 4-4-0	EL 282	16.36	Andover Jcn
09.30 BIRKENHEAD	16.37	N15 4-6-0	BM 399		
		M7 0-4-4T	LYM 362	17.00	Lymington Town (PP)
16.34 BOURNEMOUTH WEST		MN 4-6-2	NE 35	17/05	WATERLOO
15.20 WATERLOO	17.17	WC 4-6-2	BM 382	17.19	WEYMOUTH
16.38 Bournemouth Centrd	17.21	H15 4-6-0	EL 313	17.22	Eastleigh
16.30 B. West via Ringwood PP	17.35	M7 0-4-4T	BM 409		
17.05 Southampton Terminus	17.39	N15 4-6-0	EL 265	17.40	Wimborne
17.28 Lymington Pier PP	17.42	M7 0-4-4T	LYM 362		
15.30 WATERLOO		WC 4-6-2	NE 37	17.53	BOURNEMOUTH WEST.
17.05 BOURNEMOUTH WEST	17.54	LN 4-6-0	EL 253	17.57	WATERLOO
		M7 0-4-4T	LYM 362	17.57	Lymington Town (PP)
10.23 YORK	17.59	LN 4-6-0	BM 395	18.01	BOURNEMOUTH WEST.
		M7 0-4-4T	BM 409	18.06	B. West via Ringwood (PP)
17.38 Bournemouth Centrd	18.15	LN 4-6-0	EL 252	18.19	Eastleigh
17.34 Eastleigh	18.18	N15 4-6-0	EL 329		
18.15 Lymington Pier	18.28	M7 0-4-4T	LYM 362		
16.35 WEYMOUTH	18.31	WC 4-6-2	BM 383	18.34	WEYMOUTH
18.07 Southampton Terminus	18.47	T9 4-4-0	FRA 366	18.49	Bournemouth Centrd
		M7 0-4-4T	LYM 362	18.52	Lymington Town (PP)
17.35 WEYMOUTH		MN 4-6-2	NE 30	19/01	WATERLOO
18.35 Bournemouth West	19.25	N15 4-6-0	EL 266	19.27	Reading
17.30 WATERLOO	19.47	WC 4-6-2	BM 380	19.50	BOURNEMOUTH WEST.
18.50 B. West via Ringwood PP	19.56	M7 0-4-4T	BM 406		
19.46 Lymington Pier PP	19.59	M7 0-4-4T	LYM 362	20.07	Lymington Town (PP)
		M7 0-4-4T	BM 406	20.15	B. West via Ringwood (PP)
18.30 WATERLOO		WC 4-6-2	BM 386	20/22	WEYMOUTH
18.30 WEYMOUTH	20.24	WC 4-6-2	NE 33	20.26	WATERLOO
20.30 Lymington Town PP	20.40	M7 0-4-4T	LYM 362		
18.45 Portsmouth & S	20.44	Q0-6-0	BM 416	20.46	Bournemouth West
		M7 0-4-4T	LYM 362	20.50	Lymington Town (PP)
20.52 B. West via Ringwood PP	21.57	M7 0-4-4T	BM 409		
19.30 WATERLOO	21.59	WC 4-6-2	BM 385	22.03	BOURNEMOUTH WEST.
19.48 Weymouth	22.02	N15 4-6-0	BM 263	22.04	Reading
(Rear Portion of 19.30 W'loo)		M7 0-4-4T	BM 409	22.13	BOURNEMOUTH WEST.

After nearly a decade of main line work between Bournemouth and Waterloo, at the outbreak of war the Schools 4-4-0's were concentrated on the South Eastern and took no further part in South Western affairs until the last days of steam when a number, made redundant by electrification, were allocated to Nine Elms and Basingstoke.   At that time there were sufficient Pacifics for almost every service of note and only sporadic use could be found for the Schools 4-4-0's although they could generally be found in numbers on Summer Saturdays during the early 1960's when every good engine was worth its weight in gold.   30925 'Cheltenham', formerly of Bricklayers Arms, stands at New Milton with a Waterloo semi-fast whilst, lower, 30934 'St Lawrence' - another ex-Bricklayers Arms engine - passes Walkford en route for Bournemouth with another Waterloo semi-fast.

COMBINED WORKING TIME TABLE AND ENGINE WORKINGS : 1954

	1	2	3	4	5	6	7	8	9	10	11	12	13	14
Train		13.12		12.42	13.48						16.40			03.20
From		F'tham		Didcot	Reading						Yeo			Ashford
Class		Gds		Gds										Gds
Engine	700 0-6-0	H15 4-6-0	BR4 2-6-0	22xx 0-6-0	HALL 4-6-0	Q&M7	M7 0-4-4T	M7 0-4-4T	T9 4-4-0	M7 0-4-4T	14xx	T9 4-4-0	M7 0-4-4T	N15 4-6-0
Shed	BM 417	NE 65	EL 321	DID 31	OX 208	416/362	SWA 421	BM 406	SAL 444	BM 408	WEY	EL 283	EL 300	EL 264
WATERLOO														
Woking														
Basingstoke					14.35									
Worting Jcn					14/41									14/59
Micheldever					14.51									
Wallers Ash														
WINCHESTER		(09.45			15.04									15/31
Shawford		F'tham)		15/09	15.10									
Eastleigh Yard		15.09												15.53
EASTLEIGH				15.15	15.17									
EASTLEIGH		15/14		15.23	15.21								15.51	
Swaythling				15.29	(To									
St Denys		15/20	15/30	15.33	Ports)									
Bevois Park Yard			15.41											
Northam			15.34	15.37										
Southampton (T)			16D08	15.40										
SOUTHAMPTON CENT													16.02	
SOUTHAMPTON CENT												15.40	16.07	
Millbrook												15.45		
Redbridge												15.50		
Totton												15.53	16.19	
Lyndhurst Rd												16.00	(To	
Beaulieu Rd												16.06	Fawley)	
BROCKENHURST						16.01		16.10				16.19		
Lymington Jcn						16/04		16/13				16/22		
Lymington Town						16.12								
Lymington Pier						16.14								
Sway								(To				16.27		
New Milton								B'mouth)				16.34		
Hinton Admiral												16.41		
Christchurch												16.49		
Pokesdown												16.56		
Boscombe												16.59		
Central Goods														
BOURNEMOUTH CENT												17.02		
BOURNEMOUTH CENT	16.30									17.00				
BOURNEMOUTH WEST	16.38								16.52					
Branksome									16.57	17.07				
Parkstone									17.01	17.11				
POOLE									17.07	17.17				
Hamworthy Jcn									(To	17.23				
Holton Heath									S'arum)	17.30				
WAREHAM							16.57			17.38				
Worgret Jcn							17/00			17/41				
Corfe Castle							17.12			17.51				
SWANAGE							17.21			18.00				
Wool														
Moreton														
DORCHESTER														
Dorchester Jcn											17/33			
Monkton											17.35			
Wishing Well											17.41			
Upwey											17.44			
Radipole											17.48			
WEYMOUTH											17.51			

London workings, their predecessors (and some of their predecessors) were kept busy on the secondary trains and it was almost possible to travel from Southampton to Bournemouth, jumping off at each station and getting a different engine on each leg of the journey; the eleven stopping trains from Southampton being worked by three King Arthur 4-6-0's, two H15 4-6-0's, three T9 4-4-0's, a Lord Nelson 4-6-0 and a Q 0-6-0. Further variety was added by the M7 0-4-4T's that worked the 07.56 Brockenhurst – Bournemouth and the slow portion of the 19.30 ex Waterloo.

The most important operational centre between Southampton and Bournemouth was Brockenhurst where the lines to Lymington and Broadstone diverged. Under normal rules its goods role would have been a rather desultory affair, consisting of no more than a couple of branch trains a day to Lymington and the West Moors line. Because, however, the diverging West Moors line ran to Poole avoiding the Bournemouth area, it had been the traditional route of the night goods from London to Weymouth, the train calling at Brockenhurst to drop off its Bournemouth section and to attach any Wessex traffic brought in by the 21.20 and 23.35 workings from Eastleigh.

Shortly after the grouping the Southern Railway decided to withdraw the night shift from the West Moors Line and although the Nine Elms goods had to be diverted via Christchurch, it continued to avoid Bournemouth and the arrangement by which traffic was sifted at Brockenhurst was retained.

A by-product of the arrangement was the stirring sight of two H15 4-6-0's pulling out in the dead of night with their respective trains for Bournemouth and Weymouth although not everybody appreciated the dramatic side of the night shift since melding the assets of several trains in a smallish yard was not an easy task and the antics of the Yard Foreman as he harangued guards and shunters into action, simultaneously trying to establish over the telephone

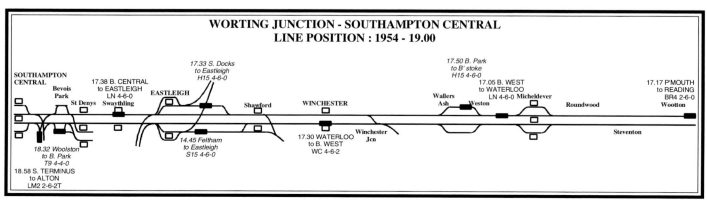

WORTING JUNCTION - SOUTHAMPTON CENTRAL
LINE POSITION : 1954 - 19.00

	17.07 Sarum	17.17 Ports	Gds	17.16 Fawley		17.16 Fawley	16.30 B'mouth					Gds	Gds		Gds
Train / From / Class															
Engine	T9 4-4-0	BR4 2-6-0	M7 0-4-4T	M7 0-4-4T	R'car	T9 4-4-0	M7 0-4-4T	T9 4-4-0	Castle	M7 0-4-4T	LN 4-6-0	700 0-6-0	T9 4-4-0	Q0-6-0	H15 4-6-0
Shed	SAL 284	EL 273	HAM 422	EL 300	WEY	EL 280	BM 409	DOR 425	OOC	LYM 362	EL 253	EL 326	FRA 365	EL 317	NE 65
WEYMOUTH					15.33			15.50	16.04						
Radipole					15.37										
Upwey					15.41										
Wishing Well					15.46										
Monkton					15.51										
Dorchester Jcn					15/54			16/05	16/19						
DORCHESTER					(To M. Ntn)			16.10	(To P'ton)				16.25		
Moreton								16.19							
Wool								16.28							
SWANAGE															
Corfe Castle			15.08												
Worgret Jcn			16/15				16/33						16/53		
WAREHAM			16.20				16.38						16/55		
Holton Heath															
Hamworthy Jcn			(To										17.11		
POOLE			B'mth)				16.50						(To		
Parkstone													B. Pk)		
Branksome															
BOURNEMOUTH WEST											17.05				
BOURNEMOUTH CENTRAL							17.00				17.13				
BOURNEMOUTH CENTRAL											17.18				
Central Goods															
Boscombe											17.22				
Pokesdown											17.25				
Christchurch											17.30				
Hinton Admiral															
New Milton											17.42				
Sway											17.49				
Lymington Pier										17.28					
Lymington Town										17.32					
Lymington Jcn							17/34			17/40	17/53				
BROCKENHURST							17.35			17.42	17.57				
Beaulieu Road															
Lyndhurst Road															
Fawley															
Totton				17.49									18.11		
Redbridge	17/47			17.52							18/11		18.15		
Millbrook				17.57											
SOUTHAMPTON CENTRAL	17.53			18.00							18.16				
SOUTHAMPTON CENTRAL	17.58			18.03							18.20		18.27		
Southampton T.													18.32		17D33
Northam															
Bevois Park Yard															18.38
St Denys	18.14			18.10											18/42
Swaythling	(To			18.14											
EASTLEIGH	P'mth)	18.11		18.19							18.31				
EASTLEIGH		18.15				18.30					18.33				18/52
Eastleigh Yard						(To A. Jn)	(To A. Jn)								19.02
Shawford		18.23													
WINCHESTER		18.32									18.47				
Weston															
Micheldever		18.49													
Worting Jcn		19/01									19/11				
Basingstoke		19.05									19.15				
Waterloo											20.23				

what was coming into the yard on the next arrival was an object lesson in man-management. It was a location where, lacking hump pilots, motive power facilities and all the other distractions that diverted attention from the prime object of goods working, a young man could receive a clear education in the very basics of goods operating.

The essence of the night operation was to regard each train as terminating in Brockenhurst and to make up new trains in station order for Wessex and Bournemouth for them to work forward. The two H15's were called back onto their trains, given the right away whilst the pilot - the Q 0-6-0 off the 00.36 ex Northam – applied itself to the remaining traffic, shunting it – again in station order – into trains for Lymington, Dorchester via Wimborne, Bournemouth direct and local stations to Bournemouth.

While this was taking place, the overnight train from Feltham arrived behind an S15 4-6-0 for a twenty-five minute stop. The Q 0-6-0 and the yard shunters would descend on the train, knock out the Bournemouth line vehicles and attach them to those already made up and, as soon as a load of fifty – the limit for this particular train - had been put together, the S15 would be waved back and sent on its way to Bournemouth.

By the time the Feltham had pulled out, the night shift was all but over and all that remained was for the Q to tidy up the yard and put together the Wimborne/Dorchester goods which it took out at 06.15.

The day shift was remarkable for the engines that came into Brockenhurst yard rather than the service itself which amounted to no more than a trio of Eastleigh – Bournemouth workings together with the Lymington branch train. The first of the series arrived just before the 06.15 and its Q drew out, and was the 04.50 ex Eastleigh; a working that ran non-stop to Brockenhurst before serving every nook and cranny via Christchurch to Bournemouth. Curiously, it was booked to an Eastleigh King Arthur 4-6-0.

The next service to arrive was the 09.15 from Eastleigh which, like the 04.50, also ran non stop to Brockenhurst where it made up its load before taking the West Moors road to Poole. Unlike the 04.50, it was worked by one of the

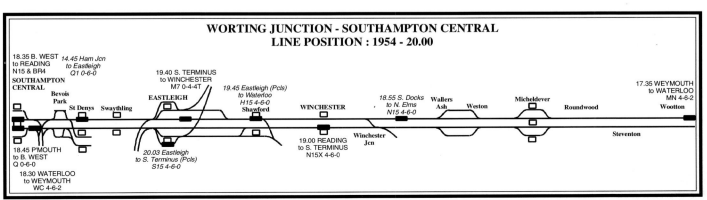

WORTING JUNCTION - SOUTHAMPTON CENTRAL
LINE POSITION : 1954 - 20.00

18.35 B. WEST to READING N15 & BR4 — 14.45 Ham Jcn to Eastleigh Q1 0-6-0 — SOUTHAMPTON CENTRAL — Bevois Park — St Denys — Swaythling — 19.40 S. TERMINUS to WINCHESTER M7 0-4-4T — EASTLEIGH — 19.45 Eastleigh (Pcls) to Waterloo H15 4-6-0 Shawford — WINCHESTER — Winchester Jcn — 18.55 S. Docks to N. Elms N15 4-6-0 — Wallers Ash — Weston — Micheldever — Roundwood — Steventon — Wootton — 17.35 WEYMOUTH to WATERLOO MN 4-6-2

18.45 P'MOUTH to B. WEST Q 0-6-0 — 18.30 WATERLOO to WEYMOUTH WC 4-6-2 — 20.03 Eastleigh to S. Terminus (Pcls) S15 4-6-0 — 19.00 READING to S. TERMINUS N15X 4-6-0

Train From Class	Gds	09.30 B'head S&D	Gds	Gds	BR4 2-6-0				16.15 W'bury		12.50 Alton Gds	16.25 B'TOL		15.20 W'loo	15.07 Scrum Gds
Engine	M7 0-4-4T	2P 4-4-0	N15 4-6-0	U 2-6-0	BR4 2-6-0	M7 0-4-4T	M7 0-4-4T	78xx 4-6-0	43xx	M7 0-4-4T	700 0-6-0	M7 0-4-4T	Hdl	WC 4-6-2	43xx 2-6-0
Shed	BM 405	TCB 59	BM 399	DOR 426	EL 321	EL 297	BM 407	CHM 22	WEY	LYM 362	EL 327	EL 329	BRD	BM 382	BTL 365
WATERLOO														15.20	
Woking															
Basingstoke		15.29													
Worting Jcn		15/35												16/14	
Micheldever															
Wallers Ash															
WINCHESTER		15.55									16.02	16.25		16.36	
Shawford											16/12	16.31			
Eastleigh Yard											16.24				
EASTLEIGH												16.38			
EASTLEIGH		16/04				16.06								16/45	16/50
Swaythling						16.12									
St Denys						16.16									17/01
Bevois Park Yard				16.12											
Northam						16.20									17.05
Southampton (T)				16.22		16.23		16.36							
SOUTHAMPTON CENT		16.12						16.41						16.53	
SOUTHAMPTON CENT		16.17						16.44						16.58	
Millbrook															
Redbridge		16/22						16.50 (To C'ham)						17/03	
Totton															
Lyndhurst Rd															
Beaulieu Rd															
BROCKENHURST		16.40									17.00			17.19	
Lymington Jcn		16/43									17/03			17/22	
Lymington Town										17.12					
Lymington Pier										17.14					
Sway															
New Milton		16.54													
Hinton Admiral															
Christchurch		17.06													
Pokesdown		17.13													
Boscombe		17.17													
Central Goods					17.28										
BOURNEMOUTH CENT		17.21												17.40	
BOURNEMOUTH CENT		17.27			17/31									17.45	
BOURNEMOUTH WEST		17.18			17.35										
Branksome		17.23												17.52	
Parkstone	17.20	17.28												17.56	
POOLE	17.30	17.38 (To T'combe)			17.51									18.02	
Hamworthy Jcn														18.08	
Holton Heath														18.14	
WAREHAM							17.45							18.21	
Worgret Jcn							17/48							18/24	
Corfe Castle															
SWANAGE															
Wool							17.53							18.30	
Moreton														18.39	
DORCHESTER														18.50	
Dorchester Jcn									18/06			18/46		18/52	
Monkton															
Wishing Well									18.13						
Upwey									18.18					19.01	
Radipole															
WEYMOUTH									18.21			18.57		19.06	

(relatively) conventional Q1 0-6-0's.

The last train of the morning, the 05.30 Eastleigh to Bournemouth, was a real gem was several points of view. It was the archetype pick-up train, calling almost everywhere and taking ten hours to cover the thirty-five mile distance, distributing empty wagons at the wayside stations for loading within the following twenty-four hours and, as its string of empties diminished, collecting loaded traffic for remarshalling at Bournemouth. It was also noteworthy on account of its motive power, being worked as far as Brockenhurst by a T9 4-4-0.

Whilst a Greyhound may not have appeared an especially appropriate engine for a stopping goods train, sixty empties could be taken by the smallest engine between Eastleigh and Totton and over this section a T9 was therefore an extravagance rather than a curiosity. In fact an M7 0-4-4T would have been adequate as far as Totton but beyond that point the 1 in 358 climb from Ashurt to Beaulieu Road limited smaller engines to 45 empties whilst a 4-4-0 could complete the section to Brockenhurst with sixty.

Beyond Brockenhurst the loading were far less generous and since the tendency was for the train's weight to increase as empties were being replaced by loaded wagons, the T9 handed over to a 700 0-6-0; the latter being able to work fifty-five wagons over the sharp dips and falls of Sway and Christchurch as opposed to thirty-six by a 4-4-0.

Up road traffic was dealt with in a more orthodox manner with none of the frantic exchanges of traffic which marked some of the down movements. Most of the trains made very brief calls at Brockenhurst – sufficient to add or subtract a few wagons from the front of each service – and the only working to spend any length of time in the yard was the 14.45 Hamworthy Junction – Eastleigh whose Q1 0-6-0 spent two hours shunting the yard and sorting waiting traffic into sections for Bevois Park, Nine Elms and Eastleigh to be picked up by the overnight services.

Given the level of activity at Brockenhurst it was surprising to find that it had no motive power resources of its own and was reliant on neighbouring locations, principally Bournemouth, for the provision of engines. Bournemouth M7 0-4-4T's had a near monopoly of services over the Old Road via Wimborne whilst the Lymington Branch was powered partly by the solitary M7 0-4-4T based at Lymington Town and partly by visiting engines. So long

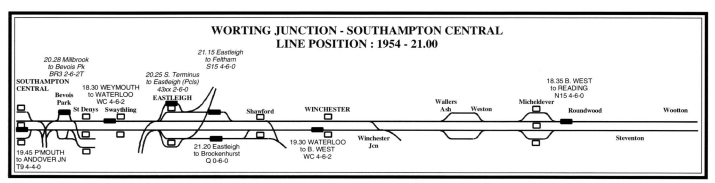

WORTING JUNCTION - SOUTHAMPTON CENTRAL
LINE POSITION : 1954 - 21.00

WEYMOUTH - BASINGSTOKE : WORKING TIMETABLE 1954

	T9 4-4-0	Castle	M7 0-4-4T	LM5 4-6-0	M7 0-4-4T	H15 4-6-0	M7 0-4-4T	LN 4-6-0	U 2-6-0	M7 0-4-4T	700 0-6-0	43xx	M7 0-4-4T	LN 4-6-0	2MT 2-6-2T
Train	16.10				10.20	17.50					18.11				
From	Brock				M'ter	B. Pk					Totton				
Class					Gds	Gds					Gds				
Engine	T9 4-4-0	Castle	M7 0-4-4T	LM5 4-6-0	M7 0-4-4T	H15 4-6-0	M7 0-4-4T	LN 4-6-0	U 2-6-0	M7 0-4-4T	700 0-6-0	43xx	M7 0-4-4T	LN 4-6-0	2MT 2-6-2T
Shed	FRA 365	OOC	BM 406	BATH 1	BM 405	NE 68	SWA 421	EL 252	DOR 427	LYM 362	EL 326	SPM	BM 407	BM 393	AJN 247
WEYMOUTH		16.10										16.30		16.40	
Radipole															
Upwey														16.48	
Wishing Well															
Monkton															
Dorchester Jcn		16/24										16/43		16/57	
DORCHESTER		(To P'ton)										(To Yeo)		17.02	
Moreton														17.11	
Wool														17.19	
SWANAGE							16.23								
Corfe Castle							16.34								
Worgret Jcn							16/43						17/20	17/24	
WAREHAM							16.45	17.17					17.22	17.28	
Holton Heath								17.26						17.34	
Hamworthy Jcn								17.32						17.40	
POOLE			17.12	17.22				17.39						17.47	
Parkstone			17.17					17.45						17.53	
Branksome			17.22		17.30			17.50						17.58	
BOURNEMOUTH WEST			17.25	17.32	17.35										
BOURNEMOUTH CENTRAL								17.55						18.03	
BOURNEMOUTH CENTRAL							17.38	17.58							
Central Goods															
Boscombe							17.42	18.02							
Pokesdown							17.45	18.05							
Christchurch							17.50	18.09							
Hinton Admiral							17.57								
New Milton							18.04								
Sway							18.10								
Lymington Pier										18.15					
Lymington Town										18.18					
Lymington Jcn							18/14			18/26					
BROCKENHURST							18.19			18.28					
Beaulieu Road							18.27								
Lyndhurst Road							18.32								
Fawley															
Totton							18.38								
Redbridge							18.40				18.50				
Millbrook							18.44				18.55				
SOUTHAMPTON CENTRAL							18.47								
SOUTHAMPTON CENTRAL							18.49								
Southampton T.	18.40														18.58
Northam	18.44														19.02
Bevois Park Yard															
St Denys	18.48						18.56								19.06
Swaything	(To P'mth)						18.59								19.09
EASTLEIGH							19.04								19.14
EASTLEIGH															19.15
Eastleigh Yard															
Shawford															19.23
WINCHESTER															19.33
Weston						19.10									(To Alton)
Micheldever															
Worting Jcn						19/35									
Basingstoke						20.18									
Waterloo															

as the timetable worked 'as booked' there was no problem but the minute anything went adrift – an engine failure, for example, just as a train was starting away from Brockenhurst – everything would fall to pieces. The price for having one engine shed twenty miles to the north and another fifteen miles to the south could at times be high.

Neither was much held in the way of carriage stock. Brockenhurst, in spite of the busy service to Lymington and Broadstone, had an allocation of one two-coach push and pull set plus another held in reserve whilst a third was based at Lymington Town. By contrast, the placid station of Christchurch on the outskirts of Bournemouth housed no less than eleven corridor vehicles: a 7-set of corridor working the school train to Brockenhurst and returning at 16.15 and a 4 set which worked down to Swanage.

The bustle of life at Brockenhurst and Bournemouth contrasted strongly with the pastoral air of the stations in between and the fifteen-odd mile stretch was an isthmus of tranquillity between two powerhouses of activity. The passenger service in both directions left little to complain about and with a population that seemed to consist very largely of retired Admirals (who carried considerable influence at Waterloo), there was no shortage of through services to and from London.

Each of the five stations had facilities for dealing with a wide range of goods traffics although the method of service was curious in that all but Hinton Admiral were served in one direction only. Both local goods trains (the 04.50 and 05.35 ex Eastleigh) called at Sway and New Milton (the 04.50 also served Hinton Admiral) before running non-stop to Bournemouth. Boscombe and Christchurch were only worked in the up direction (the 07.53 Bournemouth to Brockenhurst) with the result that in the (admittedly unlikely) event of a consignment being sent from New Milton to Christchurch, it would have to be taken initially to Bournemouth in the 04.50 ex Eastleigh to be transferred at the Central Goods to the 13.50 Bevois Park service.

The running of the local goods trains had to be fairly smart, especially in the case of the 04.50 ex Eastleigh which was allowed only five minutes to attach and detach at both Sway and Hinton Admiral.

With the intensity of the passenger service, there was always a fear that the smaller yards would not be able to fully accommodate one of the down local goods during shunting

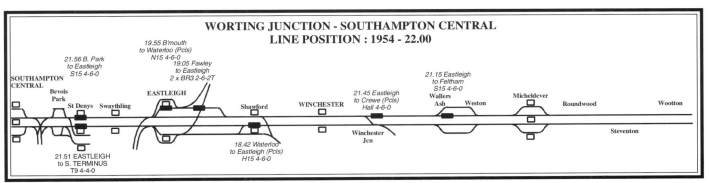

WORTING JUNCTION - SOUTHAMPTON CENTRAL
LINE POSITION : 1954 - 22.00

BROCKENHURST to BROADSTONE Via West Moors

Train	03.52	03.52		04.05	04.05				07.15	06.40	06.40	06.40		06.40
From	Sarum	Sarum		Sarum Gds	Sarum Gds	Gds			Sarum	Brock Gds	Brock Gds	Brock Gds		Brock Gds
Class														
Engine	T9 4-4-0	WC 4-6-2	T9 4-4-0	700 0-6-0	700 0-6-0	Q0-6-0	M7 0-4-4T	M7 0-4-4T	T9 4-4-0	Q0-6-0	Q0-6-0	Q0-6-0	M7 0-4-4T	Q0-6-0
Shed	SAL 444	BM 381	SAL 444	SAL 452	SAL 452	EL 317	BM 406	BM 405	SAL 443	EL 317	EL 317	EL 317	BM 408	EL 317
BROCKENHURST						06.40		07.17					08.34	
Lymington Jcn						06/44		07/20					08/37	
Holmsley						07.10				07.46			08.45	
Ringwood								07.34			08.09		09.02	
Ashley Heath								07.39					09.07	
West Moors	04.58		05/12					07.45	07.45		08.15	08.30	09.14	
Wimborne	04.50	04.58	05.17	05.22	05.50		06.45	07.53	08.21			08.54	09.25	09.32
BROADSTONE		05.04	05.23		05/58		06.50	07.59	08.27				09.30	09.42
Destination		Weymouth	Eastleigh		B.Cent		B.West	B.West	B.West				Poole	Dor

Train	09.25		08.39	08.39	08.39					07.37		17.20		
From	Sarum		Elgh Gds	Elgh Gds	Elgh Gds					Sarum		Sarum		
Class														
Engine	Q0-6-0	T9 4-4-0	Q0-6-0	M7 0-4-4T	Q1 0-6-0	Q1 0-6-0	Q1 0-6-0	M7 0-4-4T	M7 0-4-4T	700 0-6-0	M7 0-4-4T	T9 4-4-0	M7 0-4-4T	M7 0-4-4T
Shed	EL 324	EL 284	EL 324	BM 406	EL 318	EL 318	EL 318	BM 408	BM 409	SAL 451	BM 406	SAL 445	BM 409	BM 406
BROCKENHURST	09.32			11.04	11.20			12.12	14.05		16.10		18.06	20.15
Lymington Jcn	09/35			11/07	11/24			12/15	14/08		16/13		18/09	20/18
Holmsley	09.42			11.14	11.38	11.54		12.22	14.15		16.21		18.16	20.25
Ringwood	09.54			11.26				12.34	14.26		16.34		18.28	20.37
Ashley Heath	09.59			11.31				12.39	14.32		16.40		18.33	20.42
West Moors	10.04	10.13		11.36			12.20	12.45	14.37	15.05	16.47	18.12	18.38	20.47
Wimborne	10.12	10.23	11.12	11.45			12.30	12.54	14.47	15.15	16.57	18.22	18.47	20.56
BROADSTONE		10.29	11.17	11.51			12/48	12.59	14.53		17.03	18.28	18.53	21.02
Destination		B.West	B.West	B.Cent			Poole	B.West	B.West		B.West	B.West	B.West	B.West

BROADSTONE to BROCKENHURST Via West Moors

Train		06.30		07.00	07.00	07.42	08.10	08.45	08.45	08.45	08.45	10.22	12.35	08.45	13.20	14.35
From		B.Cent		Dor	B.Cent	B.West	Poole	Poole Gds	Poole Gds	Poole Gds	Poole Gds	Poole	B.West	Poole Gds	B.West	B.West
Class																
Engine	M7 0-4-4T	M7 0-4-4T	T9 4-4-0	Q0-6-0	Q0-6-0	T9 4-4-0	M7 0-4-4T	Q0-6-0	Q0-6-0	Q0-6-0	Q0-6-0	M7 0-4-4T	M7 0-4-4T	Q0-6-0	T9 4-4-0	M7 0-4-4T
Shed	BM 405	BM 408	EL 284	EL 324	EL 324	SAL 444	BM 406	BM 416	BM 416	BM 416	BM 416	BM 408	BM 409	BM 416	EL 284	BM 406
BROADSTONE		06.55	07.14	07.46			08.07	08.36	08/57			10.43	12.57		13.47	14.58
Wimborne	06.25	07.02	07.20	08.00			08.13	08.42	09.01			10.49	13.03		13.54	15.05
West Moors	06.33	07.10	07.28	08.08			08.21	08.50	09.15		10.03	10.58	13.11		14.02	15.14
Ashley Heath	06.38	07.15		08.13				08.55			10.18	11.03	13.16			15.19
Ringwood	06.44	07.22		08.18		08.30		09.00		10.42	10.52	11.10	13.22	13.33		15.26
Holmsley	06.54	07.33				08.41		09.11				11.21	13.33			15.36
Lymington Jcn	07/01	07/40				08/48		09/17				11/29	13/40	13/58		15/43
BROCKENHURST	07.03	07.42				08.50		09.19				11.31	13.42	14.02		15.45
Destination			Sarum			Sarum									Sarum	

Train	14.45	14.45	14.45	12.00	16.30	16.52	16.25	17.05		18.50	16.25	19.43	16.25	20.52	23.10
From	H.Jn	H.Jn	H.Jn	Dor	B.West	B.West	Dor	S.Term		B.West	Dor	B.West	Dor	B.West	B.West
Class	Gds	Gds	Gds	Gds	Gds			Gds		Gds				Gds	
Engine	700 0-6-0	Q1 0-6-0	Q1 0-6-0	Q0-6-0	700 0-6-0	M7 0-4-4T	T9 4-4-0	Q0-6-0	N15 4-6-0	700 0-6-0	M7 0-4-4T	Q0-6-0	T9 4-4-0	M7 0-4-4T	M7 0-4-4T
Shed	SAL 452	EL 319	EL 319	EL 319	SAL 451	BM 409	SAL 444	EL 317	EL 265	SAL 451	BM 406	EL 317	SAL 445	BM 409	BM 406
BROADSTONE		15.29		16/35	16.51	17.16	18/10	18.44		19.12		20.05		21.13	23.34
Wimborne	15.35	15.37	15.55	16.43	16.57	17.22	18.18	18.49		19.18	19.56	20.22		21.19	23.39
West Moors	15/43				17.05	17.20				18.58		20.19		21.27	
Ashley Heath					17.10					19.26				21.32	
Ringwood			16.20	16.36	17.16					19.37	20.21		20.50	21.37	
Holmsley					17.26					19.47				21.48	
Lymington Jcn				17/08	17/33					19/54			21/16	21/55	
BROCKENHURST				17.12	17.35					19.56			21.20	21.57	
Destination	Sarum			Elgh			Sarum			Sarum			Sarum	B.Park	

operations and that in consequence the down main would be blocked. As an illustration of line occupation, the 04.50 ex Eastleigh left Brockenhurst twenty-six minutes ahead of the 07.56 Bournemouth West passenger, a margin that decreased to fourteen minutes at New Milton where the goods was overtaken by three passenger services before being able to proceed. In order to minimise the risk of trouble, there was a standing order that Brockenhurst had to specifically liaise with New Milton to ensure that a down goods train could be accommodated.

Local goods traffic was rather heavier on the old road than it was on the new, the former having a large Ministry of Defence establishment at West Moors, a Ministry of Supply abattoir at Uddens Sidings, Wimborne and a Piston Ring establishment at Ringwood in addition to the ordinary flows of traffic generated

In contrast to the other regions of BR where there were no blanket restrictions on double-heading, the Southern placed so many classes under a prohibition that double-heading was almost unknown and the accompanying table shows the classes of engine affected. The only concession concerned engines running light from sheds to stations where, in a number of cases, engines could be coupled together

DOUBLE-HEADING
All mainline diesels
MN 4-6-2
BR7 4-6-2
BR6 4-6-2
LN 4-6-0
N15 4-6-0
N15X 4-6-0
S15 4-6-0
H15 4-6-0
H16 4-6-2T
B4 0-4-0T *
C14 0-4-0T *
0458 0-4-0T *
* May be coupled together

by the five branch stations.

Two through goods trains, one from Brockenhurst and the other from Eastleigh worked the line, with a third coming in from Salisbury to serve West Moors and Uddens before terminating at Wimborne and as with the trains on the main line, the goods service was by no means straightforward. Station calls were staggered between the two through trains; the 06.15 from Brockenhurst serving Ringwood, Udden's Sidings, Wimborne and Broadstone whilst the 09.15 ex Eastleigh called at Holmsley and the War Department Sidings at West Moors before continuing on to Poole. The Salisbury working served West Moors and Udden's before returning as the 15.35 back to Salisbury.

To avoid overloading the 09.15 ex Eastleigh, which was the only service to run from Brockenhurst to West Moors, the 06.15 was

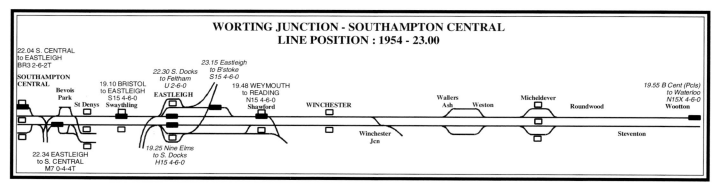

WORTING JUNCTION - SOUTHAMPTON CENTRAL
LINE POSITION : 1954 - 23.00

T9 4-4-0 30304 runs into West Moors with the 07.42 Bournemouth Central to Salisbury in 1957. Although a very light train - the booked formation was three vehicles as opposed to the two in the photograph - during the summer it was booked to be worked by a pair of T9 4-4-0's, the assisting engine being an unbalanced locomotive off a Salisbury - Bournemouth service the night previous.

permitted to maximise its load with West Moors traffic, working it to Wimborne where it was transferred to the 08.45 ex Poole.

Although the load limit from Brockenhurst to Wimborne was a generous 70 wagons (eighty from Ringwood), care also had to be taken to ensure that loadings were arranged so that only sixty would be on each service from Wimborne; the gradient towards Broadstone being rather restrictive. (Although no concession was given in respect of loadings, banking was authorised between Wimborne and Broadstone).

Whilst in passenger terms the Wimborne route had become a very inferior alternative to the new line, in the world of goods trains the old road was the line of automatic choice for special trains going beyond Poole since it avoided the congestion in and around Bournemouth whilst permitted speeds and maximum loads were identical to those of the main line. Psychology also played its part and if one had a special night train for the west under starter's orders at Eastleigh, it was certain that the driver would do everything he could – from slow running to sabotage - to get relief at Bournemouth.

Running the same train during the day and push-

ing it round the Old Road would not only see the driver taking it without complaint to Hamworthy Junction but nine times out of ten

agreeing to take it on to Dorchester.

To those of us who mixed railway operating and railway history, the treatment meted out to the Old Road was saddening especially as the pre-1914 LSWR had done its best to keep both routes on a level footing. Commerce dictated that the majority of trains had to run via Sway and Bournemouth but several second-division London expresses contained sections for Swanage or Dorchester which ran independently from Eastleigh or Southampton and were routed via the Old Road, keeping the route in contact with the main line. Two services, the 14.10 Waterloo – Weymouth and the 21.50 Dorchester Mail, ignored the New Line completely and ran complete to their destinations via Wimborne whilst the 15.00 ex Waterloo broke all the rules by running as a through train to Bournemouth West via West Moors and Broadstone. By 1917 such pleasant diversions had been sacrificed to the war effort and, apart from the Dorchester Mail which clung to the Old Road for a few years more, access to and from West Moors et al became by local train only.

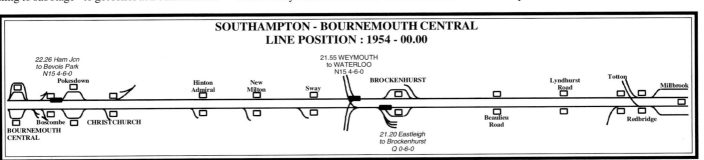

DORSET

Merchant Navy Pacifics came late to Bournemouth and prior to 1954 the only appearances in the area were on the 02.40, 08.30, 10.30 and 12.30 departures from Waterloo; all of which were Nine Elms turns. Of these, only one engine worked beyond Bournemouth; the engine of the 02.40 taking over the 10.30 and working it through to Weymouth. In 1954 the allocation of 8P locomotives was reviewed and three Merchant Navy locomotives were sent to Bournemouth, the number increasing to eight following the reintroduction of the two-hour timing in 1957. The driver of 35030 'Elder Dempster Lines' peers back for the Right Away with a Weymouth - Waterloo express.

Given Bournemouth's long established position as one of the most popular and busiest locations on the Southern Railway, it comes as a surprise to learn that its existence owed everything to the arrival of the railway and the coincidental popularity of seaside resorts. Prior to the railway age Bournemouth amounted to no more than a small group of cottages of such insignificance that some time elapsed before the railway recognised it as worth serving and indeed the first line through the locality – the 1847 extension of the London & Southampton to Dorchester – ran almost entirely inland via Brockenhurst, Ringwood, Broadstone and Poole.

In spite of being circumvented, by 1860 Bournemouth had grown sufficiently to be taken seriously by the railway; a branch being opened from Ringwood to Christchurch in 1862 and – whipped by the threat of a competing railway (the Poole & Bournemouth) authorised in 1865 – into Bournemouth itself in 1870.

The Poole & Bournemouth ran into immediate engineering problems and had to apply for several extensions of time, eventually opening in 1874 to be absorbed into the LSWR. In its short independent career the P&B conceded running powers to the Somerset and Dorset – the two companies had a common Chairman – thus setting the scene for the joint working ar-

rangement between the LSWR and the Midland which came into force in 1875.

During the post-1874 period the scope for confusion must have been unparalleled since the better trains from London conveyed two sections for Bournemouth; one being detached at Wimborne and running to Bournemouth Central (the East) whilst the other was removed from the Weymouth section at Poole for Bournemouth West. Ensuring one was in the correct portion of the train must have been, to say the least, a challenge whilst speeds were not exactly breathtaking with a timing of three hours and forty-seven minutes to Bournemouth West by the two Weymouth day expresses. (Run-

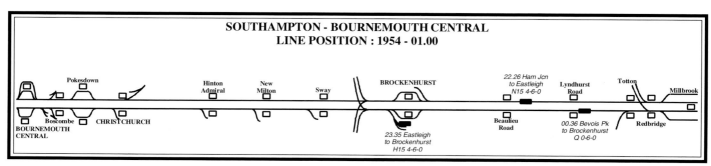

SOUTHAMPTON - BOURNEMOUTH CENTRAL
LINE POSITION : 1954 - 01.00

The allocation of U 2-6-0's at Eastleigh rose from 5 in 1954 to 17 by 1960 and led to a marked increase in their activities on the main line south of Eastleigh. 31639, above, heads the 10.29 Southampton - Bournemouth stopping train - a former T9 4-4-0 working - near New Milton in 1960. Below, the same engine is seen at Walkford with a short goods train.

	N15 4-6-0 EL 266	U 2-6-0 DOR 426	M7 0-4-4T SWA 421	U 2-6-0 DOR 426	M7 0-4-4T EL 302	N15 4-6-0 EL 265	T9 & M7 365/304	M7 0-4-4T EL 302	WC 4-6-2 NE 37	N15 4-6-0 EL 263	R'car WEY	LM5 4-6-0 BATH 4	2MT 2-6-2T AJN 247	M7 0-4-4T LYM 362
Train From		17.28 BM		17.28 BM	16.10 Alton			16.10 Alton	15.30 W'LOO	18.45 M.Ntn			17.03 Romsey	
Class		Gds		Gds								S&D		
WATERLOO									15.30					
Woking														
Basingstoke									16.33					
Worting Jcn									16/39					
Micheldever									16.51					
Wallers Ash														
WINCHESTER					16.52				17.04					
Shawford					16.57									
Eastleigh Yard														
EASTLEIGH					17.03				17.14				17.18	
EASTLEIGH								17.08	17.16				17.20	
Swaything								17.14					17.25	
St Denys								17.18					17.29	
Bevois Park Yard														
Northam								17.21					17.32	
Southampton (T)						17.05	17.15	17.24					17.35	
SOUTHAMPTON CENT						17.10	17.21	17.26						
SOUTHAMPTON CENT						17.11		17.31						
Millbrook						17.15								
Redbridge						17.19		17/36						
Totton						17.22								
Lyndhurst Rd						17.29								
Beaulieu Rd														
BROCKENHURST						17.40		17.53						17.57
Lymington Jcn						17/43		17/56						18/00
Lymington Town														18.08
Lymington Pier														18.10
Sway						17.47								
New Milton						17.53		18.05						
Hinton Admiral						17.59								
Christchurch						18.06		18.15						
Pokesdown						18.12		18.22						
Boscombe						18.14								
Central Goods														
BOURNEMOUTH CENT						18.17		18.26						
BOURNEMOUTH CENT	17.50					18.20						18.32		
BOURNEMOUTH WEST	17.58										18.40	18.40		
Branksome						18.27								
Parkstone						18.30						18.48		
POOLE		18.24				18.36					18.48	18.54		
Hamworthy Jcn		18.34		18.50		(To Wimbn)						(To Bath)		
Holton Heath														
WAREHAM			18.33	19.05										
Worgret Jcn			18/36											
Corfe Castle			18.48											
SWANAGE			18.57											
Wool														
Moreton														
DORCHESTER														
Dorchester Jcn										19/02				
Monkton										19.05				
Wishing Well										19.10				
Upwey										19.13				
Radipole										19.17				
WEYMOUTH										19.20				

ning times to Bournemouth Central were a few minutes either side of three and a half hours).

One major milestone of the times was the inauguration of an express service specifically for Bournemouth traffic; a precursor in some respects of the Royal Wessex of later years. The service ran from Bournemouth Central, leaving at 10.55 and was notable for running non-stop through the Southampton area to reach Waterloo at 14.05. Calls were made at Boscombe, Christchurch, Ringwood, Eastleigh, Winchester, Basingstoke and Vauxhall. In the reverse direction the service left Waterloo at 16.55 to reach Bournemouth Central in three hours fifteen minutes after making the same stops as the up train except that Vauxhall was excised in favour of Southampton West.

The final part of the network jigsaw was put into place in 1888 when the gap between Brockenhurst and Christchurch was filled by the present main line, from which time Bournemouth took its place as the focal point of commercial and operating interests west of Southampton. It is interesting to note that in 1873, before the railway reached Bournemouth, the balance of receipts of the LSWR – the difference between income and expenditure – lay at just over £900,000. By 1893, when the main line had been completed and allowed to gel, the balance had risen by no less than 83% to £1.6 million; an increase in which the development of Bournemouth played a major role. Bournemouth was not, of course, the only factor but the remainder of the system was largely in place by the mid-1870s whilst the benefits derived from the acquisition of Southampton Docks had not, by 1893, percolated through the system.

The 1888 line brought in its train a number of operating complications that remained in being until the end of steam operations. The preferred station in Bournemouth from most points of view was Bournemouth West but to have used it for all trains between London and Dorchester would have incurred a pointless reversal, unnecessary mileage and added time – not to mention the matter of limited accommodation in the terminus –and so the Central station became the centre of operations where engines were changed as trains split into sections for Weymouth and Bournemouth West.

With only one down platform and two for up traffic, Bournemouth Central seemed to be larger that it actually was and the rather re-

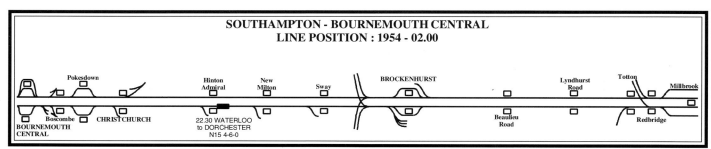

**SOUTHAMPTON - BOURNEMOUTH CENTRAL
LINE POSITION : 1954 - 02.00**

Train		16.25					16.13		18.43				16.32
From		Dorset					Ev Jn		Fawley				Bristol
Class	Gds	Gds	Gds						Pds	ECS			ECS
Engine	N15 4-6-0	H15 4-6-0	Q0-6-0	700 0-6-0	4F 0-6-0	M7 0-4-4T	MN 4-6-2	M7 0-4-4T	H15 4-6-0	M7 0-4-4T	M7 0-4-4T	U 2-6-0	M7 0-4-4T
Shed	EL 264	NE 74	EL 317	BM 417	BK 71	SWA 421	NE 30	EL 304	NE 72	LYM 362	EL 302	FRA 368	BM 407
WEYMOUTH		17.10					17.35						
Radipole													
Upwey													
Wishing Well													
Monkton													
Dorchester Jcn							17/47						
DORCHESTER		*17.30*					17.52						
Moreton													
Wool							18.05						18.10
SWANAGE						17.38							
Corfe Castle						17.52							
Worgret Jcn						18/01							18/15
WAREHAM						18.03	18.14						18.20
Holton Heath													
Hamworthy Jcn			18.00										
POOLE			(to		18.15		18.26						
Parkstone			B. Pk										
Branksome			via										
BOURNEMOUTH WEST			Wim)	18.16	18.25								
BOURNEMOUTH CENTRAL				18.24			18.36						
BOURNEMOUTH CENTRAL							18.41						
Central Goods													
Boscombe													
Pokesdown													
Christchurch													
Hinton Admiral													
New Milton													
Sway													
Lymington Pier										19.10			
Lymington Town										19.12			
Lymington Jcn							18/59						
BROCKENHURST													
Beaulieu Road													
Lyndhurst Road													
Fawley													
Totton								19.14					
Redbridge							19/11	19.17				19/37	
Millbrook								19.21					
SOUTHAMPTON CENTRAL							19.16	19.24				19.42	
SOUTHAMPTON CENTRAL							19.21	19.26				19.48	
Southampton T.	18D55						19.31				19.40		
Northam											19.44		
Bevois Park Yard													
St Denys	19/12										19.48	19.55	
Swaything											19.52	(To	
EASTLEIGH											19.57	P'mth)	
EASTLEIGH	19/22						19/31		19.45		19.59		
Eastleigh Yard													
Shawford											20.07		
WINCHESTER	19/56										20.13		
Weston													
Micheldever													
Worting Jcn	20/28						20/00		20/38				
Basingstoke	(To								20.45				
Waterloo	N. Elms)						20.50		23.10				

stricted operating facilities meant that smart working was necessary if station allowances were not to be exceeded. A typical down road operation consisted of an arrival from Waterloo which had to change engines and split into Weymouth and Bournemouth West sections; a fairly routine event from the perspective of the platform but quite complicated as seen from the signalbox where the bobby and the ASM will be setting up their stall.

" – What's happening with this one?"

'This one' is the 08.30 ex Weymouth which is passing the goods and two minutes away from coming to a stand in the down platform.

" – Inward engine light to Branksome loco."

" - Branksome loco? You sure?"

" – Branksome for the one-five back to Waterloo. Send it main line and it'll be the right way round for later on."

" – Branksome loco it is.   Where's the engine to work this forward?"

The telephone rings.

" – Dorchester 426 fireman speaking. 31623 ringing off shed for the Weymouth passenger. I've got 30111 standing behind me for the 11.25 Bournemouth West."

" – Wait signal."

" - Do you want the other fireman to come on the 'phone."

" – No. Just tell him to wait for his signal."

The 08.30 come to a stand, the shunter uncouples the engine, 35010 'Blue Star' of Nine Elms. The shed outlet dolly is pulled off and 31623 runs forward and crosses over to the down main as 2-3 is sent and received from Gas Works Junction, the signalman there being told by telephone that the engine is for Branksome loco and that both sections of the 08.30 will be right behind it. The down loop signal is cleared and 35010 moves forward along the platform, running light to Branksome. The train entering section signal is sent for the Pacific as levers are

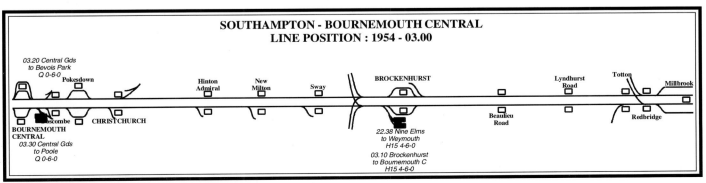

SOUTHAMPTON - BOURNEMOUTH CENTRAL
LINE POSITION : 1954 - 03.00

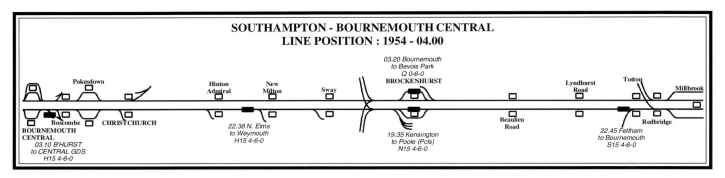

SOUTHAMPTON - BOURNEMOUTH CENTRAL
LINE POSITION : 1954 - 04.00

manipulated to set up for 31623 to take the place of 35010.

The shed outlet dolly is cleared again and, with a cheerful whistle, 30111 barks its way bunker-first off shed and stands, ready to back down into the station.

As 31623, blowing off noisily, backs down and couples up to the leading coach, the 2-1 bell code is received and acknowledged from Gas Works Junction. The signalman looks hopefully at the 2-6-0.

" – Is he ready to go?"

" – He'll be all right. Set him up."

Four beats on the bell go down the wire to Gas Works and the block indicator turns to line clear. Levers are pulled, the down main signal clears, whistles blow on the platform and, its valves still sending steam fifty feet into the air, 31623 gathers up its seven coaches and pulls away for Weymouth. Resetting levers as the train clears successive track circuits, the road is set up for 30111 which, also with valves blasting away, runs down and attaches itself to the remaining four vehicles. The signalman mutters something about firemen not daring to let engines blow off steam before the war as he dances about the frame to let 30863 'Lord Rodney' and the 11.16 Bournemouth - York express into the up platform. Bells and block indicators jangle and blink as out of section is received for the down Weymouth and line clear banged out for 30111 and its train.

The signal come off, the M7 pops its whistle and runs past on its eight minute trip to the West. The time is 11.25 and no more than a dozen minutes have passed since the signalman first mentioned the 08.30. There is no time for contemplation as the telephone rings, bringing tiding of the next move.

"Up side Inspector here. When the up Weymouth gets in at thirty-four, I don't want it shunting immediately. The rear vehicle normally goes forward on the Clapham Parcels but today they want it down the West so when the 10.29 Southampton arrives, the engine off that works the Brighton forward so cross it over to the up side, pick up the last vehicle and attach it to the Brighton. The pilot can shunt the

COMBINED WORKING TIME TABLE AND ENGINE WORKINGS : 1954

Train	10.23			15.07		16.45	14.56			17.03		17.28		
From	York			Sarum		Ports	Oxford			B'tol		BM		
Class				Gds								Gds		Gds
Engine	LN 4-6-0	M7 0-4-4T	M7 0-4-4T	43xx 2-6-0	M7 0-4-4T	U 2-6-0	BR4 2-6-0	WC 4-6-2	700 0-6-0	43xx	M7 0-4-4T	U 2-6-0	T9 4-0	N15 4-6-0
Shed	BM 395	BM 409	EL 329	BTL 365	BM 406	SAL 450	EL 277	BM 383	BM 417	WEY	BM 408	DOR 426	DOR 425	BM 401
WATERLOO								16.35						
Woking														
Basingstoke	16.44													
Worting Jcn	16/50							17/30						
Micheldever														
Wallers Ash														
WINCHESTER	17.11							17.51						
Shawford							17.48							
Eastleigh Yard														
EASTLEIGH	17.22						17.54							
EASTLEIGH	17.25		17.34					18/00						
Swaything			17.40											
St Denys			17.44		17.47									
Bevois Park Yard														
Northam				17.42										
Southampton (T)				17D50										
SOUTHAMPTON CENT	17.37		17.49		17.52			18.08						
SOUTHAMPTON CENT	17.40		17.50		17.56			18.12						
Millbrook			17.54											
Redbridge	17/45				18.02			18/17						
Totton			17.59		(To									
Lyndhurst Rd			18.06		Sarum)									
Beaulieu Rd			18.11											
BROCKENHURST	18.01	18.06	**18.18**					18.34						
Lymington Jcn	18/04	18/09						18/37						
Lymington Town														
Lymington Pier														
Sway		(To												
New Milton	18.13	Brock)												
Hinton Admiral														
Christchurch	18.23													
Pokesdown	18.30													
Boscombe	18.33													
Centrd Goods														19.15 / 19.20
BOURNEMOUTH CENT	18.36							18.55						
BOURNEMOUTH CENT	18.42							18.59	19.05					19.12
BOURNEMOUTH WEST	18.50					18.50			19.13					
Branksome						18.55								19.19
Parkstone						18.59								19.23
POOLE						19.04		19.11						19.29
Hamworthy Jcn						(To								19.35
Holton Heath						Brock)								19.40
WAREHAM								19.24				19.30	19.41	19.47
Worgret Jcn								19/27				19/33	19/45	19/50
Corfe Castle												19.43		
SWANAGE												19.52		
Wool													19.53	19.56
Moreton														20.05
DORCHESTER								19.44						20.16
Dorchester Jcn								19/46					20/02	20/18
Monkton														
Wishing Well														
Upwey													20.08	20.26
Radipole														
WEYMOUTH								19.55					20.14	20.31

The 12.30 Pullman from Waterloo enters Bournemouth Central behind the first of the Merchant Navy rebuilds, 35028 'Clan Line'. After a turnround of only two hours, during which time 35028 was serviced on Branksome loco, the engine, men and stock made their way back to London.

WEYMOUTH - BASINGSTOKE : WORKING TIMETABLE 1954

Train	14.45	17.20		15.02	18.00			18.50		16.35	15.30			15.02
From	H. Jcn	Sarum		Corfe	Brock			B'mouth		Cardiff	Bristol			Corfe
Class	Gds			Gds		Pcls						Gds		Gds
Engine	Q1 0-6-0	N15 4-6-0	T9 4-4-0	U 2-6-0	M7 0-4-4T	M7 0-4-4T	43xx 2-6-0	R'car	M7 0-4-4T	M7 0-4-4T	U 2-6-0	4F 0-6-0	BR3 2-6-2T	M7 0-4-4T
Shed	EL 319	EL 266	SAL 445	BM 414	HAM 422	BM 409	BTL 365	WEY	BM 406	LYM 362	SAL 482	BK 72	EL 320	HAM 422
WEYMOUTH			17.41					17.50						
Radipole								17.54						
Upwey			17.48					17.58						
Wishing Well								18.03						
Monkton								18.08						
Dorchester Jcn		(BR4	17/57					18/11						
DORCHESTER		pilot	18.02					(To						
Moreton		S'ton	18.11					M. Ntn)						
Wool		- EL)	18.19											
SWANAGE														
Corfe Castle														
Worgret Jcn			18/24											
WAREHAM			18.29	18.36										
Holton Heath			18.35	18.45										18.57
Hamworthy Jcn			18.41											19.06
POOLE		18.39	18.47		19.03						19.13			
Parkstone		18.45	18.53		19.08									
Branksome		18.50	18.58		19.12									
BOURNEMOUTH WEST		18.35	18.54		19.16						19.25			
BOURNEMOUTH CENTRAL		18.44	19.05											
BOURNEMOUTH CENTRAL		18.47												
Central Goods				(M7										
Boscombe		18.51		414										
Pokesdown		18.54		pilot										
Christchurch		18.59		ex										
Hinton Admiral		19.06		W'ham)										
New Milton		19.13												
Sway		19.20												
Lymington Pier										19.46				
Lymington Town										19.49				
Lymington Jcn		19/24							19/55	19/57				
BROCKENHURST	19.03	19.27							19.56	19.59				
Beaulieu Road		19.35												
Lyndhurst Road		19.41												
Fawley														
Totton		19.47												
Redbridge	19/42	19.50									20/18			
Millbrook		19.55												20.28
SOUTHAMPTON CENTRAL		19.58									20.23			
SOUTHAMPTON CENTRAL	19/52	20.03									20.28			20/33
Southampton T.						20.25								
Northam														
Bevois Park Yard														20.41
St Denys		20.10									20.37			
Swaythling		20.14									(To			
EASTLEIGH		20.19				20.38					P'mth)			
EASTLEIGH	20/09	20.24												
Eastleigh Yard	20.19													
Shawford		20.32												
WINCHESTER		20.40												
Weston														
Micheldever		20.57												
Worting Jcn		21/08												
Basingstoke		21.12												
Waterloo		(Read'g)												

BOURNEMOUTH 380 : WC 4-6-2

Arr	Station	Dep	Train
	Bournemouth Loco	05.25	
05.38	Bournemouth West	06.15	ECS
06.23	Bournemouth C	06.35	Pass
08.55	Basingstoke	10.35	Pass
11.50	Waterloo	12.24	Light
	NineElms Loco	17.01	Light
	Waterloo	17.30	Pass
20.39	Bournemouth West	21.14	Light
21.26	Bournemouth Loco		

BOURNEMOUTH 381 : WC 4-6-2

Arr	Station	Dep	Train
	Bournemouth Loco	04.08	Light
04.43	Wimborne	04.58	03.52 ex Salisbury
06.09	Weymouth		
	Weymouth Loco	07.00	
	Weymouth	07.34	Pass
10.50	Waterloo	11.14	Light
	NineElms Loco	12.59	Light
	Waterloo	13.30	Pass
17.33	Weymouth		
	Weymouth Loco	19.30	
	Weymouth	19.48	Reading
21.14	Bournemouth C		
21.30	Bournemouth Loco		

BOURNEMOUTH 382 : WC 4-6-2

Arr	Station	Dep	Train
	Bournemouth Loco	07.55	Light
	Bournemouth West	08.35	Pass
11.55	Waterloo	12.24	Light
	NineElms loco	14.50	Light
	Waterloo	15.20	Pass
19.06	Weymouth		
	Weymouth Loco	21.00	
	Weymouth	21.18	Pass
22.44	Bournemouth C		
23.00	Bournemouth Loco		

BOURNEMOUTH 383 : WC 4-6-2

Arr	Station	Dep	Train
	Bournemouth Loco	10.25	Light
	Bournemouth West	11.05	Pass
14.20	Weymouth	14.44	Light
	NineElms Loco	16.00	Light
	Waterloo	16.35	Pass
19.55	Weymouth		
	Weymouth Loco	21.30	
	Weymouth	21.55	Waterloo
23.25	Bournemouth C		
23.40	Bournemouth Loco		

BOURNEMOUTH 385 : WC 4-6-2

Arr	Station	Dep	Train
	Bournemouth Loco	09.55	
	Bournemouth C	10.23	08.00 ex Romsey
11.45	Weymouth		
	Weymouth Loco	13.00	
	Weymouth	13.25	Pass
16.50	Waterloo	17.24	Light
	NineElms Loco	18.55	Light
	Waterloo	19.30	Pass
22.38	Bournemouth West	23.00	Light
23.16	Bournemouth Loco		

BOURNEMOUTH 386 : WC 4-6-2

Arr	Station	Dep	Train
	Bournemouth Loco	08.20	
	Bournemouth C	08.59	05.40 ex Waterloo
10.07	Weymouth		
	Weymouth Loco	11.15	
	Weymouth	11.30	
14.49	Waterloo	15.14	Light
	NineElms Loco	17.55	Light
	Waterloo	18.30	Pass
20.41	Bournemouth C	20.53	18.30 ex Waterloo
21.01	Bournemouth West	21.25	Light
21.37	Bournemouth Loco		

BOURNEMOUTH 393 : LN 4-6-0

Arr	Station	Dep	Train
	Bournemouth Loco	06.40	
	Bournemouth C	07.10	Pass
08.50	Weymouth		
	Weymouth Loco	10.00	
	Weymouth	10.10	
11.34	Bournemouth C		
	Bournemouth C	12.35	
	Bournemouth C	13.01	Pass
14.21	Weymouth		
	Weymouth	16.30	
	Weymouth	16.40	Pass
18.03	Bournemouth C		
18.20	Bournemouth Loco		

BOURNEMOUTH 394 : LN 4-6-0

Arr	Station	Dep	Train
	Bournemouth Loco	05.25	
	Bournemouth C	05.54	Pass
07.17	Eastleigh		
	Eastleigh Loco	08.10	
	Eastleigh	08.35	08.00 ex Romsey
10.12	Bournemouth C		
	Bournemouth Loco	20.10	
	Bournemouth C	20.47	18.30 ex Waterloo
21.43	Weymouth	23.15	Pass
23.34	Dorchester		
23.55	Dorchester Loco		

BOURNEMOUTH 395 : LN 4-6-0

Arr	Station	Dep	Train
	Dorchester Loco	07.45	
	Dorchester	08.04	Pass
08.59	Bournemouth Loco	10.40	Light
	Bournemouth West	11.16	Pass
14.33	Oxford		
	Oxford Loco	15.00	
	Oxford	15.25	10.23 ex York
18.50	Bournemouth West	19.22	Light
19.34	Bournemouth Loco		

BOURNEMOUTH 399 : N15 4-6-0

Arr	Station	Dep	Train
	Bournemouth Loco	08.15	Light
	Bournemouth West CS	08.50	ECS
08.55	Bournemouth West	09.20	Pass
12.41	Oxford		
	Oxford Loco	14.00	
	Oxford	14.19	09.30 ex Birkenhead
17.35	Bournemouth West	18.05	Light
	Bournemouth Loco	23.00	
	Bournemouth C	23.32	21.55 ex Weymouth
00.37	S. Terminus	01.23	Portsmouth
01.37	Eastleigh	03.05	19.35 Kensington Pds
04.58	Bournemouth C		
05.15	Bournemouth Loco		

BOURNEMOUTH 401 : N15 4-6-0

Arr	Station	Dep	Train
	Bournemouth Loco	17.45	Light
17.49	Boscombe	19.15	Goods
19.20	Bournemouth C Goods	19.45	Light
20.05	Poole		
	Banking engine		
	Poole	21.45	Light
22.01	Hamworthy Jcn	22.26	Goods
02.19	Eastleigh		
	Eastleigh Loco	05.30	
	Eastleigh	06.05	Pass
08.52	Dorchester		
	Dorchester Loco	11.16	Light
11.33	Weymouth Loco	12.05	
	Weymouth Loco	12.20	Pass
13.43	Bournemouth C		
14.00	Bournemouth Loco		

BOURNEMOUTH 414 : U 2-6-0

Arr	Station	Dep	Train
	Bournemouth Loco	05.25	Light
05.35	Christchurch	06.00	Pds
09.06	Weymouth	09.30	Light
09.45	Dorchester Loco	16.40	Light
16.55	Weymouth		
19.05	Bournemouth C		
19.20	Bournemouth Loco		

> Diagrams 381 and 383 became Merchant Navy duties from early 1954 following the transfer of 35008 from Salisbury and 35011 and 35012 from Nine Elms.

stock of the Southampton and then come back for the Weymouth stock and berth it ready for the four thirty-eight."

"Right-ho, mate."

The bobby absorbs the list of instructions without batting an eyelid and if anyone sug- gested to him that such complex matters of administration ought to be communicated in a more sophisticated manner than an informal telephone

BOURNEMOUTH 404 : M7 0-4-4T

Arr	Station	Dep	Train
	Bournemouth Loco	06.20	Light
	Bournemouth C.	06.55	Pass
07.23	Broadstone	07.36	Pass
08.02	Bournemouth C.		
08.10	Bournemouth C	10.46	Loco Coal
10.54	Bournemouth C Goods		
	Yard Pilot		
	Bournemouth C Goods	13.00	Light
	Bournemouth Loco	15.15	
	Bournemouth C Goods	15.45	Goods
16.31	Poole		
	Pilot/Banker		
	Poole	02.30	
02.45	Bournemouth Loco		

BOURNEMOUTH 405 : M7 0-4-4T

Arr	Station	Dep	Train
	Bournemouth Loco	05.10	Light
05.44	Wimborne	06.25	Pass
07.03	Brockenhurst	07.17	Pass
08.22	Bournemouth West	08.40	Light
	Poole	09.30	Goods
09.58	Bournemouth C Goods	10.15	Light
	Bournemouth Loco	11.35	Light
11.51	Bournemouth West		
	Goods pilot		
	Bournemouth West	13.34	Goods
13.40	Branksome	15.30	Goods
15.36	Bournemouth West	16.38	Light
16.46	Parkstone	17.20	Goods
17.30	Poole	17.55	Light
18.10	Bournemouth Loco		

BOURNEMOUTH 406 : M7 0-4-4T

Arr	Station	Dep	Train
	Bournemouth Loco	05.10	Light
05.44	Wimborne	06.45	Pass
07.12	Bournemouth West	08.10	Pass
09.19	Brockenhurst	09.39	Pass
09.53	Lymington Pier	10.35	Pass
10.49	Brockenhurst	11.04	Pass
12.14	Bournemouth C		
12.30	Bournemouth Loco	13.50	Light
14.03	Bournemouth West	14.35	Pass
15.45	Brockenhurst	16.10	Pass
17.25	Bournemouth West	18.50	Pass
19.56	Brockenhurst	20.15	Pass
21.26	Bournemouth West	23.10	Pass
23.39	Wimborne	23.57	Light
00.30	Bournemouth Loco		

BOURNEMOUTH 407 : M7 0-4-4T

Arr	Station	Dep	Train
	Bournemouth Loco	05.15	
	Bournemouth C.	05.47	Pass
06.48	Swanage	07.15	Pass
08.12	Bournemouth West	08.30	Pass
08.57	Wareham	09.47	Pass
10.09	Swanage	11.06	
11.31	Wareham	11.57	
12.19	Swanage	12.42	
13.04	Wareham	13.30	
13.54	Swanage	14.42	
15.08	Wareham	15.37	
15.59	Swanage	17.00	
17.22	Wareham	17.45	
17.53	Wool	18.10	ECS
18.20	Bournemouth West	18.29	Pilot
19.05	Bournemouth C		
19.20	Bournemouth Loco		

BOURNEMOUTH 408 : M7 0-4-4T

Arr	Station	Dep	Train
	Bournemouth Loco	05.50	
	Bournemouth C.	06.30	Pass
07.42	Brockenhurst	07.52	Pass
08.02	Lymington T.	08.05	Pass
08.15	Brockenhurst	08.34	Pass
09.38	Poole	10.32	Pass
11.31	Brockenhurst	12.12	Pass
13.22	Bournemouth West	14.20	Pass
14.25	Bournemouth C.		
14.35	Bournemouth Loco	16.30	
	Bournemouth C.	17.00	Pass
18.00	Swanage	18.36	Pass
18.58	Wareham	19.30	Pass
19.52	Swanage	21.20	Pass
21.52	Wareham	22.04	Pass
22.26	Swanage		
	Swanage Loco		

BOURNEMOUTH 409 : M7 0-4-4T

Arr	Station	Dep	Train
	Bournemouth Loco	06.36	
	Bournemouth C.	07.00	Pass
07.38	Brockenhurst	07.56	Pass
08.44	Bournemouth West	10.12	Pass
10.20	Bournemouth C.		
	Bournemouth C.	11.00	
	Bournemouth C.	11.25	Pass
11.33	Bournemouth West	12.35	Pass
13.42	Brockenhurst	14.05	Pass
15.19	Bournemouth West	16.30	Pass
17.25	Brockenhurst	18.06	Pass
19.16	Bournemouth West	20.52	Pass
21.57	Brockenhurst	22.13	Pass
23.03	Bournemouth West	11.25	Light
23.37	Bournemouth Loco		

B'MOUTH 421 : M7 0-4-4T

Arr	Station	Dep	Train
	Swanage Loco	07.10	
	Swanage	07.36	Pass
07.59	Wareham	08.31	Pass
08.53	Swanage	08.58	Pass
09.20	Wareham	11.06	Pass
11.28	Swanage	11.36	Pass
11.58	Wareham	12.16	Pass
12.38	Swanage	13.33	Pass
13.56	Wareham	14.16	Pass
14.38	Swanage	16.23	Pass
16.45	Wareham	16.57	Pass
17.21	Swanage	17.38	Pass
18.03	Wareham	18.33	Pass
18.57	Swanage	20.10	Pass
20.32	Wareham	21.20	Pass
21.42	Swanage	22.04	Pass
22.29	Wareham	22.35	ECS
23.06	Bournemouth C		
23.20	Bournemouth Loco		

BOURNEMOUTH 410 : M7 0-4-4T

Arr	Station	Dep	Train
	Bournemouth Loco	06.13	Light
06.26	Christchurch	06.51	Pass
07.40	Wareham	08.18	Pass
08.40	Swanage	09.24	Pass
09.48	Wareham	10.16	Light
10.47	Bournemouth Loco		

BOURNEMOUTH 411 : M7 0-4-4T

Arr	Station	Dep	Train
	Bournemouth Loco	05.25	
	Bournemouth West Pilot		
20.22	Bournemouth Loco		

BOURNEMOUTH 413 : M7 0-4-4T

Arr	Station	Dep	Train
	Bournemouth Loco	07.10	
	Bournemouth C.	07.50	ECS
07.58	Christchurch	08.10	Pass
08.37	Brockenhurst	09.14	Light
10.00	Bournemouth Loco		

BOURNEMOUTH 412 : O2 0-4-4T

Arr	Station	Dep	Train
	Bournemouth Loco	05.35	
	Central Pilot		
23.55	Bournemouth Loco		

BOURNEMOUTH GOODS ENGINE DIAGRAMS

BOURNEMOUTH 415 : Q 0-6-0

Arr	Station	Dep	Train
	Bournemouth Loco	04.55	
	Bournemouth C.	05.20	Pds
05.49	Poole		
	Pilot		
	Poole	13.27	Goods
13.36	Hamworthy Jcn	13.50	Goods
13.59	Poole	14.35	Goods
15.01	Bournemouth C Gds		
15.30	Bournemouth Loco		

BOURNEMOUTH 416 : Q 0-6-0

Arr	Station	Dep	Train
	Bournemouth Loco	02.55	
	Bournemouth C Gds	03.30	Goods
03.59	Poole		
	Pilot		
	Poole	08.45	Goods
	Via W. Moors		
14.02	Brockenhurst	16.01	Pass
16.14	Lymington Pier	16.23	Light
16.25	Lymington Town	16.40	Goods
16.54	Brockenhurst		
	Pilot		
	Brockenhurst	18.40	Light
19.15	S. Terminus	19.40	Relief
19.45	Southampton C.	20.10	18.45 ex Portsmouth
21.48	Bournemouth W.	22.00	Light
22.11	Poole	23.30	Goods
23.53	Bournemouth C		
00.00	Bournemouth Loco		

BOURNEMOUTH 417 : Q 0-6-0

Arr	Station	Dep	Train
	Bournemouth Loco	06.10	
	Bournemouth C Gds	06.40	Goods
06.44	Boscombe	07.12	Light
07.15	Bournemouth C Gds	07.53	Goods
10.36	Brockenhurst	12.52	Goods (12.52 Eastleigh)
15.52	Bournemouth C Gds	16.00	Light
16.05	Bournemouth C	16.30	Pass
16.38	Bournemouth W	18.16	Pass
18.24	Bournemouth C.	19.05	16.35 ex Waterloo
19.13	Bournemouth W	19.45	Light
19.57	Bournemouth Loco		

BOURNEMOUTH 418 : G6 0-6-0T

Arr	Station	Dep	Train
	Bournemouth Loco	02.30	
02.35	Bournemouth C Gds		
	Pilot		
	Bournemouth C Gds	11.40	Loco coal
11.45	Bournemouth C Gds	12.45	
13.00	Bournemouth C Gds		
	Pilot		
	Bournemouth C Gds	00.00	
00.10	Bournemouth Loco		

BOURNEMOUTH 419 : B4 0-4-0T

Arr	Station	Dep	Train
	Bournemouth Loco	06.00	Light
06.17	Poole		
	Quay pilot		
	Poole	10.57	Goods
11.31	Broadstone	12.00	Goods
12.27	Poole	12.56	Light
13.20	Bournemouth Loco		

call, he would wonder what they were talking about.

Up expresses were even more difficult to deal with than the down because two trains had to be merged into one but for real action one had to be in the box just after midday when the up side would be combining the 12.20 Bournemouth West and 11.30 Weymouth services to form the 12.41 London whilst the down were dealing with the 09.30 Waterloo – Bournemouth West (through engine and no division) and the 10.30 from Waterloo which split into the usual Weymouth and Bournemouth sections.

Most of the Central's business was connected with express workings and their associated stopping services and local trains tended to be relatively few in number, most using the West and running to Brockenhurst via the West Moors line. Unlike later years, when the Pacifics had been rebuilt and tended to work anything and almost everything, during the greater part of the decade variety was the hallmark of Bournemouth Central and the one hundred and fourteen daily movements were handled by no less than ten different classes of locomotive. The most frequently seen engines were the light Pacifics which accounted for

BOURNEMOUTH CENTRAL : 1954

Train	Arr	Engine	Shed	Dep	Destination
21.20 Eastleigh Yard		Q0-6-0	EL 324	00/40	Dorchester
22.30 WATERLOO	02.14	N15 4-6-0	EL 253	02.28	DORCHESTER
03.30 Central Goods		Q0-6-0	BM 416	03/34	Poole
22.38 Nine Elms		H15 4-6-0	NE 73	04/24	Weymouth
19.35 Kensington Pds	04.58	N15 4-6-0	BM 399		(Fwd at 05.20)
(19.35 Kensington Pds)		Q0-6-0	BM 415	05.20	Poole
		M7 0-4-4T	BM 407	05.47	Swanage
05.17 Wimborne	05.48	T9 4-4-0	SAL 444		(Fwd at 05.54)
(05.17 Wimborne)		LN 4-6-0	BM 394	05.54	Eastleigh
02.40 WATERLOO	05.54	MN 4-6-2	NE 30		
06.00 Christchurch (Fish)		U 2-6-0	BM 414	06/08	Weymouth
06.15 ECS Bournemouth W	06.23	WC 4-6-2	BM 380		
		M7 0-4-4T	BM 408	06.30	Brockenhurst via Poole
		WC 4-6-2	BM 380	06.35	WATERLOO
		MN 4-6-2	NE 30	06.40	Bournemouth W. (Pds)
		M7 0-4-4T	BM 410	06.55	Broadstone
		M7 0-4-4T	BM 409	07.00	Brockenhurst
06.51 Christchurch	07.03	M7 0-4-4T	BM 410	07.05	Swanage
04.05 Salisbury		700 0-6-0	SAL 452	07/06	Central Goods
		LN 4-6-0	BM 393	07.10	Weymouth
07.20 Central Goods		Q1 0-6-0	EL 319	07/24	Poole
07.20 BOURNEMOUTH W.	07.28		NE 38	07.30	WATERLOO
05.28 Eastleigh (Fish)	07.35	T9 4-4-0	EL 280		(Fwd at 08.10)
07.30 Bournemouth W	07.38	H15 4-6-0	EL 313	07.41	Eastleigh
		T9 4-4-0	SAL 444	07.42	Salisbury
06.05 Eastleigh	07.45	N15 4-6-0	BM 401		(Fwd at 07.53)
		M7 0-4-4T	BM 413	07.50	Christchurch ECS
(06.05 Eastleigh)		N15 4-6-0	BM 401	07.53	Dorchester
07.36 Broadstone	08.02	M7 0-4-4T	BM 404		
(05.28 Eastleigh)		T9 4-4-0	EL 280	08.10	Bournemouth W. (Fish)
08.20 BOURNEMOUTH W	08.28	MN 4-6-2	NE 30		(Att to 08.40)
07.58 Brockenhurst	08.32	M7 0-4-4T	BM 409	08.36	Bournemouth W.
07.34 WEYMOUTH	08.35	WC 4-6-2	BM 381	08.40	WATERLOO
08.35 BOURNEMOUTH W.	08.43	WC 4-6-2	BM 382	08.46	WATERLOO
05.40 WATERLOO	08.53	LN 4-6-0	NE 31		(Fwd at 08.59)
(05.40 Waterloo)		WC 4-6-2	BM 386	08.59	WEYMOUTH
08.04 Dorchester	08.59	LN 4-6-0	BM 395		
07.46 Eastleigh	09.24	H15 4-6-0	NE 79		
09.20 BOURNEMOUTH W.	09.28	N15 4-6-0	BM 399	09.32	BIRKENHEAD
09.30 Poole Goods		M7 0-4-4T	BM 405	09/50	Central Goods
08.25 Weymouth	10.00	U 2-6-0	DOR 426		
		H15 4-6-0	NE 79	10.10	Bournemouth West (ECS)
08.00 Romsey	10.12	LN 4-6-0	BM 394		(Fwd at 10.23)
10.12 BOURNEMOUTH W.	10.20	M7 0-4-4T	BM 409		(Att to 10.33)
(08.00 Romsey)		WC 4-6-2	BM 385	10.23	Weymouth
09.20 WEYMOUTH	10.27	N15 4-6-0	EL 263		(Fwd at 10.33)
(09.20 Weymouth)		LN 4-6-0	NE 31	10.33	WATERLOO
(Loco coal)		M7 0-4-4T	BM 404	10.46	Central Goods
11.05 BOURNEMOUTH W.	11.13	WC 4-6-2	BM 383	11.17	WATERLOO
08.30 WATERLOO	11.14	MN 4-6-2	NE 32		(Fwd at 11.20)
(08.30 Waterloo)		U 2-6-0	DOR 426	11.20	WEYMOUTH
(08.30 Waterloo)		M7 0-4-4T	BM 409	11.25	Bournemouth W.
11.16 BOURNEMOUTH W.	11.24	LN 4-6-0	BM 395	11.28	YORK
10.10 Weymouth	11.34	LN 4-6-0	BM 393		
11.35 Central Goods		700 0-6-0	SAL 452	11/38	Poole
11.36 Bournemouth W.	11.44	H15 4-6-0	NE 79	11.48	Woking
10.29 S. Terminus	11.51	T9 4-4-0	EL 282		
09.40 BRIGHTON	12.09	WC 4-6-2	BTN 730		(Fwd at 12.14)
(09.40 Brighton)		T9 4-4-0	EL 282	12.14	Bournemouth W.
11.04 Brockenhurst via Poole	12.14	M7 0-4-4T	BM 406		
12.20 BOURNEMOUTH W.	12.28		EL 284		(Att to 12.41)
09.30 WATERLOO	12.30	WC 4-6-2	NE 33	12.35	BOURNEMOUTH W.
11.30 WEYMOUTH	12.35	WC 4-6-2	BM 386	12.41	WATERLOO
10.30 WATERLOO	12.43	MN 4-6-2	NE 34		(fwd at 12.49)
(10.30 Waterloo)		MN 4-6-2	NE 30	12.49	WEYMOUTH
(10.30 Waterloo)		MN 4-6-2	NE 34	12.55	BOURNEMOUTH W.
		LN 4-6-0	BM 393	13.01	Weymouth
13.05 BOURNEMOUTH W.	13.13	MN 4-6-2	NE 32	13.18	WATERLOO
11.28 Weymouth	13.24	H15 4-6-0	EL 313		
12.20 Weymouth	13.43	N15 4-6-0	BM 401		
13.50 BOURNEMOUTH W.	13.58	WC 4-6-2	BTN 730	14.01	BRIGHTON
11.30 WATERLOO	14.22	N15 4-6-0	EL 252	14.26	Weymouth
14.20 BOURNEMOUTH W.	14.28	M7 0-4-4T	BM 408		(Att to 14.40)
13.25 WEYMOUTH	14.35	WC 4-6-2	BM 385	14.40	WATERLOO
12.30 WATERLOO	14.40	MN 4-6-2	NE 35	14.44	BOURNEMOUTH W.
13.45 Hamworthy Jcn		Q0-6-0	BM 415	14/50	Central Goods
		WC 4-6-2	NE 33	14.57	Weymouth
15.05 BOURNEMOUTH W.	15.13	MN 4-6-2	NE 34	15.18	WATERLOO
13.29 Fareham	15.35	LN 4-6-0	EL 253	15.39	Bournemouth W.
14.20 Weymouth	15.50	U 2-6-0	DOR 426		(Fwd at 15.56)
15.45 Central Goods		M7 0-4-4T	BM 404	15/50	Poole
(14.20 Weymouth)		T9 4-4-0	EL 282	15.56	Andover Junction
13.30 WATERLOO	16.20	WC 4-6-2	BM 381	16.24	WEYMOUTH
(13.30 Waterloo)		700 0-6-0	BM 417	16.30	Central Goods
		H15 4-6-0	EL 313	16.38	Eastleigh
16.34 BOURNEMOUTH W.	16.42	MN 4-6-2	NE 35	16.45	WATERLOO
		M7 0-4-4T	BM 408	17.00	Swanage
15.50 WEYMOUTH	17.00	T9 4-4-0	DOR 425		(Att to 17.18)
14.10 Winchester	17.02	T9 4-4-0	EL 283		
17.05 BOURNEMOUTH W.	17.13	LN 4-6-0	EL 253	17.18	WATERLOO
09.20 BIRKENHEAD	17.21	N15 4-6-0	BM 399	17.27	BOURNEMOUTH W.
17.28 Central Goods		U 2-6-0	DOR 426	17/31	Dorchester
		LN 4-6-0	EL 252	17.38	Eastleigh
15.20 WATERLOO	17.40	WC 4-6-2	BM 382	17.45	WEYMOUTH
(15.20 Waterloo)		N15 4-6-0	EL 266	17.50	BOURNEMOUTH W.
17.17 Wareham	17.55	U 2-6-0	DOR 427	17.58	Christchurch
16.40 Weymouth	18.03	LN 4-6-0	EL 393		
17.05 S. Terminus	18.17	N15 4-6-0	EL 265	18.20	Wimborne
18.16 BOURNEMOUTH W.	18.24	700 0-6-0	BM 417		(Att to 18.41)
15.30 WATERLOO	18.26	WC 4-6-2	NE 37	18.32	BOURNEMOUTH W.
(15.30 Waterloo)		N15 4-6-0	EL 263	18.32	BOURNEMOUTH W.
17.35 WEYMOUTH	18.36	MN 4-6-2	NE 30	18.41	WATERLOO
10.23 YORK	18.36	LN 4-6-0	BM 395	18.42	BOURNEMOUTH W.
18.35 Bournemouth W.	18.44	N15 4-6-0	EL 266	18.47	Reading
16.35 WATERLOO	18.55	WC 4-6-2	BM 383	18.59	WEYMOUTH
(16.35 Waterloo)		700 0-6-0	BM 417	19.05	BOURNEMOUTH W.
17.41 Weymouth	19.05	U 2-6-0	BM 414		
		T9 4-4-0	DOR 425	19.12	Weymouth
18.07 S. Terminus	19.30	T9 4-4-0	FRA 366		
18.30 WEYMOUTH	19.44	WC 4-6-2	NE 33	19.52	WATERLOO
19.55 Bournemouth W (Pds)	20.04	N15 4-6-0	EL 265	20.30	Waterloo (Pds)
17.30 WATERLOO	20.26	WC 4-6-2	BM 380	20.31	BOURNEMOUTH W.
18.30 WATERLOO	20.41	WC 4-6-2	BM 386		(Fwd at 20.47)
(18.30 Waterloo)		LN 4-6-0	BM 394	20.47	WEYMOUTH
(18.30 Waterloo)		WC 4-6-2	BM 386	20.53	BOURNEMOUTH W.
15.02 Corfe Castle Gds		Q1 0-6-0	EL 318	20/56	Central Goods
19.48 Weymouth	21.14	WC 4-6-2	BM 381		(Fwd at 21.22)
		U 2-6-0	DOR 427	21.18	Weymouth
(19.48 ex Weymouth)		N15 4-6-0	EL 263	21.22	Reading
21.20 Central Goods		LM5 4-6-0	BATH 1	21/25	Bath
18.45 Portsmouth & S.	21.27	Q0-6-0	BM 416	21.33	Bournemouth W.
19.50 Dorchester Goods		H15 4-6-0	NE 74	21/43	Nine Elms
19.30 WATERLOO	22.24	WC 4-6-2	BM 385	22.29	BOURNEMOUTH W.
21.18 Weymouth	22.44	WC 4-6-2	BM 382		
19.30 WATERLOO	22.51	M7 0-4-4T	BM 409	22.55	BOURNEMOUTH W.
22.35 Wareham ECS	23.06	M7 0-4-4T	SWA 421		
21.55 WEYMOUTH	23.25	WC 4-6-2	BM 383		(Fwd at 23.32)
(21.55 Weymouth)		N15 4-6-0	BM 399	23.32	WATERLOO
22.26 Hamworthy Jcn		N15 4-6-0	BM 401	23/56	Eastleigh

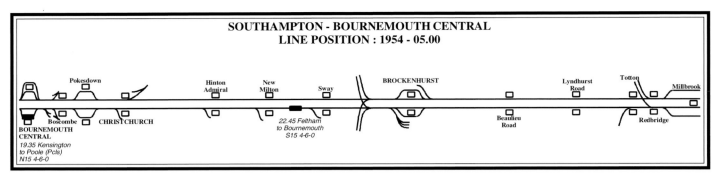

twenty-six movements – 23% - whilst Merchant Navies, Lord Nelsons and King Arthurs appeared in roughly equal numbers, each handling about a dozen workings apiece.

To native eyes the M7 0-4-4T's were a perfectly routine sight, the class handling most of the local services, but to senses brought up on affairs north of the Thames where four-coupled engines had been museum pieces for years, the presence of 0-4-4 tanks routinely bustling about with no talk of their replacement – nor any sign of it being needed – took some getting used to. Why, one repeatedly asked oneself, had the LM spent a post-war fortune on 2-6-2T's when the Southern managed so well with a class that had been working continuously since the Diamond Jubilee? Not all the Bourne-mouth M7 work was rural and one of the class played daily to the waiting gallery at Bournemouth West by arriving with the 08.30 express from Waterloo. One wonders how many laymen alighted to wonder how such a diminutive engine had managed such high speeds down Weston Bank….

The 08.30 turn was, of course, a short jaunt for the 0-4-4T between the two Bournemouth stations but later in the day the same M7 worked down to Brockenhurst with the 20.52 local from Bournemouth West via Ringwood, returning with the rear portion of the 19.30 express from Waterloo. Neither working lay in the top link of things but it was pleasing to see an M7 playing an active part in South Western express duties.

One was similarly pleased to find that one of the dozen T9 4-4-0 appearances also featured a London train; one of the Dorchester engines arriving in the Central with the five-coach 15.50 ex Weymouth: a train that ran to exactly the same timings as the Pacific-worked trains. One's pleasure at seeing 30284 (or one of its shedmates) running in on a London express was increased immeasurably by the sight of the seven-coach Bournemouth West section arriving behind an Eastleigh Lord Nelson 4-6-0; the latter working the train through to Waterloo. Observers of those times will recall that the sight of an express not worked by a Pacific was by no means a depressing prospect…

With 157 2-6-0's on its books, visitors to the area expected to see quite a sprinkling of

COMBINED WORKING TIME TABLE AND ENGINE WORKINGS : 1954

Train From	14.56 Oxford	17.05 Alton		17.45 Ports		17.10 Reading		14.45 Feltham Gds	18.42 Wool Gds	18.03 Ports	17.28 BM Gds	17.30 W'loo	13.20 S'don Gds	Gds
Engine	T9 4-4-0	BR4 2-6-0	M7 0-4-4T	M7 0-4-4T	BR4 2-6-0	T9 4-4-0	U 2-6-0	S15 4-6-0	T9 4-4-0	BR4 2-6-0	U 2-6-0	WC 4-6-2	43xx	43xx 2-6-0
Shed	FRA 366	EL 277	EL 298	LYM 362	EL 272	SAL 445	GUI 181	FEL 107	SAL 443	EL 270	DOR 426	BM 380	WBY	AJN 250
WATERLOO												17.30		
Woking														
Basingstoke								17.46				18.29		
Worting Jcn						17/52		18/00				18/35		
Micheldever						18.02						18.47		
Waller's Ash														
WINCHESTER		17.57				18.14		18/32				19.01		
Shawford		18.03				18.20								
Eastleigh Yard								18.54						
EASTLEIGH		18.09				18.26						19.11		
EASTLEIGH			18.02			18.28						19.14		
Swaythling			18.08			18.34								
St Denys			18.12		18.36	18.38				19.08				
Bevois Park Yard								_18.52_						
Northam			18.16			18.42								
Southampton (T)	18.07		18.19			18.45								19D04
SOUTHAMPTON CENT	18.12				18.41					19.13		19.23		
SOUTHAMPTON CENT	18.15				18.45							19.28		_19/35_
Millbrook	18.19													
Redbridge	18.24				18/50									_19/46_
Totton	18.27				(To Cardiff)									(To C'ham)
Lyndhurst Rd	18.34													
Beaulieu Rd	18.40													
BROCKENHURST	18.49			18.52								19.50		
Lymington Jcn	18/52			18/55								19/53		
Lymington Town				19.03										
Lymington Pier				19.05										
Sway	18.57													
New Milton	19.04											20.02		
Hinton Admiral	19.11													
Christchurch	19.18											20.12		
Pokesdown	19.24											20.19		
Boscombe	19.27											20.23		
Central Goods														
BOURNEMOUTH CENT	19.30											20.26		
BOURNEMOUTH CENT												20.31		
BOURNEMOUTH WEST										19.43		20.39		
Branksome										19.48				
Parkstone										19.51				
POOLE										19.57				
Hamworthy Jcn										(To Sarum)				
Holton Heath														
WAREHAM														
Worgret Jcn														
Corfe Castle														
SWANAGE														
Wool													_20.14_	
Moreton														
DORCHESTER													**20.50**	
Dorchester Jcn													_21/02_	
Monkton														
Wishing Well														
Upwey														
Radipole														
WEYMOUTH													_21.27_	

With four coaches from Weymouth, two from Swanage and six from Bournemouth West, the 13.25 Weymouth - Waterloo has 120 minutes to reach Waterloo and is a task best suited to a class 8 Merchant Navy.   Inevitably the light Pacifics found themselves on the two-hour trains from time to time yet, provided they were given a clear run, could manage the task without undue difficulty.   West Country 34021 'Dartmoor' starts the 13.25 from Bournemouth Central.

WEYMOUTH - BASINGSTOKE : WORKING TIMETABLE 1954														
Train								19.05		19.05		16.25		19.55
From								Fawley		Fawley		Dorset		B'mouth
Class			Gds		Milk		Gds	Gds	Pds	Gds	Pds	Gds	Gds	Pds
Engine	M7 0-4-4T	43xx	2 x BR3	WC 4-6-2	Hall	M7 0-4-4T	S15 4-6-0	M7 0-4-4T	2 x BR3	HALL 4-6-0	2 x BR3	N15 4-6-0	Q0 6-0	S15 4-6-0	N15X 4-6-0
Shed	BM 408	SPM	331/328	NE 33	OOC	LYM 362	FEL 111	EL 299	331/328	OX 208	331/328	EL 265	EL 317	FEL 106	BSK 236
WEYMOUTH			18.10	18.30	18.35									
Radipole			18.14											
Upwey			18.18											
Wishing Well														
Monkton														
Dorchester Jcn			18/28	18/43	18/49									
DORCHESTER			(To	18.48	(To									
Moreton			WBY)	18.57	Ken)									
Wool				19.05										
SWANAGE	18.36													
Corfe Castle	18.47													
Worgret Jcn	18/56			19.10										
WAREHAM	18.58			19.15										
Holton Heath														
Hamworthy Jcn				19.23										
POOLE				19.30										
Parkstone				19.35										
Branksome				19.40										
BOURNEMOUTH WEST												19.55		
BOURNEMOUTH CENTRAL				19.46								20.04		
BOURNEMOUTH CENTRAL				19.52								20.30		
Centrd Goods														
Boscombe				19.56										
Pokesdown				19.59										
Christchurch				20.04								20.41		
Hinton Admird												(Via		
New Milton				20.15								20.56	Wim)	
Sway														
Lymington Pier														
Lymington Town						20.30								
Lymington Jcn				20/23		20/38						21/04	21/16	
BROCKENHURST				20.26		20.40						21.11	21.20	
Beaulieu Road														
Lyndhurst Road														
Fawley								19.05						
Totton								20.25				21.28		
Redbridge								20/30	20/40					
Millbrook								20.35			21.03			
SOUTHAMPTON CENTRAL				20.45								21.39		
SOUTHAMPTON CENTRAL				20.50						21/13		21.42		
Southampton T.										21.05				
Northam										21.09				
Bevois Park Yard									21.22		21.36		21.56	
St Denys							20.57			21.13	21/40		22/00	
Swaythling										21.17				
EASTLEIGH				21.04					21.22			21.56		
EASTLEIGH				21.09						21.45	21/50		22/12	22.15
Eastleigh Yard								21.15			21.56		22.15	
Shawford					21.17									
WINCHESTER				21.20				21/47						22.34
Weston														
Micheldever														
Worting Jcn				21/49				22/22			22/29			22/59
Basingstoke				21.53				(To			22.34			23.05
Waterloo				22.56				Feltham)			(Crewe)			00.48

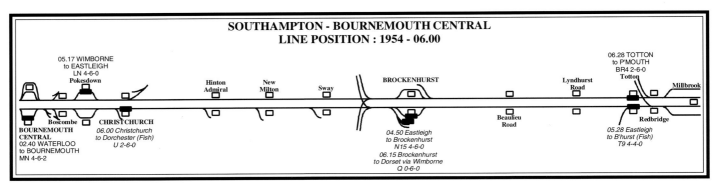

SOUTHAMPTON - BOURNEMOUTH CENTRAL
LINE POSITION : 1954 - 06.00

Southern moguls but those who came in the expectation of falling over large numbers of N's and U's were doomed to disappointment since only three examples of the type – all U class 2-6-0's – were booked to play a part in the passenger workings. Two were Dorchester engines – one of which took over the Weymouth portion of the 08.30 ex Waterloo from a Nine Elms Merchant Navy – and the third was based at Bournemouth although it spent most of the day on Dorchester shed, working down with an early morning fish train and returning with the 17.41 Weymouth stopping train. The most notable element in the Bournemouth working was the sight of the 2-6-0 being double-headed from Wareham by one of the Swanage branch M7 0-4-4T.

Double-heading was not a widespread feature of Southern life, partly because most of the engines were capable of keeping time unassisted but also because – something not widely known – double-heading was prohibited to a far greater extent on the Southern than it was elsewhere. The list of locomotives that could not be coupled to other engines (except over short stretches of line, such as from shed to station, etc) included the Merchant Navy Pacifics, Lord Nelson and King Arthur 4-6-0's and the H15, S15 and N15X mixed traffic classes.

Conveniently located for the enthusiast who could follow most of its movements from the adjacent Central station, Bournemouth MPD was a horribly cramped establishment accomodating far more movements than it had ever been designed for. Almost sixty engines a day were booked to go on and come off shed and not only was the layout reminiscent of Camden in terms of its restrictions but engine movements had to enter the running lines at the Weymouth end of the station, scuttling to and from the station over one of the busiest sections of line on the region. Some alleviation was granted by diverting engines from trains which turned round at Bournemouth West to Branksome shed – how many enthusiasts remember that Branksome daily routinely played host to Merchant Navy Pacifics – but the main shed generally teetered on the very knife edge of movement saturation.

The shed's busiest period occurred between 04.55 and 07.10 each morning when no less than twenty-four engines - one every six minutes - were booked off shed which left no scope for slack working. Any engine coming late off shed was likely to cause delays that would react on for some time but amongst the key working during the morning period was the Pacific which came off shed at 06.11 and ran light to Bournemouth West for the 07.20 Waterloo: a train well used by the business community in the area who not only demanded a high degree of punctuality but who could be pretty vociferous if they didn't get it. Guarantees of smooth working were slightly muddied by the fact that the engine concerned belonged not to Bournemouth but to Nine Elms thanks to a cockeyed working in which the up morning express was the back working of the previous day's 15.30 ex Waterloo. If London decided to send a dud, it was up to Bournemouth to either straighten the engine out or replace it with one of their own.

Important as the 07.20 was, by a short

Train	Arr	Engine	Shed	Dep	Load	Destination
TRAIN WORKING : BOURNEMOUTH WEST (1954)						
Light ex Bournemouth loco	05.38	M7 0-4-4T	BM 411			Station Pilot
Light ex Bournemouth loco	05.38	WC 4-6-2	BM 380			
		WC 4-6-2	BM 380	06.15	10	B. Central ECS/Waterloo
Light ex Bournemouth loco	06.24	WC 4-6-2	NE 38			
Light ex Branksome loco	06.30	4F 0-6-0	BK 71			
		4F 0-6-0	BK 71	06.48	3	Bath
Light ex Bournemouth loco	06.43	H15 4-6-0	EL 313			
06.40 B. Central (Pcls)	06.48	MN 4-6-2	NE 30		8	
06.45 Wimborne Motor	07.12	M7 0-4-4T	BM 406		2	
		WC 4-6-2	NE 38	07.20	11	WATERLOO
		H15 4-6-0	EL 313	07.30	6	Eastleigh
02.40 Bath (Mail & Goods)	07.44	LM5 4-6-0	BATH 1		4	
		LM5 4-6-0	BATH 1	08.00		Light to Branksome loco
		M7 0-4-4T	BM 406	08.10	2	Brockenhurst Motor via Poole
07.15 Swanage	08.12	M7 0-4-4T	BM 407		2	
05.28 Eastleigh (Fish)	08.18	T9 4-4-0	EL 280		5	
		MN 4-6-2	NE 30	08.20	6	WATERLOO
07.17 Brockenhurst Motor via Poole	08.22	M7 0-4-4T	BM 405		2	
		M7 0-4-4T	BM 407	08.30	2	Wareham Motor
Light ex Branksome loco	08.30	4F 0-6-0	BK 72			
Light ex Bournemouth loco	08.33	N15 4-6-0	BM 399			
		WC 4-6-2	BM 382	08.35	11	WATERLOO
		M7 0-4-4T	BM 405	08.40		Light to Poole
07.56 Brockenhurst via B. Central	08.44	M7 0-4-4T	BM 409		3	
		4F 0-6-0	BK 72	08.48	3	Bath
07.15 Salisbury	08.50	T9 4-4-0	SAL 443		3	
		T9 4-4-0	SAL 443	09.00		Light to Branksome loco
		N15 4-6-0	BM 399	09.20	12	BIRKENHEAD
Light ex Branksome Loco	09.25	LM5 4-6-0	BATH 1			
		T9 4-4-0	EL 280	09.28		Light to Southampton
07.28 Templecombe	09.30	2P 4-4-0	TCB 57		3	
		LM5 4-6-0	BATH 1	09.45	12	MANCHESTER
		2P 4-4-0	TCB 57	09.50		Light to Branksome loco
Light ex Branksome loco	09.55	T9 4-4-0	SAL 443			
		M7 0-4-4T	BM 409	10.12	6	WATERLOO
		T9 4-4-0	SAL 443	10.14		Light to B. Central Loco
10.10 B. Central ECS	10.19	H15 4-6-0	NE 79		10	
Light ex Bournemouth loco	10.38	WC 4-6-2	BM 383			
Light ex Bournemouth loco	10.50	LN 4-6-0	BM 395			
09.25 Salisbury	10.57	T9 4-4-0	EL 284		3	
06.05 Bristol	11.03	LM5 4-6-0	BATH 4		4	
		WC 4-6-2	BM 383	11.05	12	WATERLOO
		LN 4-6-0	BM 395	11.16	9	YORK
		LM5 4-6-0	BATH 4	11.20		Light to Branksome loco
Light ex Branksome loco	11.30	2P 4-4-0	TC 57			
08.30 WATERLOO	11.33	M7 0-4-4T	BM 409		4	
		H15 4-6-0	NE 79	11.36	10	Woking
		2P 4-4-0	TC 57	11.40	4	Bristol
11.12 Wimborne	11.41	Q 0-6-0	EL 324		3	
		Q 0-6-0	EL 324	11.56		Light to Bournemouth loco
		T9 4-4-0	EL 284	12.20	6	WATERLOO
09.40 BRIGHTON	12.22	T9 4-4-0	EL 282		5	
		M7 0-4-4T	BM 409	12.35	3	Brockenhurst Motor via Poole
Light ex Branksome loco	12.40	4F 0-6-0	BK 74			
09.30 WATERLOO	12.43	WC 4-6-2	NE 33		6	
Light ex Branksome loco	12.45	MN 4-6-2	NE 32			
Light ex Bournemouth Central	12.47	T9 4-4-0	EL 284			
09.05 Bristol	12.53	LM5 4-6-0	BATH 2		4	
		4F 0-6-0	BK 74	12.55		Bath
10.30 WATERLOO	13.03	MN 4-6-2	NE 34		6	
		MN 4-6-2	NE 32	13.05	6	WATERLOO
		LM5 4-6-0	BATH 2	13.10		Light to Branksome loco
		T9 4-4-0	EL 284	13.20	3	Salisbury
12.12 Brockenhurst Motor via Poole	13.22	M7 0-4-4T	BM 408		2	
		WC 4-6-2	NE 33	13.25		Light to Bournemouth loco
		MN 4-6-2	NE 34	13.30		Light to Branksome loco
Light ex Branksome loco	13.30	WC 4-6-2	BTN 730			
		WC 4-6-2	BTN 730	13.50	5	BRIGHTON
		T9 4-4-0	EL 282	13.59		Light to Bournemouth loco

whisker it was subordinate to the 07.34 Weymouth – Waterloo Royal Wessex which also conveyed a highly vocal complement of passengers. The Wessex however was a true Bournemouth turn, which allowed the engine used to be selected from the shed's allocation. The light Pacific concerned worked through from Weymouth to London and back again with the 13.30 ex Waterloo.

The problem with the Royal Wessex, as with most of the other Weymouth trains, was that the operating authorities insisted on the engine working through from Weymouth to London even though there were no (Southern) motive power facilities at Weymouth other than the GWR turntable. The traditional shed for working the Weymouth end of the line was Dorchester but this was out of the question since Waterloo was determined to close it.

The Wessex engine therefore had to be provided by Bournemouth which meant turning a West Country Pacific out at four in the morning to run light to Wimborne to take over the 03.52 ex Salisbury from a Salisbury T9 4-4-0.

The pick of Bournemouth's crop was usually selected for the diagram but even so a close watch would be kept on the progress of the up Wessex with a spare engine being kept up one's sleeve in case a change had to be made in the

five minutes the train spent at Bournemouth Central.

Merchant Navy engines were – during the period under review – uncommon at Bournemouth Central shed and the only example of the class booked to make a visit was the Nine Elms engine which ran down with the 02.40 News from Waterloo and continued West, after a little local work in the Bournemouth area, with the 10.30 from Waterloo. All other class 8 Pacifics were dealt with at Branksome and it was not until the class came to be allocated to Bournemouth from 1954 - following complaints about the running of the Royal Wessex – that they became familiar sights on Central shed.

Many large and medium-sized towns possessed two main line stations but Bournemouth was unusual, if not unique, in that both were operated by the same company. Had the lines in the area been planned rather evolving piece by piece, it is probable that Bournemouth West would never have seen the light of day although, in view of the large number of terminating trains, a single Bournemouth station would have needed to have been far larger than the three-platformed Central.

Although appearances suggested otherwise, the West was far more of an operating centre than the Central, dealing with a very large number of terminating services, their associated engine movements and a considerable volume of stock servicing. While Bournemouth Central had the respectable allocation of twenty-six passenger coaches and three parcels vehicles, the West had (at the quieter times of year) no less than one hundred and thirty vehicles on its books: a feature of operations that alone called for the services of about thirty members of staff.

Another feature that added to the complexities of the West was the fact it was a dead-end station with all the light engine movements that the obligatory reversals entailed. The Brockenhurst motor trains assisted a little but, on average, a service arrived or departed from the West every twelve minutes during the seventeen hours the station was open; a statistic that made it the busiest steam-worked terminus on the Southern. Multiple-units predominated at Waterloo, Brighton and Portsmouth, et al, and were therefore vastly easier to operate.

The variety of engines to be seen at Bournemouth West in the course of an ordinary day was startling even by the reckoning of 1954 with no particular class, other than the M7 0-4-4T's, predominating. Local Bulleid Pacifics had the lion's share of the Waterloo workings, especially the morning and evening business trains, with a number of Nine Elms engines – including several Merchant Navy's – appearing on the London-based services such as the Bournemouth Belle. Interestingly the proportion of Bulleid Pacifics seen at Bournemouth West was the same as that of LMS engines arriving off the Somerset & Dorset whilst the most numerous engine – a quarter of the total - was the M7 0-4-4T which monopolised the motor trains to Brockenhurst via Poole together with most of the other local services. One of

Train	On Shed	Engine	Shed	Off Shed	Train
16.12 Evercreech Jcn - B.West	18.55	4F 0-6-0	BK 71	06.20	06.48 B.West - Bath
15.30 Bristol - B.West	19.50	4F 0-6-0	BK 72	08.20	08.48 B.West - Bath
02.40 Bath Mail - B.West	08.10	LM5 4-6-0	BATH 1	09.25	10.45 B.West - Manchester
07.28 Templecombe - B.West	10.00	2P 4-4-0	TCB 57	11.20	11.40 B.West - Bristol
18.00 Bristol - B.West	23.05	4F 0-6-0	BK 74	12.30	12.55 B.West - Bath
08.30 Waterloo - B.West	11.24	MN 4-6-2	NE 32	12.35	13.05 B.West - Waterloo
09.40 Brighton - B.Central	12.18	WC 4-6-2	BTN 730	13.20	13.50 B.West - Brighton
10.30 Waterloo - B.West	13.40	MN 4-6-2	NE 34	14.35	15.05 B.West - Waterloo
09.05 Bristol - B.West	13.20	LM5 4-6-0	BATH 2	15.15	15.35 B.West - Bristol
12.30 Waterloo - B.West	15.30	MN 4-6-2	NE 35	15.50	16.34 B.West - Waterloo
13.29 Fareham - B.West	16.10	LN 4-6-0	EL 253	16.35	17.05 B.West - Waterloo
12.23 Templecombe - B.West	14.40	2P 4-4-0	TCB 59	16.50	17.18 B.West - Templecombe
06.05 Bristol - B.West	11.30	LM5 4-6-0	BATH 4	18.20	18.40 B.West - Bath
17.20 Salisbury - B.West	18.45	T9 4-4-0	SAL 445	19.30	19.43 B.West - Salisbury
10.20 Manchester - B.West	18.00	LM5 4-6-0	BATH 1	20.30	21.20 B.Cent Gds - Bath

TRAIN WORKING : BOURNEMOUTH WEST (1954)

Train	Arr	Engine	Shed	Dep	Load	Destination
Light ex Bournemouth loco	14.03	M7 0-4-4T	BM 406			
12.23 Templecombe	14.05	2P 4-4-0	TCB 59		3	
		M7 0-4-4T	BM 408	14.20	6	Waterloo
		2P 4-4-0	TCB 59	14.30		Light to Branksome loco
12.58 Salisbury	14.31	T9 4-4-0	SAL 444		3	
11.30 WATERLOO	14.34	LN 4-6-0	EL 252		6	
		M7 0-4-4T	BM 406	14.35	2	Brockenhurst Motor via Poole
Light ex Branksome loco	14.45	MN 4-6-2	NE 34			
12.30 WATERLOO PULLMAN	14.52	MN 4-6-2	NE 35		11	
		LN 4-6-0	EL 252	15.00		Light to Bournemouth loco
		MN 4-6-2	NE 34	15.05	11	WATERLOO
		MN 4-6-2	NE 35	15.15		Light to Branksome loco
14.05 Brockenhurst Motor via Poole	15.19	M7 0-4-4T	BM 409		2	
Light ex Branksome loco	15.22	LM5 4-6-0	BATH 2			
		LM5 4-6-0	BATH 2	15.35	4	Bristol
13.29 Fareham	15.47	LN 4-6-0	EL 253		3	
		LN 4-6-0	EL 253	16.00		Light to Branksome loco
Light ex Branksome loco	16.00	MN 4-6-2	NE 35			
		M7 0-4-4T	BM 409	16.30	4	Brockenhurst Motor via Poole
		MN 4-6-2	NE 35	16.34	11	WATERLOO PULLMAN
13.30 WATERLOO	16.38	700 0-6-0	BM 417		6	
Light ex Branksome loco	16.45	LN 4-6-0	EL 253			
		T9 4-4-0	SAL 444	16.52	5	Salisbury
Light ex Branksome loco	17.00	2P 4-4-0	TC 59			
		LN 4-6-0	EL 253	17.05	7	WATERLOO
		2P 4-4-0	TC 59	17.18	5	Templecombe
16.10 Brockenhurst Motor via Poole	17.25	M7 0-4-4T	BM 406		2	
10.20 MANCHESTER	17.32	LM5 4-6-0	BATH 1		12	
09.30 BIRKENHEAD	17.35	N15 4-6-0	BM 399		12	
		LM5 4-6-0	BATH 1	17.50		Light to Branksome loco
15.20 WATERLOO	17.58	N15 4-6-0	EL 266		6	
		N15 4-6-0	BM 399	18.05		Light to Bournemouth loco
		700 0-6-0	BM 417	18.16	6	WATERLOO
16.12 Evercreech Jcn	18.25	4F 0-6-0	BK 71		3	
Light ex Branksome loco	18.30	LM5 4-6-0	BATH 4			
		N15 4-6-0	EL 266	18.35	8	Reading
		LM5 4-6-0	BATH 4	18.40	6	Bath
15.30 WATERLOO	18.40	N15 4-6-0	EL 263		10	
		4F 0-6-0	BK 71	18.45		Light to Branksome loco
		M7 0-4-4T	BM 406	18.50	2	Brockenhurst Motor via Poole
10.28 YORK	18.50	LN 4-6-0	BM 395		9	
17.20 Salisbury	18.54	T9 4-4-0	SAL 445		6	
		N15 4-6-0	EL 263	19.00	6	Bournemouth Car Sidings
		T9 4-4-0	SAL 445	19.05		Light to Branksome loco
16.35 WATERLOO	19.13	700 0-6-0	BM 417		6	
18.06 Brockenhurst Motor via Poole	19.16	M7 0-4-4T	BM 409		2	
		LN 4-6-0	BM 395	19.22		Light to Bournemouth loco
15.30 Bristol	19.25	4F 0-6-0	BK 72		4	
Light ex Wimborne	19.32	N15 4-6-0	EL 265			
Light ex Branksome loco	19.35	T9 4-4-0	SAL 445			
		4F 0-6-0	BK 72	19.40		Light to Branksome loco
		T9 4-4-0	SAL 445	19.43	4	Salisbury
		700 0-6-0	BM 417	19.45		Light to Bournemouth loco
		N15 4-6-0	EL 265	19.55	4	Waterloo (Pds)
		M7 0-4-4T	BM 411	20.10		Light to Bournemouth loco
Station Pilot						
17.30 WATERLOO	20.39	WC 4-6-2	BM 380		10	
		M7 0-4-4T	BM 409	20.52	2	Brockenhurst Motor via Poole
18.30 WATERLOO	21.01	WC 4-6-2	BM 386		6	
		WC 4-6-2	BM 380	21.14		Light to Bournemouth loco
		WC 4-6-2	BM 386	21.25		Light to Bournemouth loco
20.15 Brockenhurst Motor via Poole	21.26	M7 0-4-4T	BM 409		2	
18.45 Portsmouth & S	21.48	Q 0-6-0	BM 416		3	
		Q 0-6-0	BM 416	22.00		Light to Poole
18.00 Bristol	22.35	4F 0-6-0	BK 74		4	
19.30 WATERLOO	22.38	WC 4-6-2	BM 385		6	
		4F 0-6-0	BK 74	22.55		Light to Branksome loco
		WC 4-6-2	BM 385	23.00		Light to Bournemouth loco
19.30 WATERLOO	23.03	M7 0-4-4T	BM 409		2	
		M7 0-4-4T	BM 406	23.10	2	Wimborne Motor
		M7 0-4-4T	BM 409	23.25		Light to Bournemouth loco

Unrebuilt West Country 34006 'Bude' accelerates the 15.50 Weymouth - Waterloo away from Christchurch towards its next stop, New Milton, during the winter of 1961.   A few years earlier the 15.50 had been the last main line express to be booked to a T9 4-4-0.

N 2-6-0 31412 attacks the 1 in 266 bank out of Hinton Admiral with the 17.40 Bournemouth Central - Eastleigh stopping train.

At the end of the war the Southern Railway had possessed 347 pre-grouping 4-4-0 locomotives spread over 18 classes. Within ten years the number had been reduced to six classes of 93 engines, a contraction that owed something to the smaller BR Standard locomotives such as the 4MT 2-6-0's of which twenty were working from Eastleigh by early 1955. In addition to their duties over the Southern District of the LSWR, two of the Eastleigh BR moguls were outbased at Branksome for working across the Somerset & Dorset. 76010 waits for its next duty at Bournemouth Central.

the class spent the day as station pilot – a duty which mainly consisted of drawing stock from the platforms to the adjacent carriage sidings and vice versa – whilst others played a part in the running of the London trains: the 10.12 and 14.20 departures and the 08.30 ex Waterloo being worked as far to and from the Central by 0-4-4T's. Those who hankered for a longer session behind an M7 could ride down to Brockenhurst in the 20.52 via West Moors and come back in the rear two coaches of the 19.30 ex Waterloo.

Although it was considered a wonderful thing in the age of the Pacific to see an M7 pulling away with a London train (even if it was only taking it a couple of miles), the exercise was not without its risks since it meant relying on four driving wheels to get a train moving up a gradient of 1 in 90. There was not much of a margin on a wet or greasy rail and to minimise any risks one tried to ensure that the station pilot drew the stock into the station (as opposed to propelling it from the sidings) and was in a position to assist at the critical moment of starting. Assisting in rear was prohibited out of Bournemouth West but the pilot was allowed to bank departing trains as far as the end of the platform or – and this was an especially valuable concession – as far as the signalbox if the train consisted of ten or more coaches and started from either platform four or five.

As it happened the diagrammers were careful to see that the London trains booked to the M7's were on the light side and the heaviest was the six-coach 10.12. The class did of course handle empty stock workings of up to twelve vehicles – the maximum allowed into the West – although they always had the benefit of as-

sistance from the train engine to the platform end. It was permitted for trains of stock to be propelled from the sidings to the station but not vice versa and outgoing empty stock always had to be drawn out of the platform.

As always with a terminal station, the secret of fluid working lay in clearing the platforms quickly and soon as a train arrived, the pilot was brought onto the rear to be ready to draw the stock out as soon as the last passenger had alighted. The train engine would assist the stock as far as the end of the platform and, when the signal had been restored and pulled off again, be sent light to Bournemouth Central shed if it was a local engine finishing its diagram or to Branksome shed if it was a foreigner. None of these movements were given a precise time and it was accepted that engines and stock would be cleared as quickly as possible after arriving.

In the reverse direction the stock would generally be propelled from the sidings into the platform either by the train engine or the station pilot but if there were any misgivings about the ability of the train engine to get up the bank then orders would be given to have the stock drawn into the station so that the pilot could give a push upon starting. When this happened provision had to be made for any other trains of stock that became engineless because the station pilot was trapped in the platform.

Local trains were operated by push and pull working – an M7 0-4-4T and two vehicles – and simply changed direction without any shunting being needed. This was of considerable assistance at the West although the full economics of push and pull operation did not apply since the Brockenhurst and Wareham services were required, like any other train, to

carry a guard. (One of these trains, the 08.30 to Wareham, was unique in that it was the only service that ran between Bournemouth West and the main line west of Poole. All other trains taking the Branksome road from the West took the Broadstone route from Poole).

M7's were not the strangest sight to be seen on London trains at the West and a particularly gladdening spectacle for those with a detestation of anything that had happened since 1923 was the departure of the 18.16 to Waterloo behind a 700 0-6-0. Their happiness was completed an hour later when the same engine returned on no less a train than the Royal Wessex itself although, with a tractive effort very similar to that of a King Arthur, no one should really have been surprised.

One thing that was a little depressing was the near absence of N15 4-6-0's at the West especially as their appearance more than any other engine encapsulated the pre-war Southern Railway. Not so very many years earlier the class had been responsible for almost every train of any importance on the system whilst now, with the class still intact, the only major working left to them at Bournemouth West was the through service to Birkenhead which one of the Bournemouth allocation worked as far as Oxford. On an ordinary winter's day only four of the class were diagrammed to make an appearance although during the summer many of the engines normally relegated to night parcels and goods services would see the light of day on Saturday holiday workings.

Unlike Bournemouth Central, the West did not have an engine shed specifically devoted to its needs although Branksome, a mile and a quarter towards Poole, came very close to fulfilling the role.

COMBINED WORKING TIME TABLE AND ENGINE WORKINGS : 1954

Train	18.02	14.45				18.45	18.30	18.30		18.30		18.00		
From	Alton	Feltham				Ports	W'loo	W'loo		W'loo		P'ton		
Class		Gds												Gds
Engine	M7 0-4-4T	S15 4-6-0	S15 4-6-0	M7 0-4-4T	M7 0-4-4T	BR4 2-6-0	WC 4-6-2	LN 4-6-0	M7 0-4-4T	WC 4-6-2	Castle	M7 0-4-4T	U 2-6-0	LMS 4-6-0
Shed	EL 299	FEL 107	FEL 106	LYM 362	BM 406	EL 271	BM 386	BM 394	BM 409	BM 386	OOC	SWA 421	DOR 427	BATH 1
WATERLOO							18.30							
Woking														
Basingstoke		19.00												
Worting Jcn		19/06					19/26							
Micheldever														
Walers Ash														
WINCHESTER	19.05		19.26											
Shawford	19.11		19.32											
Eastleigh Yard		19.21												
EASTLEIGH	19.18		19.39											
EASTLEIGH	19.20	19/25					19/53							
Swaything	19.26													
St Denys	19.30	19/35				19.47								
Bevois Park Yard														
Northam	19.34	19.40												
Southampton (T)	19.37	20D25												
SOUTHAMPTON CENT						19.52	20.01							
SOUTHAMPTON CENT							20.05							
Millbrook														
Redbridge														
Totton														
Lyndhurst Rd														
Beaulieu Rd														
BROCKENHURST				20.07	20.15									
Lymington Jcn				20/10	20/18	20/24								
Lymington Town				20.18										
Lymington Pier				20.20										
Sway					(To									
New Milton					B'mouth)									
Hinton Admiral														
Christchurch														
Pokesdown														
Boscombe														
Centrd Goods														21.20
BOURNEMOUTH CENT							20.41							
BOURNEMOUTH CENT						20.47				20.53				21/25
BOURNEMOUTH WEST									20.52	21.01				
Branksome									20.56			21.25		21/34
Parkstone									21.00			21.29		
POOLE						20.59		21.05				21.37		22.20
Hamworthy Jcn								(To				21.43		
Holton Heath								Brock)				21.49		(To
WAREHAM						21.12					21.20	21.56		Bath)
Worgret Jcn						21/15					21/23	21/59		
Corfe Castle											21.33			
SWANAGE											21.42			
Wool												22.06		
Moreton												22.16		
DORCHESTER						21.32						22.27		
Dorchester Jcn						21/34				21/57		22/29		
Monkton														
Wishing Well														
Upwey												22.37		
Radipole														
WEYMOUTH						21.43				22.08		22.42		

Contrary to general belief, Branksome did not exist solely for the benefit of the Somerset & Dorset any more than it survived in the twilight operational haze that many, who should know better, have ascribed to it. Its reportedly tenuous grasp on life arose from the fact that it did not have a permanent allocation of locomotives even though an allocation of engines was largely incidental to the status of any depot. An engine shed existed by virtue of having train crews and mechanical staff attached to it and the number which continued to operate long after the enthusiast press had written them off as closed is legion.

All the same it has to be admitted that with a booking of fifteen engines a day, Branksome was by no means the busiest location on the system although the engines that the shed dealt were enough to prevent Bournemouth (71B) from seizing up with congestion and this alone justified its existence.

Locomotives using the shed fell into two categories: Somerset & Dorset engines which in all but name were allocated to the shed and foreign engines which came onto the shed for servicing between turns at Bournemouth West.

The former consisted of three LMS 4F 0-6-0's which stabled overnight for the 06.48, 08.48 and 12.55 S&D workings to Bath. The trio were actually allocated to Bath and were replaced in diagram by Green Park or Templecombe as and when it became necessary. During the course of the day further

BANKING ARRANGEMENTS : POOLE (1954)				
Train	Train Engine	Poole Dep	Banker	Assist to
04.05 Salisbury - Bournemouth C. Goods	700 0-6-0 (452)	06.42	Q 0-6-0 (416)	Branksome
13.45 Hamworthy Jcn - Bournemouth C. Goods	Q 0-6-0 (415)	14.30	Q1 0-6-0 (318)	Branksome
15.08 Corfe Castle - Bournemouth C. Goods	Q1 0-6-0 (318)	20.36	N15 4-6-0 (401)	Branksome
19.50 Dorchester - Nine Elms	H15 4-6-0 (74)	21.25	N15 4-6-0 (401)	Branksome
22.26 Hamworthy Jcn - Eastleigh	N15 4-6-0 (401)	23.30	Q1 0-6-0 (416)	B'mouth Central

S&D locomotives – four LMS 4-6-0's, and a pair of 2P 4-4-0's – visited the shed for coal and water before returning north and not infrequently 7F 2-8-0's could be seen when either standing in for one of the usual engines or working special trains.

Most enthusiasts who paid a visit to Branksome did so on Sundays and it is their reported observations that have rather distorted the shed's history. Had they changed their visits to a weekday, a very different bill of fare would have been digested since between 11.30 and 16.30 the shed had to prepare the engines of no less than four Waterloo expresses – including the Bournemouth Belle – together with the through train for Brighton. This brought onto the depot three Merchant Navy Pacifics, a West Country and a Lord Nelson 4-6-0 and no establishment responsible for the servicing of such engines could be described as insignificant.

Generally a town of around ninety thousand inhabitants would have merited all the railway facilities one could think of and in this respect it was Poole's (population: 87,000) misfortune to be situated only five miles down the line from Bournemouth. As though trying to redress an imbalance, Poole station consisted of no more than an up and a down platform through which something like one hundred and seventy daily movements – a train every nine minutes - were channelled. Apart from the summer week-ends when the stock of two long distance excursion were stored in the yard, passenger terminal work was al-

WEYMOUTH – BASINGSTOKE : WORKING TIMETABLE 1954														
Train	s		15.02				16.25	21.56		20.52	19.48		21.56	
From	Corfe		Corfe				Dorset	B Park		B'mouth	W'mouth		B Park	
Class	Gds	Gds	Gds				Gds	Gds					Gds	Gds
Engine	Q1 0-6-0	H15 4-6-0	Q1 0-6-0	BR3 2-6-2T	BR4 2-6-0	M7 0-4-4T	Q0-6-0	S15 4-6-0	Hdl	M7 0-4-4T	WC 4-6-2	N15 4-6-0	S15 4-6-0	U 2-6-0
Shed	EL 318	NE 74	EL 318	EL 330	EL 270	SWA 421	FEL 106	FEL 106	BRD	BM 409	BM 381	EL 263	FEL 106	GUI 181
WEYMOUTH									19.30		19.48			
Radipole														
Upwey											19.55			
Wishing Well														
Monkton														
Dorchester Jcn									19/44		20/04			
DORCHESTER		19.50							(To B'tol)		20.15			
Moreton											20.24			
Wool											20.31			
SWANAGE						20.10								
Corfe Castle						20.21								
Worgret Jcn		20/15				20/30					20/37			
WAREHAM		20/19				21.32					20.41			
Holton Heath											20.47			
Hamworthy Jcn	19.56	20.28									20.52			
POOLE	20.05		20.36								20.59			
Parkstone											21.04			
Branksome											21.09			
BOURNEMOUTH WEST														
BOURNEMOUTH CENTRAL											21.14			
BOURNEMOUTH CENTRAL			20/56							(To Read'g)	21.22			
Central Goods			21.02											
Boscombe											21.26			
Pokesdown											21.29			
Christchurch											21.35			
Hinton Admiral											21.42			
New Milton											21.50			
Sway											21.57			
Lymington Pier														
Lymington Town														
Lymington Jcn										21/55	22/00			
BROCKENHURST						21.40				21.57	22.04			
Beaulieu Road											22.12			
Lyndhurst Road											22.18			
Fawley														
Totton											22.24			
Redbridge							22/06							
Millbrook														
SOUTHAMPTON CENTRAL											22.31			
SOUTHAMPTON CENTRAL				22.04			22/12				22.35			
Southampton T.					22.00									22D30
Northam					22.04									
Bevois Park Yard							22.22							
St Denys				22.11	22.13						22.42			22/47
Swaythling				22.15	(To									
EASTLEIGH				22.20	P'mth)						22.49			
EASTLEIGH											22.52			23/00
Eastleigh Yard								22.30						
Shawford											23.00			
WINCHESTER								22/52			23.07			23/25
Weston								23.04					23.22	
Micheldever											23.23			
Worting Jcn											23/34		23/50	00/03
Basingstoke											23.39			
Waterloo											(Reading)		(To Feltham)	(To Feltham)

most unknown and for most of the year was confined to the 06.10 Christchurch and 19.35 Kensington parcels train which ran from Poole to Weymouth as one train. Later in the morning the Q 0-6-0 which spent several hours shunting in the goods yard would amble across to the up platform and remove the rearmost vanfit from the 09.23 Salisbury - Bournemouth West and attach it, half an hour later, to the rear of the 10.10 Weymouth – Bournemouth Central. It seemed a rather complicated way of getting a vehicle from Salisbury to Bournemouth Central but it was preferable to running it via Bournemouth West with the shunting movements that would have been involved.

Poole was unique in that it was the only location on the main line (Weymouth being Great Western) where banking was permitted; the reason being descent from Branksome which averaged 1 in 113 for the two miles between Parkstone and Poole.

Banking was authorised for both passenger and goods trains although the limit for the former was on the generous side; a small engine such as an M7 0-4-4T or 2P 4-4-0 being allowed to take up to six vehicles unassisted. For larger engines the allowance increased in the proportion of one coach per engine classification rising to thirteen vehicles for a Merchant Navy Pacific. Unfortunately there was never

any question of an up train being able to rush the bank since, apart from the 30 mph restriction at the country end of the station, there was a long standing instruction that required all trains to come to a stand in Poole station.

Although authority for assistance was given from Poole to Bournemouth, in practice the assisting engine only ran the full course if Branksome box was closed. In normal circumstances the engine was not coupled to the train and dropped off at Branksome where it crossed over to the down line to return light to Poole. When Branksome box was switched out, the procedure became rather complicated. An engine banking a non-stop train from Poole to

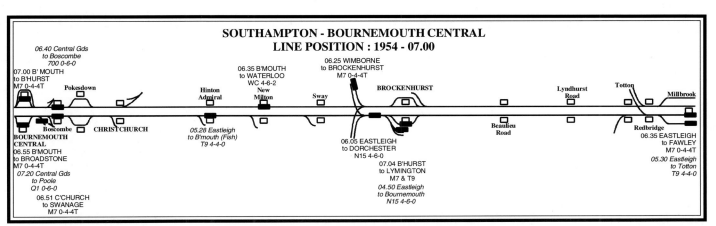

SOUTHAMPTON - BOURNEMOUTH CENTRAL
LINE POSITION : 1954 - 07.00

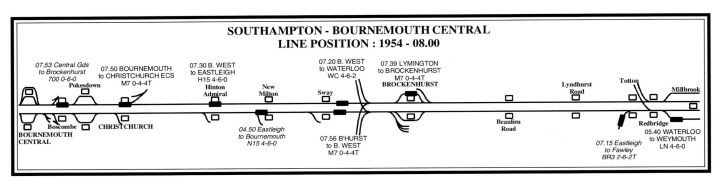

SOUTHAMPTON - BOURNEMOUTH CENTRAL
LINE POSITION : 1954 - 08.00

Bournemouth had to be coupled and piped to the train unless the train happened to stop at Parkstone but not Branksome, the brake and coupling being disconnected during the Parkstone stop so that the engine could continue pushing as far as Branksome.

When arranging heavy special trains on the up road from Poole one quickly learned to stop them either at both stations or none and the easiest way of failing a member of staff at his annual rules examination was to ask him: "… you have a ten coach train with a class 2 engine running as a special from Swanage to Bournemouth West calling at Parkstone only. Describe everything that happens." (If by some quirk, the candidate dealt with everything correctly – probably because

S W A N A G E : 1954					
Train	Arr	Engine	Shed	Dep	Destination
05.47 Bournemouth Central Motor	06.48	M7 0-4-4T	BM 407	07.15	Bournemouth West
		M7 0-4-4T	SWA 421	07.36	WATERLOO
06.51 Christchurch	08.40	M7 0-4-4T	BM 410		
08.31 Wareham Motor	08.53	M7 0-4-4T	SWA 421	08.58	Wareham Motor
		M7 0-4-4T	BM 410	09.24	WATERLOO
06.55 Hamworthy Jcn Goods	09.52	M7 0-4-4T	H. Jn 422		
09.47 Wareham Motor	10.09	M7 0-4-4T	BM 407		
		M7 0-4-4T	H. Jn 422	10.45	Wareham Goods
		M7 0-4-4T	BM 407	11.06	Wareham Motor
11.06 Wareham Motor	11.28	M7 0-4-4T	SWA 421	11.36	Wareham Motor
11.57 Wareham Motor	12.19	M7 0-4-4T	BM 407		
12.16 Wareham Motor	12.38	M7 0-4-4T	SWA 421		
		M7 0-4-4T	BM 407	12.42	Wareham Motor
		M7 0-4-4T	SWA 421	13.33	WATERLOO
10.30 WATERLOO	13.54	M7 0-4-4T	BM 407		
14.16 Wareham Motor	14.38	M7 0-4-4T	SWA 421		
		M7 0-4-4T	BM 407	14.42	Wareham Motor
15.37 Wareham Motor	15.59	M7 0-4-4T	BM 407		
		M7 0-4-4T	SWA 421	16.23	Wareham Motor
		M7 0-4-4T	BM 407	17.00	Wareham Motor
16.57 Wareham Motor	17.21	M7 0-4-4T	SWA 421	17.38	Wareham Motor
17.00 Bournemouth Central Motor	18.00	M7 0-4-4T	BM 408	18.36	Wareham Motor
18.33 Wareham Motor	18.57	M7 0-4-4T	SWA 421		
16.35 WATERLOO	19.52	M7 0-4-4T	BM 408		
		M7 0-4-4T	SWA 421	20.10	Wareham Motor
		M7 0-4-4T	BM 408	21.20	Wareham Motor
18.30 WATERLOO	21.42	M7 0-4-4T	SWA 421		
		M7 0-4-4T	SWA 421	22.04	Wareham Motor
22.04 Wareham Motor	22.26	M7 0-4-4T	BM 408		

Southampton with seventy and only by adding an engine to the rear could anything like a full load be taken over the section in between. (The rule for banked loadings between Poole and Branksome was to aggregate the normal loadings for the train and banking engine. Thus a train worked by a U class 2-6-0 (32 wagons) and banked by an M7 0-4-4T (18 wagons) could take fifty vehicles up the bank).

Having no motive power depot, Poole lacked the facility of permanently based banking engines and arrangements for each up goods train via Bournemouth had to be made well in advance of its arrival at Poole. To assist in this respect and at the

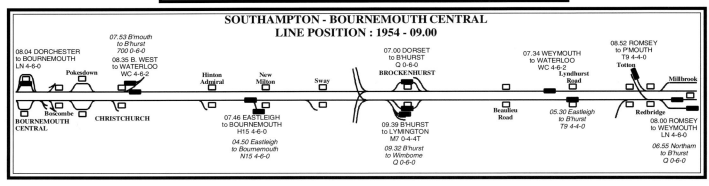

SOUTHAMPTON - BOURNEMOUTH CENTRAL
LINE POSITION : 1954 - 09.00

he'd been asked the same question innumerable times before – the examiner would wipe the smile from his face by repeating the question but imposing a blanket of fog or falling snow when even more variations on the theme were brought into play…).

While the great majority of passenger trains operated within the limits and rarely needed

assistance, goods trains were a different matter since the maximum load that could be worked from Poole to Branksome unassisted was a paltry thirty-five wagons. This was a serious operating handicap since main line services could run into Poole with up to fifty vehicles behind the engine and leave Bournemouth for

same time to guard against any overloaded passenger trains, as many engines as possible with spare time in their diagrams were directed towards Poole with the result that the bankers were represented by a curious assortment of engines ranging from an M7 0-4-4T which put in quite a long evening shift (16.31 to 02.30) to

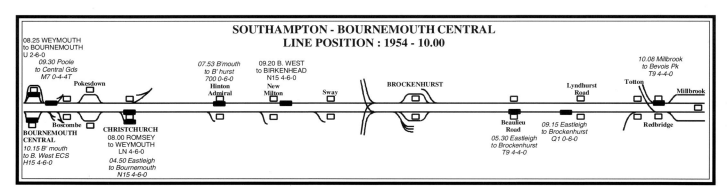

SOUTHAMPTON - BOURNEMOUTH CENTRAL
LINE POSITION : 1954 - 10.00

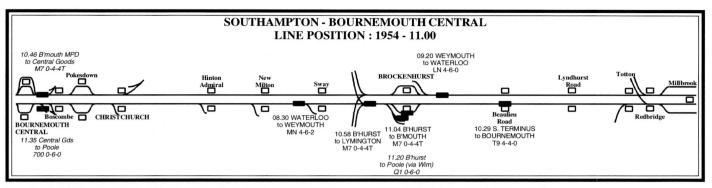

SOUTHAMPTON - BOURNEMOUTH CENTRAL
LINE POSITION : 1954 - 11.00

	BOURNEMOUTH CENTRAL LOCO				
Inward Working	**On Shed**	**Loco**	**Diagram**	**Off Shed**	**Working**
Centrd Goods Pilot	00.10	G6 0-6-0T	BM 418	02.30	Centrd Goods Pilot
Light ex B. West (11.12 ex Wimborne)	12.08	Q0-6-0	EL 325	02.50	03.20 Bevois Park Goods
22.26 Hamworthy Jcn Banker	00.00	Q0-6-0	BM 416	02.55	03.30 Poole Goods
19.48 Weymouth	21.25	WC 4-6-2	BM 381	04.08	Light to Wimborne for 04.58 Weymouth
14.35 Poole Goods	15.15	Q0-6-0	BM 415	04.55	05.20 Poole Pds
Light ex Poole (17.20 ex Parkstone)	18.10	M7 0-4-4T	BM 405	05.10	Light to Wimborne for 06.25 Brockenhurst
Light ex Wimborne (23.10 ex B. West)	00.30	M7 0-4-4T	BM 406	05.10	Light to Wimborne for 06.45 B. West
18.29 Wareham (Pilot)	19.15	M7 0-4-4T	BM 407	05.15	05.47 Swanage
Light ex B. West (17.30 Waterloo)	21.26	WC 4-6-2	BM 380	05.25	Light to B. West for 06.15 Waterloo
Light ex B. West (10.16 ex York)	19.34	LN 4-6-0	BM 394	05.25	05.54 Eastleigh
B. West Pilot	20.22	M7 0-4-4T	BM 411	05.25	B. West Pilot
17.41 Weymouth	19.20	U 2-6-0	BM 414	05.25	Light to Christchurch for 06.00 Weymouth Pds
Centrd Goods Pilot	23.55	O2 0-4-4T	BM 412	05.35	Centrd Goods Pilot
14.10 Winchester	17.25	T9 4-4-0	EL 284	05.50	Light to Broadstone for 07.14 Salisbury
22.35 Wareham ECS	23.15	M7 0-4-4T	BM 408	05.50	06.30 Brockenhurst
Poole Quay Pilot	13.20	B4 0-4-0T	BM 419	06.00	Light to Poole
18.07 ex Southampton Terminus	19.50	T9 4-4-0	FRA 403	06.10	Light to Brockenhurst for 07.04 Lymington
Light ex B. West (19.05 ex Centrd)	19.57	700 0-6-0	BM 417	06.10	06.40 Boscombe Goods
15.30 ex Waterloo	18.35	WC 4-6-2	N. ELMS 38	06.11	07.20 B. West - Waterloo
11.44 Hamworthy Jcn ECS	10.47	M7 0-4-4T	BM 410	06.13	Light to Christchurch for 06.51 Wareham
Light ex Poole	02.45	M7 0-4-4T	BM 404	06.20	06.55 Broadstone
23.35 Eastleigh Goods	04.10	H15 4-6-0	EL 313	06.30	Light to B. West for 07.30 Eastleigh
Light ex B. West (22.13 ex Brockenhurst)	23.37	M7 0-4-4T	BM 409	06.35	07.00 Brockenhurst
16.40 Weymouth	18.20	LN 4-6-0	BM 393	06.40	07.10 Weymouth
15.08 Corfe Castle Goods	21.20	Q1 0-6-0	EL 319	06.50	07.20 Parkstone Goods
Light ex Brockenhurst (08.10 ex Christchurch)	10.00	M7 0-4-4T	BM 413	07.10	07.50 Christchurch
05.17 ex Wimborne	05.55	T9 4-4-0	SAL 443	07.10	07.42 Salisbury
21.18 Weymouth	22.54	WC 4-6-2	BM 382	07.55	Light to B. West for 08.35 Waterloo
19.35 Kensington Pds	05.10	N15 4-6-0	BM 399	08.15	Light to B. West (09.20 Birkenhead)
Light ex B. West (18.30 Waterloo)	21.37	WC 4-6-2	BM 386	08.20	08.59 Weymouth (05.40 ex Waterloo)
Light ex B. West (19.30 Waterloo)	23.16	WC 4-6-2	BM 385	09.55	10.23 Weymouth (08.00 ex Romsey)
05.40 Waterloo	09.05	LN 4-6-0	N. ELMS 31	10.10	10.33 Waterloo (09.20 ex Weymouth)
21.55 Weymouth	23.40	WC 4-6-2	BM 383	10.25	Light to B. West for 11.05 Waterloo
10.12 ex B. West	10.30	M7 0-4-4T	BM 409	11.00	11.25 B. West
08.25 Weymouth	10.20	U 2-6-0	DOR 426	11.00	11.20 Weymouth (08.30 ex Waterloo)
04.05 Salisbury Goods	07.25	700 0-6-0	SAL 452	11.05	11.35 Poole Goods
Light ex Centrd Goods (09.30 Poole Goods)	10.10	M7 0-4-4T	BM 405	11.35	B. West Pilot
08.20 Bournemouth W. - B. Centrd	08.40	MN 4-6-2	N. ELMS 30	12.15	12.49 Weymouth (10.30 ex Waterloo)
10.10 Weymouth	11.45	LN 4-6-0	BM 393	12.35	13.01 Weymouth
Centrd Goods Pilot	11.45	G6 0-6-0T	BM 418	12.45	Centrd Goods Pilot
23.14 Feltham Goods	05.40	S15 4-6-0	FEL 111	13.15	13.45 Goods to Bevois Park
11.04 ex Brockenhurst	12.25	M7 0-4-4T	BM 406	13.50	Light to B. West for 14.35 Brockenhurst
Light ex B. West (09.30 ex Waterloo)	13.39	WC 4-6-2	N. ELMS 33	14.15	14.57 Weymouth
Light ex Centrd Goods	13.10	M7 0-4-4T	BM 404	15.15	15.45 Poole Goods
Light ex B. West (09.40 Brighton)	14.11	T9 4-4-0	EL 282	15.40	15.56 Eastleigh (14.20 ex Weymouth)
11.28 Eastleigh	13.50	H15 4-6-0	EL 313	16.15	16.38 Eastleigh
09.05 Light ex B. West	09.15	T9 4-4-0	SAL 444	16.15	Light to West for 16.52 Salisbury
14.20 ex B. West	14.35	M7 0-4-4T	BM 408	16.30	17.00 Swanage
14.20 ex Weymouth	15.55	U 2-6-0	DOR 426	16.55	17.28 Dorchester Goods
04.50 Eastleigh Goods	10.45	N15 4-6-0	EL 266	17.30	17.50 B. West
12.20 ex Weymouth	14.05	N15 4-6-0	BM 401	17.45	Light to Boscombe for 19.15 B'mouth Goods
09.20 Weymouth	10.27	N15 4-6-0	EL 263	18.00	18.32 B. West (15.30 ex Waterloo)
Light ex B. West (09.55 ex Eastleigh)	11.42	N15 4-6-0	EL 259	18.40	Light to B. West for 19.28 Eastleigh
15.50 ex Weymouth	16.00	T9 4-4-0	DOR 425	18.45	19.12 Weymouth
08.00 Romsey	10.20	LN 4-6-0	BM 394	20.10	20.47 Weymouth (18.30 ex Waterloo)
Light ex Christchurch (17.17 ex Wareham)	19.10	U 2-6-0	DOR 427	20.55	21.18 Weymouth
Light ex B. West (09.30 Birkenhead)	18.17	N15 4-6-0	BM 399	23.00	23.32 Southampton T. (21.55 ex Weymouth)

a King Arthur N15 4-6-0 which came light from Bournemouth Central to push the Corfe Castle and Dorchester goods trains up to Branksome before dropping down to Hamworthy Junction and working the 22.26 Goods to Eastleigh. As with passenger trains, banking assistance was normally given as far as Branksome where the banker would leave the train before being crossed over to the down main. In the (rather unlikely) event of Branksome box being switched out, the banker had to be coupled to the rear of the train, assistance being given to Bournemouth Central passenger station.

Hamworthy Junction two miles west of Poole was of historic importance being the point of convergence of the old route from London via West Moors and the new via Bournemouth. The connection to Broadstone had been reduced to a single line, sufficient for the handful of services using it, and it was unfortunate that the singling had somehow created an impression that Hamworthy Junction had seen its best days. In fact the main line junction had very little bearing on Hamworthy's fortunes which relied not on

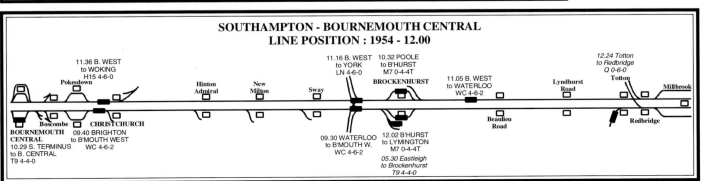

SOUTHAMPTON - BOURNEMOUTH CENTRAL
LINE POSITION : 1954 - 12.00

Train	18.45		19.00	19.00		19.25		19.45	11.45	17.55				20.38
From	Ports		Reading	Reading		Alton		Ports	St. Jn	Didcot				Alton
Class		Pds							Gds				Gds	
Engine	Q0-6-0	S15 4-6-0	N15X 4-6-0	BR4 2-6-0	M7 0-4-4T	M7 0-4-4T	M7 0-4-4T	T9 4-0	28xx	BR4 2-6-0	WC 4-6-2	M7 0-4-4T	Q0-6-0	2MT 2-6-2T
Shed	BM 416	FEL 106	BSK 236	EL 270	LYM 362	EL 297	BM 408	BM 403	EJN	EL 278	BM 385	BM 409	EL 324	AJN 247
WATERLOO											19.30			
Woking														
Basingstoke			19.35								20.28			
Worting Jcn			19/41								20/34			
Micheldever			19.51								20.46			
Walers Ash														
WINCHESTER			20.03			20.13					20.59			21.20
Shawford			20.09			20.18				20.33				21.27
Eastleigh Yard													21.20	
EASTLEIGH			20.16			20.25				20.39	21.09			21.34
EASTLEIGH		20.03		20.22						20.42	21.13		21/25	
Swaything				20.28						20.48				
St Denys				20.32				20.50		20.52	21.21		21/35	
Bevois Park Yard														
Northam				20.36						20.56				
Southampton (T)		20.17		20.39						20.59				
SOUTHAMPTON CENT								20.55			21.26			
SOUTHAMPTON CENT	20.10							21.01			21.31		21/45	
Millbrook	20.14							21.05						
Redbridge	20.19							21.10						
Totton	20.22							(To A. Jcn)			21.39		21.55	
Lyndhurst Rd	20.29										21.46			
Beaulieu Rd	20.37										21.52			
BROCKENHURST	20.46				20.50						22.03	22.13		
Lymington Jcn	20/49				20/53						22/06	22/16		
Lymington Town					21.00									
Lymington Pier														
Sway	20.54											22.20		
New Milton	21.01											22.27		
Hinton Admiral	21.08											22.33		
Christchurch	21.15											22.39		
Pokesdown	21.21											22.45		
Boscombe	21.24											22.48		
Centrd Goods														
BOURNEMOUTH CENT	21.27										22.24	22.51		
BOURNEMOUTH CENT	21.33										22.29	22.55		
BOURNEMOUTH WEST	21.41										22.38	23.03		
Branksome														
Parkstone														
POOLE														
Hamworthy Jcn														
Holton Heath														
WAREHAM							22.04							
Worgret Jcn							22/07							
Corfe Castle							22.17							
SWANAGE							22.26							
Wool														
Moreton														
DORCHESTER														
Dorchester Jcn									22/09					
Monkton														
Wishing Well														
Upwey														
Radipole														
WEYMOUTH									22.34					

the interconnection of main line services but on the two-mile branch to Hamworthy Goods and the eight establishments it served.

Having never had a passenger service, the Hamworthy branch was probably one of the least-known byways in Southern England, yet for those who took the trouble to pay a visit it was by no means devoid of operating interest. Diverging on the down side of the main line, it followed a semi-circle in its two mile length to terminate a stones' throw from Poole. (In the course of its routine the branch engine - a B4 0-4-0T - spent the latter part of its shift shunting at Poole Harbour and was therefore one of the few engines in the Southern fleet to box the compass within the span of a day's work).

There was no signalbox at the far end of the branch and trains were met by a shunter who directed them as needs dictated.

Although the branch could be used by a fairly wide range of engines, the sidings at the bottom could not and main line locomotives venturing down the line – which they did on a daily basis - could only proceed as far as the entrance to the sidings from which point a B4 0-4-0T took over.

Had the 0-4-0T been able to work all the traffic produced unaided to Hamworthy Junction, the line would have been worked on the one-engine-in-steam principle but because a Q1 0-6-0 worked down the branch each day, a system had to be devised which allowed two en-

gines to work simultaneously at the unsignalled end of the line whilst avoiding the expense of full-blown signalling.

The method employed was the 'no signalman key token' system in which the driver of a train leaving Hamworthy Junction was given a token and told whether or not a second train would be admitted to the branch.

If no other train was involved then the driver of the B4 simply retained the token until he returned to the Junction. If, on the other hand, he was to be followed by a second train then, when he arrived at Hamworthy Goods, the token would be handed to the shunter who would place it into his token machine, permitting the Hamworthy Junction signalman to withdraw a

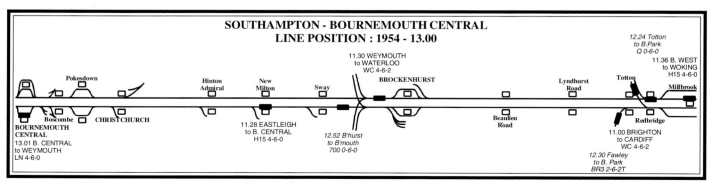

SOUTHAMPTON - BOURNEMOUTH CENTRAL
LINE POSITION : 1954 - 13.00

Train		20.15	19.50		19.50	20.22	19.50		19.10	19.50		18.00			19.50
From		Brock	Dorset		Dorset	Scrum	Dorset		Bristol	Dorset		Bristol			Dorset
Class	Gds		Gds		Gds		Gds			Gds	Gds			Gds	Gds
Engine	Hdl	M7 0-4-4T	H15 4-6-0	M7 0-4-4T	H15 4-6-0	T9 4-4-0	H15 4-6-0	14xx	S15 4-6-0	H15 4-6-0	S15 4-6-0	4F 0-6-0	WC 4-6-2	N15 4-6-0	H15 4-6-0
Shed	SPM	BM 406	NE 74	BM 408	NE 74	SAL 444	NE 74	WEY	SAL 467	NE 74	EL 310	BK 74	BM 382	BM 401	NE 74
WEYMOUTH	19.55							20.35					21.18		
Radipole								20.39							
Upwey								20.43							
Wishing Well															
Monkton															
Dorchester Jcn	20/11							20/52					21/31		
DORCHESTER	(To							(To					21.36		
Moreton	B'tol)							Yeo)					21.45		
Wool													21.53		
SWANAGE				21.20											
Corfe Castle				21.34											
Worgret Jcn				21/43									21/58		
WAREHAM				21.45									22.06		
Holton Heath													22.12		
Hamworthy Jcn			21.08										22.18	22.26	
POOLE		21.11	21.15		21.25	21.35						22.19	22.27	22.34	
Parkstone		21.17				21.41						22.25	22.33		
Branksome		21.22				21.46						22.31	22.38		
BOURNEMOUTH WEST		21.26										22.35			
BOURNEMOUTH CENTRAL						21.51							22.44		
BOURNEMOUTH CENTRAL					21/43										
Central Goods					21.47		22.09								
Boscombe							22.13		22.21						
Pokesdown															
Christchurch															
Hinton Admiral															
New Milton															
Sway															
Lymington Pier															
Lymington Town															
Lymington Jcn										22/47					
BROCKENHURST										22.51					23.20
Beaulieu Road															
Lyndhurst Road															
Fawley															
Totton															
Redbridge									22/43						23/43
Millbrook															
SOUTHAMPTON CENTRAL									22.48						
SOUTHAMPTON CENTRAL									22.50						23/50
Southampton T.															
Northam															
Bevois Park Yard															23.58
St Denys									22.57						(To
Swaything									23.01						N. Elms)
EASTLEIGH									23.06						
EASTLEIGH															
Eastleigh Yard											23.15				
Shawford															
WINCHESTER											23.48				
Weston															
Micheldever															
Worting Jcn											00/29				
Basingstoke											00.42				
Waterloo															

second token for the following train.

The normal sequence of daily events was kicked off by the B4 0-4-0T which worked the 07.40 trip from the Junction, taking down traffic that had arrived off the overnight train from Nine Elms. On reaching Hamworthy the token was given to the shunter in order to allow allowing a Q1 0-6-0, which had spent an hour shunting out wagons that had arrived from Dorchester and Poole, to work down from the Junction and join the B4 at the lower end.

The reason for having a second engine was because of the volume of traffic produced at Hamworthy Goods and where the B4 was only permitted to take twenty-nine vehicles up to the Junction, the much larger Q1 was allowed fifty. The maximum that could be worked on any single service was sixty-seven vehicles and this was achieved by using the 0-6-0 as the train and the 0-4-0T as its banker.

The combined working worked back to Hamworthy Junction at midday after which the Q1 returned to its home station with the 14.45 Eastleigh goods via Wimborne. The B4 went back to Hamworthy goods with an afternoon trip of empties and returned at 16.00 with loaded traffic for the night trains to Eastleigh and Nine Elms.

In addition to shunting out traffic for the Hamworthy Goods branch, Hamworthy Junction was also the junction point for the Swanage branch so far as goods traffic was concerned, an arrangement that achieved both an economy of scale and an M7 0-4-4T.

The latter was one of the few members of the class that never performed any passenger work; its regular working being to marshal together the Wareham and Swanage vehicles that arrived off the 22.38 ex Nine Elms and work them to Swanage, returning with the afternoon goods from Corfe Castle. It was really a working suited to an 0-6-0 but an 0-4-4T was a use-ful port in a storm should an engine be hurriedly needed for the busy Swanage branch passenger service. So far as goods loadings were concerned, the M7 could handle fifty wagons as far as Wareham and twenty-two over the Swanage Branch.

There was nothing very remarkable in the working except one element that sticks in the author's mind as being the most absurd piece of diagramming he came across in the whole of his railway career.

The Swanage goods left Hamworthy Junction at 06.30 and the first set of men were relieved at Swanage by a local set of men who worked it to Wareham with the branch goods and back to Corfe Castle in readiness for the 15.08 Hamworthy Junction train. A second set of Hamworthy Junction men signed on and travelled passenger to Wareham on the 13.01 ex Bournemouth to pick up the branch train as far as Corfe Castle. The branch train, the 14.16

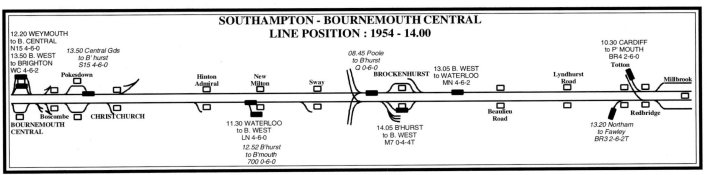

SOUTHAMPTON - BOURNEMOUTH CENTRAL
LINE POSITION : 1954 - 14.00

Although Weymouth was the point of termination for most services, the final seven miles were run over GW metals; the South Western finishing at Dorchester. Station facilities at the latter were far smaller than one might have expected of a county town although the curious arrangement which obliged all up trains to run past and back into the station provided an element of interest to what otherwise might have been a run-of-the-mill rural station. The fireman of West Country Pacific 34006 'Bude' lays it on with a trowel as his train pulls away from the up platform with the 07.34 Weymouth - Waterloo 'Royal Wessex' in June 1956. This working was a long duty, the engine starting with the 03.52 Salisbury - Weymouth (which it took over at Wimborne), the Wessex to London and the 13.30 Waterloo - Weymouth, the day concluding with the 19.48 Weymouth to Reading as far as Bournemouth.

In the lower view, N15 4-6-0 30742 'Camelot' of Bournemouth arrives in the down platform with the 06.05 Eastleigh - Dorchester and uncouples prior to running light to the loco. The G6 0-6-0T station pilot will be coupling up at the far end, ready to shunt the stock which will stay at Dorchester for almost twenty-four hours before forming the 08.04 to Bournemouth Central. In a couple of hours time the N15 will run light to Weymouth to work the 12.20 stopping train to Bournemouth Central.

POOLE : 1954

Train	Arr	Engine	Shed	Dep	Destination
23.57 light ex Wimborne		M7 0-4-4T	BM 406	00/15	Light to Bournemouth
21.20 Eastleigh Goods	01.01	Q0-6-0	EL 324	01.30	Dorchester
(Banker)		M7 0-4-4T	BM 404	02.30	Light to Bournemouth
22.30 WATERLOO	02.39	N15 4-6-0	EL 253	02.41	DORCHESTER
03.30 Bournemouth Goods	03.55	Q0-6-0	BM 416		(Bank to 08.45)
Light ex Bournemouth MPD		WC 4-6-2	BM 381	04/20	Wimborne
22.38 Nine Elms	04.44	H15 4-6-0	NE 73	05.15	Weymouth
2 LE's Light ex Bournemouth	05.29	2 X M7	405/406	05.30	Wimborne
05.17 Wimborne	05.31	T9 4-4-0	SAL 444	05.32	Eastleigh
19.35 Kensington Pds	05.49	Q0-6-0	BM 415		(Att to 06.50)
05.47 Bournemouth C.	06.02	M7 0-4-4T	BM 407	06.04	Swanage
04.05 Salisbury Goods	06.08	700 0-6-0	SAL 452		(Fwd at 06.42)
Light ex Bournemouth MPD	06.17	B4 0-4-0T	BM 419		Quay pilot
06.00 Christchurch (Fish)	06.27	U 2-6-0	BM 414		(Fwd at 06.50)
(04.05 Salisbury Goods)		700 0-6-0	SAL 452	06.42	Bournemouth Goods
06.30 Bournemouth C. (PP)	06.45	M7 0-4-4T	BM 408	06.46	Brockenhurst
(06.00 Christchurch)		U 2-6-0	BM 414	06.50	Weymouth (Fish)
06.45 Wimborne (PP)	06.58	M7 0-4-4T	BM 406	06.59	Bournemouth W.
06.48 Bournemouth W.	06.58	4F 0-6-0	BK 71	07.00	Bath
02.40 Bath Goods	07.09	WC 4-6-2	BATH 1		(Fwd at 07.27)
06.55 Bournemouth C.	07.10	M7 0-4-4T	BM 407	07.11	Broadstone
06.51 Christchurch	07.20	M7 0-4-4T	BM 410	07.22	Swanage
(02.40 Bath Goods)		WC 4-6-2	BATH 1	07.27	Bournemouth West
07.10 Bournemouth C	07.28	LN 4-6-0	BM 393	07.34	Weymouth
07.36 Broadstone	07.45	M7 0-4-4T	BM 404	07.46	Bournemouth C.
07.15 Swanage	07.54	M7 0-4-4T	BM 407	07.57	Bournemouth W.
07.42 Bournemouth C	07.57	T9 4-4-0	SAL 444	07.58	Salisbury
07.17 Brockenhurst (PP)	08.07	M7 0-4-4T	BM 405	08.08	Bournemouth W.
06.05 Eastleigh	08.08	N15 4-6-0	BM 401	08.10	Dorchester
07.20 Bournemouth Gds	08.14	Q1 0-6-0	EL 319		
07.34 WEYMOUTH	08.25	WC 4-6-2	BM 381	08.26	WATERLOO
08.10 Bournemouth W. (PP)	08.23	M7 0-4-4T	BM 406	08.28	Brockenhurst
07.15 Salisbury	08.35	T9 4-4-0	SAL 443	08.37	Bournemouth W.
08.30 Bournemouth W (PP)	08.41	M7 0-4-4T	BM 407	08.42	Wareham
08.04 Dorchester	08.44	LN 4-6-0	BM 395	08.45	Bournemouth C.
		Q0-6-0	BM 416	08.45	Brockenhurst Goods
		Q1 0-6-0	EL 319	08.50	Hamworthy Jcn Goods
Light ex B. West	08.50	M7 0-4-4T	BM 405		(For 09.30 goods)
08.48 Bournemouth	09.01	4F 0-6-0	BK 72	09.03	Bath
07.28 Templecombe	09.10	2P 4-4-0	TCB 57	09.14	Bournemouth W.
05.40 WATERLOO	09.14	WC 4-6-2	BM 386	09.17	WEYMOUTH
		M7 0-4-4T	BM 405	09.30	Bournemouth Goods
08.34 Brockenhurst (PP)	09.38	M7 0-4-4T	BM 408		(Forms 10.32)
08.25 Weymouth	09.43	U 2-6-0	DOR 426	09.44	Bournemouth C.
09.45 BOURNEMOUTH W	09.53	WC 4-6-2	BATH 1	09.55	MANCHESTER
09.20 WEYMOUTH	10.14	N15 4-6-0	EL 263	10.16	WATERLOO
10.14 Bournemouth W.	10.19	T9 4-4-0	SAL 443	10.20	Salisbury
(PP)		M7 0-4-4T	BM 408	10.32	Brockenhurst
09.25 Salisbury	10.36	T9 4-4-0	EL 284	10.40	Bournemouth W.
08.00 Romsey	10.39	WC 4-6-2	BM 385	10.42	Weymouth
06.05 Bristol	10.44	LM5 4-6-0	BATH 4	10.47	Bournemouth W.
		B4 0-4-0T	BM 419	10.57	Broadstone Goods
10.10 Weymouth	11.15	LN 4-6-0	BM 393	11.17	Bournemouth C.
11.12 Wimborne	11.26	Q0-6-0	EL 324	11.27	Bournemouth W.
08.30 WATERLOO	11.31	U 2-6-0	DOR 426	11.34	WEYMOUTH
11.40 Bournemouth W.	11.48	2P 4-4-0	TCB 57	11.50	Bristol
11.04 Brockenhurst (PP)	11.58	M7 0-4-4T	BM 406	11.59	Bournemouth C.
11.30 WEYMOUTH	12.23	WC 4-6-2	BM 386	12.25	WATERLOO
12.00 Broadstone Goods	12.27	B4 0-4-0T	BM 419		
09.05 Bristol	12.40	WC 4-6-2	BATH 2	12.43	Bournemouth W.
12.35 Bournemouth W (PP)	12.47	M7 0-4-4T	BM 409	12.49	Brockenhurst
		B4 0-4-0T	BM 419	12.56	Light to Bournemouth
03.20 Eastleigh Goods	12.58	Q1 0-6-0	EL 318		(Via W. Moors)
10.30 WATERLOO	13.00	MN 4-6-2	NE 30	13.02	WEYMOUTH

Train	Arr	Engine	Shed	Dep	Destination
12.12 Brockenhurst (PP)	13.07	M7 0-4-4T	BM 408	13.08	Bournemouth W.
12.55 Bournemouth W.	13.08	4F 0-6-0	BK 74	13.09	Bath
13.01 Bournemouth C.	13.16	LN 4-6-0	BM 393	13.19	Weymouth
11.35 Bournemouth Goods	13.24	700 0-6-0	SAL 452		
(Banker)		Q0-6-0	BM 415	13.27	Hamworthy Jcn Goods
12.20 Weymouth	13.25	N15 4-6-0	BM 401	13.28	Bournemouth C.
13.20 Bournemouth W.	13.33	T9 4-4-0	EL 284	13.37	Salisbury
12.23 Templecombe	13.49	2P 4-4-0	TCB 59	13.51	Bournemouth W.
13.45 Hamworthy Jcn Goods	13.54	Q0-6-0	BM 415		(Fwd at 14.30)
		700 0-6-0	SAL 452	13.55	Light to Wimborn
13.25 WEYMOUTH	14.23	WC 4-6-2	BM 385	14.25	WATERLOO
(13.45 H. Jcn Goods)		Q0-6-0	BM 415	14.30	Bournemouth Goods
14.35 Bournemouth W. (PP)	14.46	M7 0-4-4T	BM 406	14.49	Brockenhurst
14.05 Brockenhurst (PP)	15.02	M7 0-4-4T	BM 409	15.04	Bournemouth W.
14.57 Bournemouth C.	15.12	WC 4-6-2	NE 33	15.14	Weymouth
14.20 Weymouth	15.31	U 2-6-0	DOR 426	15.35	Andover Junction
15.35 Bournemouth W.	15.47	WC 4-6-2	BATH 2	15.49	Bristol
15.45 Bournemouth Goods	16.31	M7 0-4-4T	BM 404		(Bank until 02.30)
13.30 WATERLOO	16.34	WC 4-6-2	BM 381	16.36	WEYMOUTH
16.30 Bournemouth W. (PP)	16.42	M7 0-4-4T	BM 409	16.43	Brockenhurst
15.50 WEYMOUTH	16.48	T9 4-4-0	DOR 425	16.50	WATERLOO
16.52 Bournemouth W.	17.05	T9 4-4-0	SAL 444	17.07	Salisbury
16.10 Brockenhurst (PP)	17.11	M7 0-4-4T	BM 406	17.12	Bournemouth W.
17.00 Bournemouth C. (PP)	17.15	M7 0-4-4T	BM 408	17.17	Swanage
10.20 MANCHESTER	17.20	WC 4-6-2	BATH 1	17.22	BOURNEMOUTH W.
17.20 Parkston Goods	17.30	M7 0-4-4T	BM 405		
17.18 Bournemouth W.	17.33	2P 4-4-0	TCB 59	17.38	Templecombe
17.17 Wareham	17.37	U 2-6-0	DOR 427	17.39	Christchurch
16.40 Weymouth	17.45	LN 4-6-0	BM 393	17.47	Bournemouth C.
17.28 Bournemouth Goods	17.51	U 2-6-0	DOR 426		(Fwd at 18.24)
		M7 0-4-4T	BM 405	17.55	Light to Bournemouth
15.20 WATERLOO	18.00	WC 4-6-2	BM 382	18.02	WEYMOUTH
16.13 Evercreech Jcn	18.13	4F 0-6-0	BK 71	18.15	Bournemouth W.
(B'mouth Goods)		U 2-6-0	DOR 426	18.24	Dorchester
17.35 WEYMOUTH	18.24	MN 4-6-2	NE 30	18.26	WATERLOO
17.05 Southampton (T)	18.34	N15 4-6-0	EL 265	18.36	Wimborne
17.20 Salisbury	18.37	T9 4-4-0	SAL 445	18.39	Bournemouth W.
17.41 Weymouth	18.45	U & M7	414/407	18.47	Bournemouth C.
18.40 Bournemouth W.	18.52	LM5 4-6-0	BATH 4	18.54	Bath
18.06 Brockenhurst (PP)	19.01	M7 0-4-4T	BM 409	19.03	Bournemouth W.
18.50 Bournemouth W. (PP)	19.03	M7 0-4-4T	BM 406	19.04	Brockenhurst
16.35 WATERLOO	19.09	WC 4-6-2	BM 383	19.11	WEYMOUTH
15.30 Bristol	19.10	4F 0-6-0	BK 72	19.13	Bournemouth W.
19.12 Bournemouth C.	19.27	T9 4-4-0	DOR 425	19.29	Weymouth
18.30 WEYMOUTH	19.27	WC 4-6-2	NE 33	19.30	WATERLOO
19.43 Bournemouth W.	19.55	T9 4-4-0	SAL 445	19.57	Salisbury
Light ex Bournemouth C	20.05	N15 4-6-0	BM 401		Banker to 21.25
15.02 Corfe Castle Goods	20.05	Q1 0-6-0	EL 318	20.36	Bournemouth Goods
18.30 WATERLOO	20.57	LN 4-6-0	BM 394	20.59	WEYMOUTH
19.48 Weymouth	20.57	WC 4-6-2	BM 381	20.59	Reading
20.52 Bournemouth W.	21.04	M7 0-4-4T	BM 409	21.05	Brockenhurst
20.15 Brockenhurst (PP)	21.09	M7 0-4-4T	BM 406	21.11	Bournemouth W.
19.50 Dorchester Goods	21.15	H15 4-6-0	NE 74	21.25	Nine Elms
21.18 Bournemouth C.	21.33	U 2-6-0	DOR 427	21.37	Weymouth
21.20 Bournemouth Goods	21.45	WC 4-6-2	BATH 1		(Fwd at 22.20)
Light ex Bournemouth W.	22.11	Q0-6-0	BM 416		Bank 23.30 Goods
18.00 Bristol	22.17	4F 0-6-0	BK 74	22.19	Bournemouth W.
(Bournemouth Goods)		WC 4-6-2	BATH 1	22.20	Bath
21.18 Weymouth	22.23	WC 4-6-2	BM 382	22.27	Bournemouth C.
22.26 Hamworthy Jcn (Goods)	22.34	N15 4-6-0	BM 401		(Fwd at 23.30)
22.04 Swanage ECS	22.50	M7 0-4-4T	SWA 421	22.51	Bournemouth C.
21.55 WEYMOUTH	23.05	WC 4-6-2	BM 383	23.09	WATERLOO
23.10 Bournemouth W. (PP)	23.22	M7 0-4-4T	BM 406	23.24	Wimborne
(22.26 Hamworthy Jcn (Goods))		N15 4-6-0	BM 401	23.30	Eastleigh

Wareham – Swanage, had been brought in to Wareham by a set of Swanage men who could have worked the train back to Swanage before signing off but, for reasons known only to the diagramming clerk, were booked to travel passenger in the train while the Hamworthy Junction men took over the engine for the short distance to Corfe Castle; the final stage of the trip being worked by the Swanage men off the Hamworthy Junction M7. Thus in the short space of five miles an M7 was manned by no less than three sets of men when one could have sufficed.

If the Bournemouth West and the West Moors lines were secondary routes rather than branches then the only branch line west of Bournemouth was the eleven mile line from Wareham to Swanage whose popularity as a resort led to it having quite a respectable service of trains which included a number of through workings to and from Waterloo. Swanage also had its own motive power facilities although the M7 0-4-4T based at the shed rotated daily with another of the same class from Bournemouth and minimised thus the need for extensive maintenance equipment on the branch.

For most of the day the branch service consisted of push & pull trains – an M7 0-4-4T coupled to a two-coach push and pull set - between Swanage and Wareham but since a number of trains comprised conventional stock for the London workings, the working could not operate with a single engine shuttling up and down the branch and more than one set was needed to work the sixteen daily departures.

Altogether no less than four engines – all M7 0-4-4T's - were required to work the line on an ordinary weekday. The first part of the

DORCHESTER ENGINE DIAGRAMS

DORCHESTER 425 : T9 4-4-0

Arr	Station	Dep	Train
	Dorchester Loco	14.43	Light
15.00	Weymouth	15.50	Pass
17.50	Bournemouth C		
	Bournemouth Loco	18.45	
	Bournemouth C	19.12	
20.31	Weymouth	21.00	Light
21.16	Dorchester Loco		

DORCHESTER 426 : U 2-6-0

Arr	Station	Dep	Train
	Dorchester Loco	06.20	
	Dorchester	06.45	Pass
07.02	Weymouth	08.25	Pass
10.00	Bournemouth C		
	Bournemouth Loco	11.00	
	Bournemouth C	11.20	08.30 ex Waterloo
12.24	Weymouth		
	Weymouth Loco	14.00	
	Weymouth	14.20	Andover Jcn
15.50	Bournemouth C		
	Bournemouth Loco	16.55	
	Bournemouth C Goods	17.28	Goods
20.50	Dorchester West	21.30	Light
21.33	Dorchester Loco		

DORCHESTER 427 : U 2-6-0

Arr	Station	Dep	Train
	Dorchester Loco	11.50	
	Dorchester	12.10	Goods
15.47	Hamworthy Jcn	16.27	Light
16.35	Wareham	17.17	Pass
18.09	Christchurch	19.02	Light
19.10	Bournemouth Loco	20.55	
	Bournemouth C	21.18	Pass
22.42	Weymouth	23.30	Light
23.45	Dorchester Loco		

DORCHESTER 428 : G6 0-6-0T

Arr	Station	Dep	Train
	Dorchester Loco	03.20	
	Station pilot		
19.53	Dorchester Loco		

Although through running between Waterloo and Weymouth was limited to a handful of services, it was enough to ensure that the twelve Bournemouth West portions were worked from the Central by a fascinating variety of motive power. Merchant Navy Pacifics arrived with the 10.30 and 12.30 expresses from Waterloo, Light Pacifics the 09.30, 17.30, 18.30 and 19.30, a Lord Nelson on the 11.30, King Arthur 4-6-0's with the 15.20 and 15.30, an M7 0-4-4T on the 08.30 and - best of all - a 700 0-6-0 on the 13.30 and 16.35. West Country Pacific 34042 'Dorchester' draws to a stand in the West with the 09.30 ex Waterloo.

day was shared between a Bournemouth engine which arrived with an early push & pull from Bournemouth Central and the Swanage-based locomotive which took the Royal Wessex coaches as far as Wareham. A third 0-4-4T arrived with a local from Christchurch and left with the 09.24 Waterloo whilst a fourth ran in on an evening through working from Bournemouth to replace the engine that had spent the previous night at Swanage.

The branch could probably have been worked more economically – on paper at least – with a single engine but this would have stifled the running of through trains from beyond Wareham whilst the Southern disliked having to make its passengers change trains any more than was absolutely necessary.

If the branch engine diagrams were more complex than might have been supposed, the carriage workings were equally sophisticated. The branch trains were worked by a pair of two-coach push & pull sets, based at Swanage and Bournemouth respectively, with a third set being held at Swanage to strengthen the 16.23

Swanage – Wareham and 16.57 Wareham – Swanage school trains and to provide extra capacity that may have been needed on any other services.

Three two-coach sets of conventional stock were also based at Swanage to work the 07.38, 09.24 and 13.33 through trains to Waterloo although all of these services ran with more than just the London coaches. The 07.38 'Royal Wessex' included a push & pull set as far as Wareham as did the 13.33 whilst the 09.20 consisted of six vehicles: the two Swanage-based

DORCHESTER SOUTH : 1954											
Train	Arr	Engine	Shed	Dep	Destination	Train	Arr	Engine	Shed	Dep	Destination
21.20 Eastleigh Goods	02.47	Q0-6-0	EL 324			*13.01 Bournemouth C.*	14.04	LN 4-6-0	BM 393	14.06	Weymouth
22.30 WATERLOO	03.16	N15 4-6-0	EL 263					T9 4-4-0	DOR 425	14.43	Light to Weymouth
03.52 Salisbury	05.55	WC 4-6-2	BM 381	05.58	Weymouth	*14.20 Weymouth*	14.39	U 2-6-0	DOR 426	14.44	Andover Jcn
22.38 Nine Elms Gds	06.38	H15 4-6-0	NE 73		*(Fwd at 07.15)*			H15 4-6-0	NE 74	15.52	Light to Weymouth
		U 2-6-0	DOR 426	06.45	Weymouth	14.57 Bournemouth C	15.58	WC 4-6-2	NE 33	16.00	Weymouth
		Q0-6-0	EL 324	07.00	Brockenhurst	15.50 WEYMOUTH	16.08	T9 4-4-0	DOR 425	16.10	WATERLOO
(22.38 Nine Elms Gds)		N15 4-6-0	EL 263	07.15	Weymouth Goods			Q0-6-0	EL 317	16.25	Bevois Pk Goods
07.34 WEYMOUTH	07.49	WC 4-6-2	BM 381	07.50	WATERLOO			U 2-6-0	BM 414	16.40	Light to Weymouth
06.00 Christchurch (Fish)	08.02	U 2-6-0	BM 414		*(Fwd at 08.50)*	16.40 Weymouth	17.00	LN 4-6-0	BM 393	17.02	Bournemouth C
		LN 4-6-0	BM 395	08.04	Bournemouth C	13.30 WATERLOO	17.15	WC 4-6-2	BM 381	17.18	WEYMOUTH
07.10 Bournemouth C	08.27	LN 4-6-0	BM 393	08.33	Weymouth	*17.10 Weymouth Goods*	17.30	H15 4-6-0	NE 74		*(Fwd at 19.50)*
08.25 Weymouth	08.47	U 2-6-0	DOR 426		*(Fwd at 08.56)*	17.35 WEYMOUTH	17.50	MN 4-6-2	NE 30	17.52	WATERLOO
(06.00 Christchurch (Fish))		U 2-6-0	BM 414	08.50	Weymouth	17.41 Weymouth	18.00	U 2-6-0	BM 414	18.02	Bournemouth C
(08.25 Weymouth)	08.52	N15 4-6-0	BM 401			18.30 WEYMOUTH	18.46	WC 4-6-2	NE 33	18.48	WATERLOO
		U 2-6-0	DOR 426	08.56	Bournemouth C	15.20 WATERLOO	18.48	WC 4-6-2	BM 382	18.50	WEYMOUTH
09.20 WEYMOUTH	09.36	N15 4-6-0	EL 263	09.40	WATERLOO	16.35 WATERLOO	19.42	WC 4-6-2	BM 383	19.44	WEYMOUTH
Light ex Weymouth	09.45	U 2-6-0	BM 414			*(17.10 Weymouth Goods)*		H15 4-6-0	NE 74		Nine Elms Gds
05.40 WATERLOO	09.53	WC 4-6-2	BM 386	09.56	WEYMOUTH	19.48 Weymouth	20.07	WC 4-6-2	BM 381	20.15	Reading
10.10 Weymouth	10.29	LN 4-6-0	BM 393	10.31	Bournemouth C	19.12 Bournemouth C	20.14	T9 4-4-0	DOR 425	20.16	Weymouth
		U 2-6-0	BM 401	11.16	Light to Weymouth	*17.28 Bournemouth Gds*	20.50	U 2-6-0	DOR 426		
08.00 Romsey	11.28	WC 4-6-2	BM 385	11.30	Weymouth	*Light ex Weymouth*	21.16	T9 4-4-0	DOR 425		
11.30 WEYMOUTH	11.46	WC 4-6-2	BM 386	11.48	WATERLOO	18.30 WATERLOO	21.30	LN 4-6-0	BM 394	21.32	WEYMOUTH
		U 2-6-0	DOR 427	12.00	Poole Goods	21.18 Weymouth	21.34	WC 4-6-2	BM 382	21.36	Bournemouth C
08.30 WATERLOO	12.09	U 2-6-0	DOR 426	12.13	WEYMOUTH	21.55 WEYMOUTH	22.11	WC 4-6-2	BM 383	22.20	WATERLOO
12.20 Weymouth	12.39	N15 4-6-0	BM 401	12.41	Bournemouth C	21.18 Bournemouth C	22.25	U 2-6-0	DOR 427	22.27	WEYMOUTH
06.15 Brockenhurst Goods	12.56	Q0-6-0	EL 317			23.15 Weymouth	23.34	LN 4-6-0	BM 394		
10.30 WATERLOO	13.34	MN 4-6-2	NE 30	13.36	WEYMOUTH	*Light ex Weymouth*	23.46	U 2-6-0	DOR 427		
13.25 WEYMOUTH	13.41	WC 4-6-2	BM 385	13.44	WATERLOO						

90

coaches plus four vehicles which arrived with the 06.51 from Christchurch.

DORCHESTER LOCO : 1954					
Train	On Shed	Engine	Diagram	Off Shed	Train
Station Pilot	19.53	G6 0-6-0T	DOR 428	03.20	Station Pilot
23.50 Brockenhurst - Dorset Goods	02.52	Q0-6-0	EL 324	06.15	07.00 Dorset - Brockenhurst Passenger
17.28 Bournemouth - Dorset Goods	21.33	U 2-6-0	DOR 426	06.20	06.45 Dorset - Weymouth
22.30 Waterloo Mail	03.26	N15 4-6-0	EL 263	07.05	07.15 (22.38 N. Elms) Dorset - Weymouth Goods
23.15 Weymouth - Dorset passenger	23.55	LN 4-6-0	BM 395	07.45	08.04 Bournemouth C
06.05 Eastleigh - Dorset	09.00	N15 4-6-0	BM 401	11.16	Light to Weymouth @ 12.20 Bournemouth
21.18 B'mouth - Dorset and home light	23.46	U 2-6-0	DOR 427	11.50	12.10 Dorset - Hamworthy Jcn Goods
19.12 B'mouth - Weymouth and home light	21.16	T9 4-4-0	DOR 425	14.43	Light to Weymouth @ 15.50 Waterloo (B'mouth)
22.38 Nine Elms Goods	07.15	H15 4-6-0	NE 74	15.52	Light to Weymouth @ 17.10 Nine Elms Goods
06.15 Brockenhurst - Dorset Goods	13.23	Q0-6-0	EL 317	16.10	16.25 Bevois Park Goods
06.00 Christchurch - Weymouth and Light	09.45	U 2-6-0	BM 414	16.40	Light to Weymouth @ 17.41 Bournemouth

Goods traffic, which was heavier than might have been expected on a seaside line, was handled by the Hamworthy Junction M7 which arrived in Swanage with the down morning goods just before ten in the morning, returning as far as Wareham an hour later. In the afternoon the same engine worked down to Corfe Castle with empties from Wareham and formed the 15.08 through goods to Hamworthy Junction; the load comprising very largely of traffic collected from Furzebrook and Norden between Corfe Castle and Worgret Junction. The maximum load allowed an M7 0-4-4T over most of the branch was twenty-two wagons with an extra eight being granted between Furzebrook and Wareham.

Seen at the right time – the quiet time – of year it was a pleasant little branch that had slept unaltered for years; the relatively complex engine workings and goods service giving just the right amount of operational interest although the lack of variety in engines could become a little trying after a while.

The extent to which the Swanage branch changed on Summer Saturdays has been exaggerated at times since the number of departures was very little different from that of an ordinary weekday whilst many of the Wareham locals continued to run as two-coach motor workings. What was different was the nature of the London trains, three of which expanded into full-blown services with Pacific haulage throughout. The fourth was the four-coach 07.38 Royal Wessex, which ran much as it did on a weekday.

At Dorchester one reached the English equivalent of Kinnaber Junction and where in the case of the latter the LNER and LMS reached Aberdeen over a common pair of metals, so LSWR services from Bournemouth and Waterloo completed their journeys by running over the Great Western Castle Cary – Weymouth line. It was unfortunate that the junction between the two routes was made to the west of the town and, but for a half-mile's realignment, Dorchester might have had a joint station serving both Paddington and Waterloo.

As it was the LSWR station of Dorchester South was noted more for its geometry than its size, consisting of no more than an up and a down platform; the former differing from the conventional in that it – known as the Up Bay - lay on a siding which made a trailing junction with the up main line. All up trains therefore had to run past the station, reverse and set back into the platform; a curious procedure that was allowed to persist until the end of steam operations.

Although the majority of SR passenger services started or ended their journeys at Weymouth, goods workings tended to respect the regional boundary, only the through train from Nine Elms continuing on to Weymouth because of the Portland stone traffic.

Dorchester (GWR) West Yard had a service of five trains in each direction on the Weymouth – Westbury axis and there existed a respectable volume of exchange traffic between the two systems. Under normal circumstances these flows of traffic operated from the Great Western to the Southern in the morning and vice versa in the evening, engines hauling to Dorchester Junction in both cases before propelling their trains into the South or West yard. Some care had to be exercised by the Southern when arranging ad hoc exchange transfers since the Great Western yard was only given a pilot engine for the early turn – and even that disappeared halfway through the shift to work the 10.45 Weymouth – Dorchester passenger – which made it difficult for the West to accept

WEYMOUTH : 1954											
Train	Arr	Engine	Shed	Dep	Destination	Train	Arr	Engine	Shed	Dep	Destination
		Hdl 4-6-0	SPM ex 06.47	01.50	Bristol (Goods)			GW Railcar	WEY off 13.01	14.05	Dorchester W.
		43xx 2-6-0	WBY ex 23.15	03.30	Westbury (Goods)			U 2-6-0	DOR 426	14.20	Andover Jcn
22.40 Bristol Goods	03.30	Hdl 4-6-0	SPM for 19.55			13.01 Bournemouth C	14.21	LN 4-6-0	BM 393		
		Hdl 4-6-0	SPM ex 16.50	04.30	Bristol (Goods)			43xx 2-6-0	WBY ex 11.24	14.33	Westbury
		GW Railcar		05.45	Castle Cary	Light ex Dorchester loco	15.00	T9 4-4-0	DOR 425		
				06.00	Dorchester W.	10.30 PADDINGTON	15.07	Hdl 4-6-0	Wey for 07.17		
03.52 Salisbury	06.09	WC 4-6-2	BM 381			15.00 Dorchester W.	15.20	GW Railcar	WEY @ 15.33		
01.20 Bristol Goods	06.47	Hdl 4-6-0	SPM for 01.50					GW Railcar	WEY off 15.20	15.33	Maiden Newton
06.45 Dorchester S	07.02	U 2-6-0	DOR 426					Castle 4-6-0	OCC ex 12.16	15.40*	PADDINGTON
		43xx 2-6-0	WEY ex 15.07	07.17	Chippenham			T9 4-4-0	DOR 425	15.50	WATERLOO
22.38 Nine Elms	07.44	N15 4-6-0	EL 263			08.40 Bristol Goods	16.00	43xx 2-6-0	SPM for 16.30		(Next day)
		WC 4-6-2	BM 381	07.34	WATERLOO	Light ex Dorchester loco	16.07	H15 4-6-0	NE 74		
22.40 Paddington (Goods)	08.19	Hdl 4-6-0	WEY ex 18.57	08.15	BRISTOL			Castle 4-6-0	OCC ex 22.08	16.10	PADDINGTON
		Hdl 4-6-0	OCC for 18.35			14.57 Bournemouth C	16.18	WC 4-6-2	NE 33		
		U 2-6-0	DOR 426	08.25	Bournemouth C	14.25 Easton	16.18	57xx	Dor 428		
		14xx 0-4-2T		08.40	Maiden Newton			43xx 2-6-0	SPM off 16.00	16.30	Yeovil
07.10 Bournemouth C	08.50	LN 4-6-0	BM 393			12.30 PADDINGTON	16.33	Hdl 4-6-0	OCC for 09.00		
07.27 Castle Cary	08.56	GW Railcar	WEY @ 09.28					LN 4-6-0	BM 393	16.40	Bournemouth C
		57xx 0-6-0T	DOR 428a	08.58	Portland	12.55 Yeovil Goods	16.50	Hdl 4-6-0	SPM for 04.30		
		Castle 4-6-0	OCC ex 16.33	09.00	PADDINGTON	Light ex Dorchester loco	16.55	U 2-6-0	BM 414		
06.00 Christchurch (Fish)	09.08	U 2-6-0	BM 414			16.30 Maiden Newton	17.07	GW Railcar	WEY @ 17.50		
		N15 4-6-0	EL 263	09.20	WATERLOO	11.55 WOLVERHAMPTON	17.14	Hdl 4-6-0	SRD for 09.33		
05.45 Bristol	09.25	Hdl 4-6-0	BRD for 12.35					H15 4-6-0	NE 74	17.15	Dorchester Goods
		GW Railcar	WEY off 08.56	09.28	Yeovil	13.30 WATERLOO	17.33	WC 4-6-2	BM 381		
		U 2-6-0	BM 414	09.30	Light Dorchester Loco			MN 4-6-2	NE 30	17.35	WATERLOO
		43xx 2-6-0	WBY off 21.27	09.37	Yeovil (Goods)			U 2-6-0	BM 414	17.41	Bournemouth C
09.25 Maiden Newton	10.00	14xx 0-4-2T						GW Railcar	WEY off 17.50	17.50	Maiden Newton
05.40 WATERLOO	10.07	WC 4-6-2	BM 386			16.40 Yeovil	17.51	14xx 0-4-2T	WEY @ 20.35		
		LN 4-6-0	BM 393	10.10	Bournemouth C			43xx 2-6-0	SPM off 20.14	18.10	Westbury
07.30 Westbury (Milk)	10.18	43xx 2-6-0	WBY for 11.40			16.15 Westbury	18.21	54xx 0-6-0T	WBY for 12.50		
		57xx 0-6-0T	WEY	10.20	Upwey (Goods)			WC 4-6-2	NE 33	18.30	WATERLOO
Light ex Dorchester W.	10.26	14xx 0-4-2T	WEY @ 10.45					Hdl 4-6-0	OCC ex 08.19	18.35	Paddington (Milk)
		Hdl 4-6-0	SRD ex 17.14	10.33	WOLVERHAMPTON	16.25 BRISTOL	18.57	Hdl 4-6-0	WEY for 08.15		
08.05 BRISTOL	10.35	Hdl 4-6-0	BRD for 13.40			15.20 WATERLOO	19.06	WC 4-6-2	BM 382		
		14xx 0-4-2T	LE ex Dor W.	10.45	Dorchester W.	18.45 Maiden Newton	19.20	GW Railcar	WEY @ 22.00		
10.25 Portland	11.06	57xx 0-6-0T	DOR 428a					Hdl 4-6-0	BRD ex 12.09	19.30	BRISTOL
09.27 Westbury	11.24	43xx 2-6-0	WBY for 14.33					WC 4-6-2	BM 381	19.48	Reading GW
		WC 4-6-2	BM 386	11.30	WATERLOO			Hdl 4-6-0	SPM off 03.30	19.55	Reading (Goods)
		57xx 0-6-0T	DOR 428a	11.30	Easton	16.35 WATERLOO	19.55	WC 4-6-2	BM 383		
Light ex Dorchester loco	11.33	N15 4-6-0	BM 401			17.02 Bristol	20.14	43xx 2-6-0	SPM for 18.10		
		43xx 2-6-0	WBY ex 10.18	11.40	Westbury	19.12 Bournemouth C	20.31	T9 4-4-0	DOR 425		
08.00 Romsey	11.45	WC 4-6-2	BM 385					14xx 0-4-2T	WEY off 17.51	20.35	Yeovil
11.49 Upwey	11.58	57xx 0-6-0T	WEY					T9 4-4-0	DOR 425	21.00	Light Dorchester Loco
08.30 WESTON-S-MARE	12.09	Hdl 4-6-0	BRD for 19.30					WC 4-6-2	BM 382	21.18	Bournemouth C
08.20 PADDINGTON	12.16*	Castle 4-6-0	OCC for 15.40			13.20 Swindon Goods	21.27	43xx 2-6-0	WBY for 09.37		
		N15 4-6-0	BM 401	12.20	Bournemouth C	18.30 WATERLOO	21.43	LN 4-6-0	BM 394		
08.30 WATERLOO	12.24	U 2-6-0	DOR 426					WC 4-6-2	BM 383	21.55	WATERLOO
		Hdl 4-6-0	BRD ex 09.25	12.35	BRISTOL			GW Railcar	WEY off 19.20	22.00	Maiden Newton
		54xx 0-6-0T	WBY off 18.21	12.50	Yeovil	18.00 PADDINGTON	22.08	Castle 4-6-0	OCC @ 16.10		
11.50 Yeovil	13.01	GW Railcar	WEY @ 14.05			23.15 S. Tunnel Jcn	22.34	28xx 2-8-0	E. Jcn for 13.48		
		WC 4-6-2	BM 385	13.25	WATERLOO	21.13 Bournemouth C	22.42	U 2-6-0	DOR 427		
Light ex Dorchester West	13.30	14xx 0-4-2T						LN 4-6-0	BM 394	23.15	Dorchester S
		Hdl 4-6-0	BRD off 10.35	13.40	WESTBURY	20.50 Westbury	23.15	43xx/14xx	WBY for 03.30		
10.30 WATERLOO	13.47	MN 4-6-2	NE 30					U 2-6-0	DOR 427	23.30	Light Dorchester Loco
		28xx 2-8-0	E. Jcn off 22.34	13.48	Ebbw Jcn (Goods)	22.58 Maiden Newton	23.34	GW Railcar	WEY @ 05.45		

traffic at other times. To prevent the South from becoming congested with immovable traffic for the West, up to twenty-two wagons could be tripped to one of the sidings at Dorchester Junction and left there until the Western found the facilities for dealing with them.

Apart from allowing Southern engines to use the turntable, the Western did not share its motive power facilities at Weymouth and as a result Dorchester shed survived as a base for the Western end of the LSWR even though the introduction of through engine working between Weymouth and Waterloo from 1953 gave Bournemouth shed the lion's share of the time-table. The small amount of residual traffic that could not be covered by Bournemouth plus a degree of contingency work – banking and special trains –was retained by Dorchester; the shed's most exalted duty being a T9 booking on the15.50 Weymouth – Waterloo as far as Bournemouth. In the opposite direction one of the shed's U 2-6-0's took over the 08.30 ex Waterloo from a Nine Elms Merchant Navy at Bournemouth Central.

Dorchester shed's moment of prominence came during the summer of 1954 when a pair of Lord Nelson 4-6-0's – still a power to be reckoned with in SR circles – were allocated to the shed. To say the move electrified the enthusiast world would be an understatement, especially as it was generally believed that plans were afoot to close the shed. In fact the transfer of the two engines was little more than a smokescreen to alleviate tensions locally. A number of the stopping trains between Weymouth and Bournemouth were operated by a pair of Bournemouth Lord Nelson's, each spending an alternate night on Dorchester shed, and the transfer was little more than a paper exercise designed to suggest that the shed was not being run down quite as quickly as it in fact was.

The turn-round of coaching stock was almost entirely done at Weymouth and all that remained for Dorchester to deal with was the down Night Mail from Waterloo plus a pair of local coaching sets. The Mail terminated at Dorchester at 03.16, the four postal vehicles being recessed until the return working to London at 22.20 whilst its 3-coach L set was attached to another L set to form the 06.45 to Weymouth. After the closure of the motive power depot in 1955, the Down train was extended to Weymouth since the engine had to be serviced there; the 3-L set working through although the postal vehicles continued to terminate at Dorchester as before.

The other stock based on Dorchester included a Second Corridor (SK) which worked up to Weymouth in the 06.45 and then returned to Dorchester as a school traffic strengthener in the 08.25 Weymouth to Bournemouth Central. The remainder of the 06.45 consisted of a 3-L set which arrived in the 23.15 ex Weymouth. A second 3-L set spent almost twenty-four hours at Dorchester; the vehicles arriving with the 06.03 ex Eastleigh recessing until returning north with the 08.04 to Bournemouth Central.

It was a characteristic of the Southern that the further one travelled from Waterloo, the less substantial ones train became and the Royal Wessex, for example, which pulled out of London with no less than thirteen coaches behind the engine, had only five left upon leaving Wareham; a reduction accompanied by a growing air of shabbiness that had certainly been absent the other side of Bournemouth. This sense of desolation reached its nadir at Dorchester after which the train simply became an interloper over the Great Western Railway.

Although a lawyer could have given an exact opinion, to the spectator it was not always obvious which railway company Weymouth belonged to. The LSWR and the Great Western started running to Weymouth from their respective Dorchester stations on the same day in 1857 and while the promoter of the line had been the Wiltshire, Somerset & Weymouth (absorbed by the GWR six years before the line was completed to Weymouth), services to Waterloo were provided on a more generous and faster scale than those to Paddington.

On the other hand the Great Western was more interested in freight traffic and of the seven main line goods departures from Weymouth, only one was worked by the Southern. The reason for this was that the Southern could only lay claim to traffic that originated on the jointly owned branch to Portland (worked for alternate five-year stretches by each company).

The same rules did not apply to passenger traffic and not only could one find Paddington and Waterloo trains standing on adjacent tracks but passengers wanting a late evening train to Reading were often surprised to find that their train was a Southern service via Basingstoke but with through coaches. As a coincidental quid pro quo, the 19.12 Bournemouth – Weymouth and its 21.18 return working were formed of a GW 3-coach set which came onto the system with a morning Newbury – Southampton service and returned, after a day's roaming on the Southern, in the 10.10 Southampton – Cheltenham.

Such arrangements were the result of opportunism rather than cooperation and should not be allowed to disguise the fact that during the 1950's the Southern and the Western were as operationally separate as they had been a hundred years earlier. It would not have been the most difficult task in the world to have had the engines of both companies based at Weymouth shed but the Southern preferred to centralise on Bournemouth as far as possible with Dorchester looking after the handful of local workings that could not be covered from afar. The thoroughness of this policy can be judged by the fact that not one SR engine stabled overnight at Weymouth. The U 2-6-0 which brought in the last arrival of the day ran light to Dorchester shed at 23.30 for servicing and no other Southern engine was seen at Weymouth until the arrival of the Salisbury News at 06.09.

The strategy was not as absurd as many people thought at the time since main line engines based at Bournemouth not only performed a return trip between Weymouth and Waterloo but also ran an extra sixty miles getting to and from Bournemouth and thus raised their mileage to a far higher daily figure than would have been the case had the engines been based at Weymouth.

		WEYMOUTH CARRIAGE WORKINGS : 1954		
Train	Arr	Incoming Stock (Outward Working)	Dep	Destination
		GW Railcar	05.45	Castle Cary
		Auto	06.00	Dorchester W.
03.52 Sdisbury	06.09	5: News van (08.25).. News van (11.30), 3-L (12.10).		
06.45 Dorchester S	07.02	6: 3-L (19.48) 3- L (13.25).		
		BSK, CK, BSK	07.17	Chippenham
		5: BSK, CK, CK, SK, BSK	07.34	WATERLOO
		BSK, CK, BSK, slip	08.15	BRISTOL
		8: BG, SK, 5-H, SK	08.25	Bournemouth C
		Auto	08.40	Maiden Newton
07.10 Bournemouth C	08.50	3: 3-M (10.10)		
07.27 Castle Cary	08.56	GW Railcar (09.28)		
		7: BSK, SK, CK, RC, CK, SK, BSK	09.00	PADDINGTON
06.00 Christchurch (Fish)	09.08	Fish Vans (17.10 Goods)		
		5: SK, CK, 3-L	09.20	WATERLOO
05.45 Bristol	09.25	BG ex York (14.33),		
		GW Railcar	09.28	Yeovil
09.25 Maiden Newton	10.00	Autocar (10.45)		
05.40 WATERLOO	10.07	8: SK (13.25), 5-H, PMV (18.30), PMV : 19.48)		
		4: 3-M, S(Non-Corr)	10.10	Bournemouth C
		6: BSK, SK, SK, CK, SK, BSK	10.33	WOLVERHAMPTON
08.05 BRISTOL	10.35	6: BSK, SK, SK, CK, SK, SK, BSK (12.35)		
		Autocar	10.45	Dorchester W.
09.27 Westbury	11.24	3: BSK, CK, BSK (13.40)		
		6: BG, SK, 3-L SK	11.30	WATERLOO
		BSK, CK, BSK	11.40	Westbury
08.00 Romsey	11.45	7: SK, 3-L (14.20), 3-L (21.55)		
08.30 WESTON-S-MARE	12.09	6: BSK, SK, CK, SK, BSK (19.30))		
08.20 PADDINGTON	12.16*	12: BCK, RC, BSK, FK, CK, SK, SK, SK, SK, BSK, BG. (15.40)		
		3: 3-L	12.20	Bournemouth C
08.30 WATERLOO	12.24	7: SK (15.50), SK, 5-H (17.35)		
		6: BSK, SK, CK, SK, BSK	12.35	BRISTOL
		Autocar	12.50	Yeovil
11.50 Yeovil	13.01	GW Railcar (14.05)		
		4: SK, 3-L	13.25	WATERLOO
		3: BSK, CK, BSK	13.40	WESTBURY
10.30 WATERLOO	13.47	4: 3L, SK (15.50)		
		GW Railcar	14.05	Dorchester W.
		4: SK, 3-L	14.20	Andover Jcn
13.01 Bournemouth C	14.21	4: S (non-corr), 3-M (16.40)		
		BSK, CK, BSK (14.33)	14.33	Westbury
10.30 PADDINGTON	15.02	BG ex Paddington (18.10), Slip (07.17), BSK, CK, BSK (08.15)		
15.00 Dorchester W.	15.20	GW Railcar (15.33)		
		GW Railcar	15.33	Maiden Newton
		12: BCK, RC, BG, BSK, SK, SK, SK, SK, CK, FK, BSK	15.40*	PADDINGTON
		5: 3-L, SK, SK	15.50	WATERLOO
		5 :BSK, CK, SK, SK, BSK	16.10	PADDINGTON
14.57 Bournemouth C	16.18	6: 5H (17.41), PMV (As Required)		
			16.30	Yeovil
12.30 PADDINGTON	16.33	5: BSK, SK, SK, CK, BSK (16.10)		
		4: S(non-corr), 3M	16.40	Bournemouth C
16.30 Maiden Newton	17.07	GW Railcar (17.50)		
11.55 WOLVERHAMPTON	17.14	6: BSK, SK, SK, CK, SK, BSK (10.33)		
13.30 WATERLOO	17.33	5: SK , 3-L, SK (11.30)		
		6: SK, 5-H	17.35	WATERLOO
		6: 5-H, BG	17.41	Bournemouth C
		GW Railcar	17.50	Maiden Newton
16.40 Yeovil	17.51	1 Autocar (20.35)		
			18.10	Westbury
16.15 Westbury	18.21	BSK, CK, BSK (11.40)		
		6: 5-H, PMV	18.30	WATERLOO
15.30 PADDINGTON	18.57	Slip (08.15), BSK, CK, BSK (08.15)		
15.20 WATERLOO	19.06	5: SK, SK (08.25), 3-L (11.30)		
18.45 Maiden Newton	19.20	GW Railcar (22.00)		
		6: BSK, SK, SK, CK, SK, BSK	19.30	BRISTOL
		6: SK, 3-L, PMV, PMV	19.48	Reading GW
16.35 WATERLOO	19.55	5: BSK, CK, CK, SK, BSK (07.34)		
17.02 Bristol	20.14	BSK, CK, BSK (14.33)		
19.12 Bournemouth C	20.31	3 GW set (21.18)		
		Autocar	20.35	Yeovil
		3-GW set	21.18	Bournemouth C
18.30 WATERLOO	21.43	4: 3-L (23.15), CK (0917)		
		3: 3-L	21.55	WATERLOO
		GW Railcar	22.00	Maiden Newton
18.00 PADDINGTON	22.08	7: BSK, SK, CK, RC, CK, SK, BSK (09.00)		
21.13 Bournemouth C	22.42	3: 3-L (23.15)		
		3: 3-L	23.15	Dorchester S
20.50 Westbury	23.15	BG ex Paddington (11.40). Autocar (06.00), BSK, CK, SK (07.17)		
22.58 Maiden Newton	23.34	GW Railcar		

The proliferation of Pacifics resulted in Nine Elms' King Arthur 4-6-0's becoming a rare sight in Dorset except on Summer Saturdays when no less than five appearances were booked; four being on services between Waterloo and Bournemouth West. Bogiè-tendered 30778 'Sir Pelleas' prepares the stock of the 12.10 Bournemouth West to Waterloo in August 1957, having worked down earlier in the day with the 07.52 ex Waterloo.

A summer Saturday in June 1956 and the day has not got off to a good start. The Bournemouth King Arthur for the 06.05 Eastleigh - Dorchester has had to be replaced by an Eastleigh H15 4-6-0 whilst the back working - on Saturdays - is the 12.20 Weymouth to Waterloo - non stop from Bournemouth Central. As 30476 pauses at Wareham with the down train, someone will be scratching his head and wondering whether to risk letting the engine complete the diagram or to replace it at Bournemouth on the way up. The dilemma is not made any easier by the fact that the engine is 'return as required' from Nine Elms which means it could end the weekend anywhere between Esher and Exeter. The Swanage motor train sits in the bay platform on the down side.

COMBINED WORKING TIME TABLE AND ENGINE WORKINGS : 1954

Train	21.20	19.00	20.50			23.13				21.38		19.25	16.55	21.20	
From	ELGH	DID	WBY			Dor W.				Ports		N. Elms	Worcs	ELGH	
Class	Gds	Gds					Pds	Pds				Gds	Pds	Gds	Gds
Engine	Q0-6-0	Q1 0-6-0	43xx	M7 0-4-4T	T9 4-4-0	R'car	H15 4-6-0	N15 4-6-0	T9 4-4-0	LN 4-6-0	M7 0-4-4T	H15 4-6-0	N15 4-6-0	Q0-6-0	H15 4-6-0
Shed	EL 324	EL 314	WEY	BM 406	EL 282	WEY	NE 64	EL 263	EL 286	NE 31	EL 302	EL 311	EL 266	EL 324	EL 313
WATERLOO							18.42			21.00					
Woking															
Basingstoke							21.07			21.59			23.10		
Worting Jcn							21.15			22/05		22/11	23/16		
Micheldever															
Wallers Ash															
WINCHESTER							21.55			22.27	22.34	22/39	23.50		
Shawford		21/20									22.40				
Eastleigh Yard		21.43													23.35
EASTLEIGH							22.05				22.47		00.05		
EASTLEIGH				21.51				22.20		22/36	22.49	23/00			23/42
Swaything				21.57							22.55				
St Denys				22.01					22.36		22.58	23/10			23/53
Bevois Park Yard															
Northam				22.05											
Southampton (T)				22.08				22.32	22.41	22.47		23D23			
SOUTHAMPTON CENT											23.03				
SOUTHAMPTON CENT															00/04
Millbrook															
Redbridge															00/13
Totton	22.05														
Lyndhurst Rd															
Beaulieu Rd															
BROCKENHURST	22.35													23.59	00.39
Lymington Jcn														00/03	
Lymington Town															
Lymington Pier															
Sway															
New Milton															
Hinton Admiral															
Christchurch															
Pokesdown															
Boscombe															
Central Goods															
BOURNEMOUTH CENT															
BOURNEMOUTH CENT															00/40
BOURNEMOUTH WEST				23.10											
Branksome				23.15											
Parkstone				23.18											
POOLE				23.24										01.01	
Hamworthy Jcn				(To											
Holton Heath				W'brn)											
WAREHAM															
Worgret Jcn															
Corfe Castle															
SWANAGE															
Wool															
Moreton															
DORCHESTER															
Dorchester Jcn			23/03			23/14									
Monkton						23.18									
Wishing Well						23.24									
Upwey						23.28									
Radipole						23.33									
WEYMOUTH			23.15			23.38									

Similar economies were derived from train crews and of the twenty-one departures, twelve were worked by Bournemouth men with the balance being handled by Dorchester shed. Whilst the Southern maximised the productivity of its main line engines in this way, it was difficult if not impossible to cater for the vagaries of special traffic from Bournemouth – a large school party from the Channel Islands would call for an engine, and possibly a pilot as well, at very short notice – and therefore minimal facilities were retained at Dorchester; sufficient to prevent a sudden demand from catching the system unawares.

From the spectators point of view the contrast between the Great Western and the Southern could hardly have been more marked; the former operating a series of adequate if unadventurous services to Westbury and beyond with Hall 4-6-0's on the better Bristol trains and 43xx 2-6-0's on almost everything else. Such a degree of standardisation was unknown in Southern circles and of the twenty-one SR locomotives arriving in Weymouth daily, no less than half a dozen different classes - pretty well encapsulating the motive power history of the line - were represented.

The largest of the Southern visitors was the Nine Elms Merchant Navy Pacific which brought in the 10.30 ex Waterloo (which it took over at Bournemouth) and returned with the 17.35, the engine working through to London. Several other Bulleid Pacifics made daily visits but all were West Country class locomotives, the Class 8P engines being uncommon until Bournemouth shed received a trio with which to work the up Royal Wessex and the 16.35 ex Waterloo.

London-based engines tended to be infrequent visitors to Weymouth and apart from the Merchant Navy Pacific mentioned above, the only other examples were the light Pacific which powered the 18.30 to Waterloo and the H15 4-6-0 of the 17.15 Nine Elms Goods.

It is interesting to note that not one of the three London engines worked directly from London to Weymouth. The Pacifics left Waterloo with the 02.40 and 09.30 Bournemouth trains respectively whilst the H15 spent the day on Dorchester loco before proceeding light in the evening to Weymouth.

The middle period of engine design was represented by the Moguls – five trains ran in behind U 2-6-0's – and Lord Nelson 4-6-0's with the older days being kept alive by a pair of King Arthur's and a Dorchester T9 which worked a trip to Bournemouth and back.

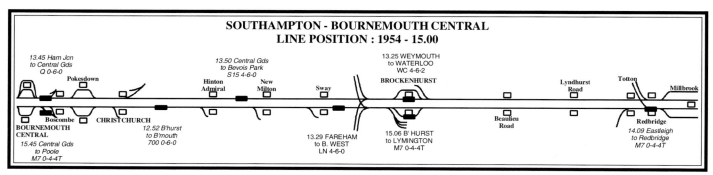

SOUTHAMPTON - BOURNEMOUTH CENTRAL
LINE POSITION : 1954 - 15.00

Train From					21.55 W'mouth	21.55 W'mouth	22.26 H.Jcn Gds	19.50 Dorset Gds	Gds	Gds	Gds	22.26 H.Jcn Gds		00.50 Fratton Gds
Class		ECS												
Engine	R'car	M7 0-4-4T	M7 0-4-4T	WC 4-6-2	N15 4-6-0	LN 4-6-0	LN 4-6-0	N15 4-6-0	H15 4-6-0	H15 4-6-0	T9 4-4-0	N15 4-6-0	N15 4-6-0	S15 4-6-0
Shed	WEY	SWA 421	SWA 421	BM 383	BM 399	NE 31	BM 394	BM 401	NE 74	NE 64	EL 282	BM 401	BM 399	FEL 101
WEYMOUTH	21.30			21.55			23.15							
Radipole	21.34													
Upwey	21.38						23.22							
Wishing Well	21.44													
Monkton	21.50													
Dorchester Jcn	21.53				22.08		23/31							
DORCHESTER	(To Dor W)			22.20			23.34							
Moreton				22.29										
Wool				22.37										
SWANAGE		22.04												
Corfe Castle		22.13												
Worgret Jcn		22/26		22/42										
WAREHAM		22.29	22.35	22.48										
Holton Heath				22.54										
Hamworthy Jcn				22.59										
POOLE			22.51	23.09				23.30						
Parkstone				23.15										
Branksome				23.19										
BOURNEMOUTH WEST														
BOURNEMOUTH CENTRAL			23.06	23.25	23.32									
BOURNEMOUTH CENTRAL								23/56						
Central Goods														
Boscombe				(To W'loo)										
Pokesdown														
Christchurch					23.41									
Hinton Admiral														
New Milton					23.52									
Sway														
Lymington Pier														
Lymington Town														
Lymington Jcn					00/00		00/26							
BROCKENHURST					00.06		00.30						00.45	
Beaulieu Road														
Lyndhurst Road														
Fawley														
Totton														
Redbridge					00/21								01/13	
Millbrook														
SOUTHAMPTON CENTRAL					00.25									
SOUTHAMPTON CENTRAL					00.32								01/21	
Southampton T.					00.37	01.10					00.01		01.23	
Northam														
Bevois Park Yard									00.12		00.45	01.30		
St Denys					(To W'loo)						00.50		01/27	
Swaything														
EASTLEIGH						01.20							01.37	01.53
EASTLEIGH						01.31		00/27			01/00		01.46	02.05
Eastleigh Yard										00.45	01.05			
Shawford														
WINCHESTER						01.47		00/50	01/17				(To Ports)	02/36
Weston														
Micheldever														
Worting Jcn						02/11		01/25	01/52					03/14
Basingstoke						02.15		(To N. Elms)	(To Feltham)					
Waterloo						03.53								(To Feltham)

Having no motive power facilities – so far as the Southern was concerned – the failure of an engine immediately prior to departure was a prospect that tended to haunt the operating staff since any replacement would almost certainly have to come from Dorchester, seven miles and about half an hour away and to minimise the effects of such an eventuality the diagrammers tried to arrange matters so that there were always two SR engines in the station at the same time; one acting as a standby for the other.

Another reason for having two SR engines on hand was because Great Western engines, which used 26" of vacuum, could not be used to double-head Southern trains on the climb to Bincombe and any assistance therefore had to be provided by a second SR locomotive.

The rules governing the assistance of trains up the 1 in 50 climb out of Weymouth varied according to circumstance but for express services running non-stop to Dorchester, the assisting engine pushed in the rear as far as Bincombe Tunnel, the engine not being coupled to the train. This was a considerable concession – if the two became separated between Weymouth and Bincombe the result was two trains in a block section – and it was not a routine that Waterloo was especially happy about. There were also complications where stopping trains were concerned and in the case of assisted trains that called at Broadwey, the banker had to be coupled and piped to the rear coach at Weymouth, the banker being uncoupled during the station stop at Broadwey.

In theory an SR train that was to be banked

'uncoupled' could be assisted by any type of locomotive but in practice, since a special stop might be called for Broadwey or it might be decided that assistance had to go beyond Bincombe to Dorchester in which case the double-heading had to be used, each company looked after its own train. (Arranging for trains to be double-headed was not made any easier by the restrictions imposed by the Southern and of the engines that ran into Weymouth, the Merchant Navy, King Arthur and Lord Nelson were prohibited from double-heading).

The allowances for unassisted loads were on the generous side and with a Class 7 engine being allowed to take nine coaches unaided, the only service to be regularly banked on an ordinary weekday was the 15.40 Weymouth Quay to Paddington boat train, all other services com-

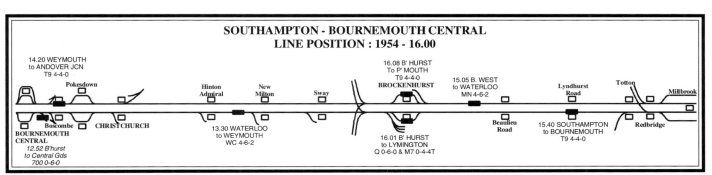

SOUTHAMPTON - BOURNEMOUTH CENTRAL
LINE POSITION : 1954 - 16.00

Bournemouth had to husband its allocation of M7 0-4-4T's with care since twelve of the fourteen were in daily traffic - nine at Bournemouth and one each at Lymington, Hamworthy Junction and Swanage - leaving little scope for repairs or maintenance. One also had to be careful in distinguishing the motor-fitted examples from others in the class: putting a non push and pull-fitted into a motor train diagram tended to raise the temperature iun operating circles. In spite of such problems 30036 is seen covering the station pilot at Bournemouth Central: a duty usually worked by an 02 0-4-4T.

ing within the limits of their train engines. The service that had to be particularly watched was the 15.50 to Waterloo which, booked for five vehicles, was at the very limit for its T9 4-4-0. An interesting feature of the single engine loadings was that the limit for a Lord Nelson 4-6-0 was greater than that for a light Pacific; nine and eight coaches being allowed respectively. Another anomaly was that a Hall 4-6-0 was allowed to work eight vehicles unassisted to Dorchester West yet only seven to Dorchester South.

On summer Saturdays when the usual four or five coach SR workings expanded to ten or more vehicles - well over the single engine load - assistance at one end of the train or the other became the norm rather than the exception.

Confusing though they may have been at time, the arrangements necessary for getting a Waterloo or Paddington express away from Weymouth were as nothing compared to the machinations needed for the military leave trains that originated on the Easton branch and called for no less than four engines in the first half-dozen miles of travelling.

Regular branch services had been rather a curious affair since a reversal at the junction opposite the motive power depot was needed for trains operating to and from Weymouth; a difficulty that had long been resolved by limiting the service to the branch and terminating trains at Melcombe Regis, a quarter of a mile

the Portland side of Weymouth Junction. Melcombe Regis was also used as an overspill for Summer Saturday main line trains; a role that continued for some years after the branch passenger service ceased in March 1952.

One of the features of the Portland branch

was that it was operated (nominally) by the Southern albeit under GW regulations using a 57xx 0-6-0T and Weymouth GW men. The Southern succeeded in retaining possession of the line after nationalisation because the afternoon goods from Easton 'became' the 17.10 goods to Dorchester and the 19.50 on to Nine Elms. Included in Dorchester shed's working

requirements was the need (purely nominal) to provide a 57xx 0-6-0T for the Easton goods trips. In fact the nearest the engine ever actually got to Dorchester was Broadwey after banking the 17.10 goods out of Weymouth.

After 1952 the only regular services over the branch were the goods trips to Portland and Easton although occasionally military leave trains of up to twelve coaches would be run from Portland to various parts of the country. When these services operated two 57xx engines had to be provided, the stock being double-headed from Weymouth to Portland goods yard where the 57xx's were repositioned - one at each end of the train – for the shunt into the Admiralty sidings. The pair of 0-6-0's would then double-head the loaded service as far as Weymouth Junction where they would give way to a main line engine. If the service was a Great Western working, a pilot engine would be attached for the run as far as Yeovil otherwise a banker would be provided, uncoupled, to assist in rear to Bincombe Tunnel.

The leave trains were not frequent and operated only two or three times a year. The branch however was maintained only to the level needed for the daily goods trips and when arrangements were being made to operate a leave train, the local permanent way inspector was required to furnish a certificate confirming that the branch was sufficiently healthy for the two engines and twelve coaches.

All change. In 1959 the Great Western route from Castle Cary to Weymouth was transferred to the Southern; the main result being the allocation of SR engines to Weymouth shed. Services over the Castle Cary line continued to be worked by GW motive power although the through services to Paddington were withdrawn with the Bristol workings being taken over by multiple-unit. The layout of Weymouth Town station was curious and consisted of three island platforms dealing, respectively with arrivals, departures and excursions. Rumours abounded of remodelling schemes but little happened until 1957 when a fourth island - perhaps twice the length of the existing platforms - was constructed on the down side. Whether the value of the development was ever realised is doubtful since a short time after completion the Paddington services were under review whilst multiple-units were operating many of the Bristol trains. The Christmas rush of 1961 brought a few faces out of retirement, one of which was N15 4-6-0 30774 'Sir Lavaine', seen at Weymouth on 8th December prior to departure with a Bournemouth working. A GW multiple-unit for Bristol stands at the excursion platform on the right.

57xx 0-6-0 Pannier tanks 3737 and 3759 shunt at Weymouth Junction.

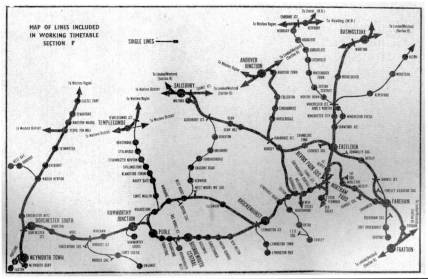

The network following the absorption of the Great Western's Dorset lines, as depicted in the goods working timetable. Except for the distinction between single and double tracks, all lines were shown as being equal; the Old Road between Brockenhurst and Wimborne carrying as much weight as the trunk line via Christchurch.

Similarly the two routes between Fareham and Romsey as shown as being of equal importance whilst in fact most passenger services were routed through Southampton with goods services running via Eastleigh.

LOCOMOTIVE ALLOCATIONS : DECEMBER 1953

EASTLEIGH

0F : 0-6-0 Dsl	**3MT 2-6-2T (1952)**
15230	82012
15231	82014
15233	82015
0F : C14 0-4-0T (1906)	82016
30588	**4F : Q 0-6-0 (1938)**
30589	30530
0F : P 0-6-0T (1909)	30531
31558	30532
0P : O2 0-4-4T (1889)	30535
30177	30536
30225	30542
30233	30543
1F : B4 0-4-0T (1891)	**4MT 2-6-0 (1952)**
30082	76005
30083	76006
30096	76007
2F : '0395' 0-6-0 (1881)	76008
30566	76009
2F : E1 0-6-0T (1874)	76010
32113	76011
32151	76012
2F : E4 0-6-2T (1897)	76013
32491	76014
32492	76015
32510	76016
32556	76017
32557	76018
32558	76019
32559	76025
32562	76026
32563	76027
32579	76028
2P : L12 4-4-0 (1904/15)	76029
30434	**4P : U 2-6-0 (1928)**
2P : M7 0-4-4T (1897)	31613
30030	31620
30031	31621
30032	31626
30033	31801
30127	**5F : H15 4-6-0 (1914)**
30133	30473
30322	30474
30323	30475
30324	30476
30375	30477
30376	30478
30479	30483
2P : M7(P&P) 0-4-4T (1897)	30489
30029	30491
30125	**5F : Q1 0-6-0 (1942)**
30328	33017
30379	33020
30480	33021
30481	33023
2MT 2-6-2T (1946)	33025
41293	33038
41304	**5P : N15 4-6-0 (1918)**
41305	30456
3F : '700' 0-6-0 (1897)	30457
30306	30755
30316	30780
30700	30781
3F : E6 0-6-2T (1911)	30784
32409	30785
3P : D15 4-4-0 (1912)	30786
30464	30787
30467	30788
3P : L1 4-4-0 (1926)	30789
31786	30790
31787	**6F : Z 0-8-0T (1929)**
31788	30952
31789	30955
3P : T9 4-4-0 (1899)	30956
30117	**7P : LN 4-6-0 (1926)**
30120	30850
30282	30851
30283	30852
30285	30853
30287	30854
30288	30855
30289	30856
30300	30857
30310	

SOUTHAMPTON

2F : E1 0-6-0T (1874)
32606
32689
3F : U.S.A. 0-6-0T (1942)
30061
30062
30063
30064
30065
30066
30067
30068
30069
30070
30071
30072
30073
30074

BOURNEMOUTH

0P : O2 0-4-4T (1889)
30212
30223
1F : B4 0-4-0T (1891)
30086
30087
30093
2F : G6 0-6-0T (1894)
30260
2P : M7 0-4-4T (1897)
30112
30318
30319
2P : M7(P&P) 0-4-4T (1897)
30056
30057
30058
30059
30060
30105
30106
30107
30111
30128
3F : '700' 0-6-0 (1897)
30695
3P : T9 4-4-0 (1899/1922)
30727
30728
30729
4F : Q 0-6-0 (1938)
30541
30548
30549
5P : N15 4-6-0 (1918)
30736
30737
30738
30739
30740
30741
30742
30743
30746
30750
30782
30783
7P : LN 4-6-0 (1926)
30861
30862
30863
30864
30865
7P : WC 4-6-2 (1945)
34043
34044
34093
34094
34095
34105
34106
34107
34108
34109
34110

DORCHESTER

0P : O2 0-4-4T (1889)
30179
30229
2F : G6 0-6-0T (1894)
30162
3P : T9 4-4-0 (1899/1922)
30284
4P : U 2-6-0 (1928)
31618
31622
31623
31631
31632

WEYMOUTH

1P : RAILCAR
25
1F : 1366 0-6-0T
1367
1370
1P : 14xx 0-4-2T (1932)
1403
1453
1467
2F : 74xx 0-6-0T (1936)
7408
7421
3F : 57xx 0-6-0T (1933)
3692
5781
9620
9642
4MT : 51xx 2-6-2T (1928)
4150
5190
4MT : 45xx 2-6-2T (1906)
4562
4MT : 43xx 2-6-0 (1911)
5314
5337
5384
5MT : HALL 4-6-0 (1928)
4988
5978
6902
6945
5MT : MOD-HALL 4-6-0 (1944)
6988
6993

With a tight margin ahead of the 15.20 Waterloo - Weymouth, M7 0-4-4T 30376 speeds the 16.25 Winchester - Eastleigh through the trailing connection with the DN&S at Shawford Junction on the 7th September 1957.

Lying in the shadow of the main line, the cross country route between Portsmouth and Salisbury tended to be neglected not only by the enthusiast but also, to the relief of the author who once had charge of the section, by Waterloo. Given the amount of traffic that moved over the line, this neglect was something of a mystery and the author put it down to the difficulties involved in paying a visit: through services from London being all but non-existent.

The line in fact needed no interference from outside: it looked after itself with an esprit de corps that had no parallel in the railway service; a state of affairs assisted by the fact that most of Fareham's staff had joined the railway from the Royal Navy; most of them giving the impression that the transfer from sea to steam involved few changes of habit. "I'll tell the signalman to bring the empty stock into the starboard side…", one of the Inspectors was heard to suggest to the SM.

The SM was just about the only member of staff who had no direct Royal Navy back-ground but, having a distant relative who had earned a posthumous VC at Jutland, was granted an honorary commission for the duration of his tenure and – so long as he played by the rules - treated with most of the perks of rank. At ten o'clock every morning, for example, there would be a loud knock on his office door and the Inspector, cap under his arm, would come to attention and salute smartly.

"Permission to serve tea, sir?"

"Carry on, No.1."

The response would be followed by a pa-

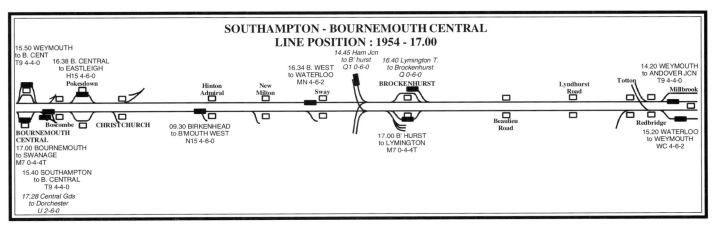

SOUTHAMPTON - BOURNEMOUTH CENTRAL
LINE POSITION : 1954 - 17.00

Up to 1953 the Great Western worked all services over the Didcot, Newbury & Southampton; a responsibility that required an engine shed at Winchester (Chesil) where two GW locomotives stayed overnight. From 1953 the line virtually became a joint operation with the Southern operating ten of the eighteen in daily trains. The result was an interesting, if widely spaced, array of locomotives which ranged from T9 4-4-0's to modern BR 4MT 2-6-0's. The most impressive sight may have been the Q1 0-6-0 which made two fast return trips between Southampton and Didcot with the express goods service introduced by the Southern in 1953. 22xx 0-6-0 3206 flies the Great Western flag at Winchester (Chesil) as it departs with the 12.42 Didcot - Southampton Terminus on 21st July 1956.

rade of staff sufficient to marshal a fair sized goods train; one carrying a tray, another with a saucer of chocolate biscuits, etc. The only drawback to what was an otherwise exemplary display of seagoing efficiency was that by the time the staff about-turned and been marched out, the tea had gone cold.

The Station Inspectors got themselves as close to the role of Master at Arms as the railway would permit and for the most part the district ran pretty smoothly.

It was unfortunate therefore that the Old Man should choose a Sunday evening, when Fareham was operated by a skeleton staff of one porter, to travel by train from Portsmouth to Southampton.

"Who was supposed to be on duty last night?" he enquired of the Inspector the following morning.

"The Frenchie, sir."

This was an individual who had fled from France to England during the war and who had never been allowed to forget his roots.

"Well he wasn't here. The station was unstaffed when I went through last night. Send him a letter and get an explanation out of him. Then I'll Form One him."

A week or so later the Old Man realized that no reply had been forthcoming and asked the Inspector to chase the matter up.

"It's like this, sir. The letter to the Frenchie was written in English but he says he can only read French."

"Then how does he issue excess tickets if he can't read or write English?"

"He says he looks to see what someone has written before and then copies it."

"He's trying it on. I presume he can actually read English."

"Well, it's never been put to the test but if you ask me, he's just being insolent. What are you going to do about it, sir?"

The Old Man thought for a moment before replying.

"I'll tell you what I'm going to do. I'll

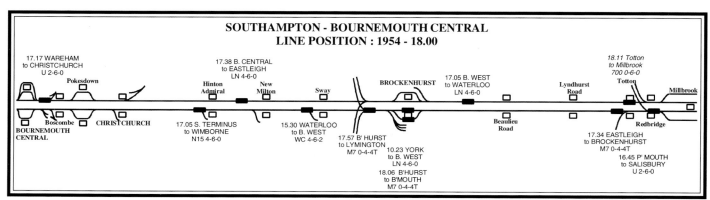

T9 4-4-0 30120 pulls away from Enborne Junction with the 14.56 Oxford - Southampton Terminus on the 14th September 1957 and with any luck will be passed by the 14.48 Eastleigh - Didcot goods before reaching the start of the single line section at Woodhay. Under normal circumstances the only T9 on the DN&S worked the 10.22 Eastleigh - Newbury, returning with the 10.50 Didcot - Eastleigh but inevitably the class found itself on BR4 workings as was the case with 30120. Care had to be taken when substituting a T9 for another engine since one of the 2-6-0 workings was extended from Didcot to Reading and the Western was not keen on receiving engines that were not formally cleared for passage. Quite a lively correspondence took place between the two systems on the question of placing the correct engines in diagrams.

Form One him in French."

"I didn't know you spoke French, sir?"

"Didn't you now...."

In his heart of hearts the Old Man wasn't too sure whether his French went much beyond asking for a cup of coffee at the Gare de Lyon but he recalled the days when, at the academy where he had expensively learn to write illegible English, he had been tutored in the language by Ken Laflin, master of the Bach fugue, medium pace spin and GWR locomotive fleet and who may well read these words. It seemed a pity that the fruits of KL's efforts should never have gone beyond the boundaries of a Parisienne station buffet.

A week or two later the Frenchman and his Union representative stood in front of the Old Man who, not without glee, realized that for the first time during a Form One hearing, the Union rep would not have a clue as to what

was taking place.

"Attention!" said the Old Man. "A quell heure terminee-vous votre travail a cet Gare deux Samedi passé?"

The Frenchman remained silent which may have been a symptom of surprise that the Old Man was able to speak French. It may also have been that the OM's French was of such an appalling standard that he had no comprehension of what it was he was being asked.

"Je m'attendre!" shouted the Old Man who knew that raising one's voice at foreigners automatically engendered understanding.

"A huit heure."

Two hours early.

"A huit heure pour travaille dans le Hart Blanc?"

"Oui, monsieur."

A little sniffing about earlier had elicited the information that the Frenchman had a part

time job as a barman.

If the Frenchman had been wearing medals and a sword, the Old Man would with a grand gesture have ripped the one off and broken the other over his knee. It was developing into genuine Dreyfuss material and the OM – not normally the most Gallic of individuals - found himself being absorbed into the part.

"Pour votre plaisir vous desolez votre chemin de fer!" he roared. "Nom du nom! Il n'y a pas le course professional. C'est vraiment ...vraiment..."

"Je le regret, Monsieur. Beaucoup."

"Regret! Ne demi pas!"

"Monsieur?"

"Alors. Le charge c'est prove. Votre service avec le companie est terminee. Alley-zi."

Dismissed with a very Gallic shake of the hand, the Frenchman grasped the sense if not

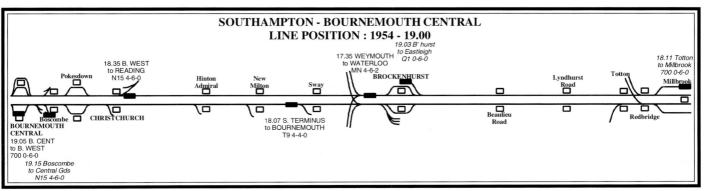

SOUTHAMPTON - BOURNEMOUTH CENTRAL
LINE POSITION : 1954 - 19.00

PORTSMOUTH - ROMSEY (SALISBURY) : WORKING TIMETABLE 1954

Station	Q1 0-6-0	U 2-6-0	S15 4-6-0	U 2-6-0	H15 4-6-0	M7 0-4-4T	BR3 2-6-2T	M7 0-4-4T	L1 4-4-0	K 2-6-0	M7 0-4-4T	M7 0-4-4T	T9 4-4-0
Train			02.10		03.58					01.45		06.10	
From			B.Park		B.Park					Brighton		Gosport	
Class	Gds		Gds	Gds	Gds	Vans	ECS		Gds	Gds		Gds	
Shed	EL 319	SAL 482	SAL 467	FRA 363	EL 311	FRA 373	EL 331	EL 300	EL 309	FRA 815	FRA 371	GOS 378	FRA 365
PORTSMOUTH H.						05.00	05.40						
Portsmouth &S						05.03	05.50						05.58
Fratton				02.00									06.02
Hilsea													
Cosham				02/12						05.55			06.11
Portchester													06.17
FAREHAM												06.21	06.24
FAREHAM				02/27						06/10	06.03		06.26
Swanwick											06.12		06.35
Bursledon											06.17		06.40
Hamble											06.21		06.44
Netley											06.25		06.48
Sholing											06.30		06.53
Woolston											06.33		06.56
Bitterne											06.37		07.00
St Denys											06.40		07.05
Southampton											06T45		07T12
Nursling													
Botley													
Eastleigh Yard	00.20			03.02				06.40	06.41				
EASTLEIGH													
EASTLEIGH	00/25	01.55	02.26		04.14			06.07	06/45				
Chandlers Ford								06.14					
ROMSEY	00/46	02/00	02/52		04/38			06.23	07.05				
Destination	Sarum	Yeovil	Bristol		Sarum			A.Jn					

Station	E4 0-6-2T	LM2 2-6-2T	LM2 2-6-2T	U 2-6-0	BR4 2-6-0	M7 0-4-4T	BR4 2-6-0	E1 0-6-0T	S15 4-6-0	N 2-6-0	N 2-6-0	T9 4-4-0	M7 0-4-4T
Train				05.58					06.15		06.40	06.55	07.52
From				Ports					Chich		Fratton	Ports	Gosport
Class	Vans							Vans	Gds	Gds	Gds		
Shed	FRA 377	EL 306	EL 306	SAL 450	EL 316	EL 298	EL 271	FRA 375	FEL 106	FRA 370	FRA 370	FRA 365	GOS 378
PORTSMOUTH H.								07.00					
Portsmouth &S	06.10				06.23		06.55						
Fratton	06.20				06.28		06.59	07.11			06.40		
Hilsea													
Cosham					06.33		07.08		07/15	06.58		07.23	
Portchester					06.45		07.15						
FAREHAM					06.52		07.21					07.39	07.53
FAREHAM		06.30			06.53		07.23		07/30				07.58
Swanwick					07.02		07.32						
Bursledon					07.07		07.37						
Hamble					07.12		07.41						
Netley					07.15		07.45						
Sholing					07.20		07.50						
Woolston					07.23		07.53						
Bitterne					07.27		07.57						
St Denys					07.32		08.00						
Southampton				07T27	07T40		08C05					08C13	
Nursling													
Botley		06.41											
Eastleigh Yard									08.04				
EASTLEIGH		06.50											
EASTLEIGH			07.30			07.56							
Chandlers Ford			07.35			08.02							
ROMSEY			07.44	07.57		08.11						08.32	
Destination				Sarum		A.Jcn							Alton

Station	H15 4-6-0	BR3 2-6-2T	T9 4-4-0	S15 4-6-0	M7 0-4-4T	N 2-6-0	N 2-6-0	T9 4-4-0	BR4 2-6-0	BR3 2-6-2T	M7 0-4-4T	U 2-6-0	43xx 2-6-0
Train				08.33	06.15	06.40	06.40						10.10
From				S'ton	Chich	Fratton	Fratton						S'ton
Class		Vans							Gds		Gds		
Shed	NE 72	EL 331	AJN 248	FEL 106	EL 304	FRA 370	FRA 370	EL 283	EL 316	EL 331	EL 300	FRA 368	AJN 249
PORTSMOUTH H.													
Portsmouth &S	07.29	07.45						08.07		08.50		09.03	
Fratton	07.33	07.50						08.11		08.54			
Hilsea													
Cosham	07.42							08.20		09.03		09.16	
Portchester	07.48							08.26		09.10			
FAREHAM	07.54							08.32		09.16		09.25	
FAREHAM	07.56				08.02	08.12		08.34		09.18		09.28	
Swanwick					08.11			08.43					
Bursledon					08.16			08.48					
Hamble					08.19	08.29	08.36	08.51					
Netley					08.23		08.40	08.54				09.41	
Sholing					08.27			08.59					
Woolston					08.30			09.01				09.47	
Bitterne					08.34			09.05					
St Denys					08.37			09.08				09.53	
Southampton				08T33	08C42			09T15				10C02	10T10
Nursling													
Botley	08.07									09.29	09.40		
Eastleigh Yard				09.00					09.30				
EASTLEIGH	08.17									09.38			
EASTLEIGH	08.26		08.55	09/05					09/35	09.44			
Chandlers Ford	08.32		09.01										
ROMSEY	08.41		09.10	09/27					09.55			10.16	10.31
Destination	Bristol		A.Jcn	Newport				S'ton	Sarum	W'ter	B.Wd	Sarum	C'ham

C: Southampton Central. T: Southampton Terminus. B: Bevois Park. D: Southampton Docks.

ROMSEY (SALISBURY) - PORTSMOUTH : WORKING TIMETABLE 1954

First Section

Train		01.23	01.35		02.10		02.40	02.40	02.10	03.30			
From	Pds	S'ton	Sarum		C.Jcn		W'loo	W'loo	C.Jcn	Sarum			
Class			Gds	Gds	Fish	Vans			Fish	Gds	Vans	Vans	Vans
Engine	U 2-6-0	T9 4-4-0	U 2-6-0	BR4 2-6-0	N15 4-6-0	M7 0-4-4T	BR3 2-6-2T	BR4 2-6-0	N15 4-6-0	Q1 0-6-0	N15 4-6-0	LM2 2-6-2T	E1 0-6-0T
Shed	FRA 363	EL 286	SAL 450	EL 316	EL 265	FRA 372	EL 331	EL 271	EL 265	GUI 214	EL 265	EL 306	FRA 375
ROMSEY			02/20							04/24			
Chandlers Ford													
EASTLEIGH										04.50			
EASTLEIGH	01.00	01.46	02/41		04.08		04.36			04.53		05.39	
Eastleigh Yard			02.46	03.45									
Botley							04/46			05/10		05.49	
Nursling													
Southampton													
St Denys													
Bitterne													
Woolston													
Sholing													
Netley													
Hamble													
Bursledon													
Swanwick													
FAREHAM	01.22	02.03			04.33		04.53			05.28		05.59	
FAREHAM	01.29	02.09		04/21			04.58	05.06	05.20	05.36			
Portchester													
Cosham	01/39	02/18		04/37			05/07	05.17	05.42	05/54			
Hilsea													
Fratton	01/47	02.21		04.49			05/09	05.24	05.50		05.59		
Portsmouth & S.	01.49	02.28				04.10		05.35			06.02		06.15
PORTSMOUTH HBR						04.13	05.18						06.20
Destination										Cicht'r			

Second Section

Train	05.15		04.25				06.28			06.28	06.57		
From	S'ton		Sarum				Totton			Totton	S'ton		
Class		Gds	Gds	Gds					ECS				ECS
Engine	T9 4-4-0	Q0-6-0	Q1 0-6-0	U 2-6-0	M7 0-4-4T	M7 0-4-4T	BR4 2-6-0	M7 0-4-4T	BR3 2-6-2T	BR4 2-6-0	M7 0-4-4T	M7 0-4-4T	BR3 2-6-2T
Shed	EL 283	EL 325	EL 319	FRA 363	EL 300	EL 304	EL 270	GOS 378	EL 331	EL 270	EL 308	EL 300	EL 331
ROMSEY			05/24		06.43								
Chandlers Ford					06.53								
EASTLEIGH					06.57								
EASTLEIGH			05/45		07.00							07.52	
Eastleigh Yard			05.50	06.00									
Botley				06/15	07.12							08.03	
Nursling													
Southampton	05T15		05B45				06C42				06T57		
St Denys	05.32						06.49				07.12		
Bitterne	05.36						06.53				07.16		
Woolston	05.41		05.58				06.57				07.20		
Sholing							07.00				07.23		
Netley	05.48						07.05				07.28		
Hamble							07.09				07.32		
Bursledon	05.53						07.12				07.36		
Swanwick	05.59						07.17				07.42		
FAREHAM	06.07				07.21		07.24				07.49	08.13	
FAREHAM	06.10			06/30				07.25		07.26			
Portchester	06.16									07.32			
Cosham	06.22			06/45						07.38			
Hilsea										07.42			
Fratton	06.33									07.48			08.13
Portsmouth & S.	06.36								07.30	07.51			08.18
PORTSMOUTH HBR									07.35				
Destination			Northam	Havant				Gosport					

Third Section

Train		07.39	07.38	06.45	06.45				08.05	07.30	07.47	07.25	
From		S'ton	Alton	A.Jcn	A.Jcn				Elgh	A.Jcn	Sarum	B.Park	
Class	Gds						Vans	Gds	Gds				Gds
Engine	Q0-6-0	BR4 2-6-0	700 0-6-0	T9 4-4-0	M7 0-4-4T	LM2 2-6-2T	700 0-6-0	BR4 2-6-0	BR4 2-6-0	43xx 2-6-0	T9 4-4-0	Q0-6-0	43xx 2-6-0
Shed	EL 312	EL 272	GUI 217	AJN 248	FRA 371	EL 306	GUI 217	EL 315	EL 315	AJN 249	SAL 445	EL 312	AJN 249
ROMSEY				07.32	08.00						08.08	08.19	08.32
Chandlers Ford					08.10								08.42
EASTLEIGH					08.15								08.47
EASTLEIGH													
Eastleigh Yard									08.05				
Botley									08.23	08.38			
Nursling													
Southampton	07B25	07T39		07T58		08T02					08C41		
St Denys	07/28	07.47				08.10					08.48		
Bitterne		07.51				08.14							
Woolston	07.38	07.57				08.18					08.54	09.07	
Sholing		08.00				08.21							
Netley		08.06				08.26					09.00	09.15	
Hamble		08.09				08.29							
Bursledon		08.13				08.32					09.08		
Swanwick		08.19				08.38							
FAREHAM		08.25	08.37			08.44				08.46	09.13		
FAREHAM		08.27				08.45		08.50			09.15		
Portchester		08.33				08.51							
Cosham		08.39				08.57					09.25		
Hilsea		08.43											
Fratton		08.48				09.06					09.34		
Portsmouth & S.		08.51				09.09					09.36		
PORTSMOUTH HBR													
Destination	F'ham				Weymouth		Gosport	Gosport	Gosport			F'ham	

C: Southampton Central. T: Southampton Terminus. B: Bevois Park. D: Southampton Docks.

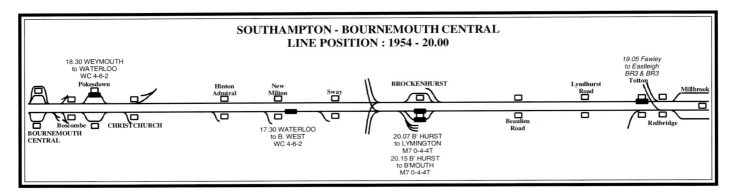

SOUTHAMPTON - BOURNEMOUTH CENTRAL
LINE POSITION : 1954 - 20.00

18.30 WEYMOUTH to WATERLOO WC 4-6-2 / Pokesdown / Hinton Admiral / New Milton / Sway / BROCKENHURST / Lyndhurst Road / 19.05 Fawley to Eastleigh BR3 & BR3 Totton / Millbrook / Boscombe / CHRISTCHURCH / BOURNEMOUTH CENTRAL / 17.30 WATERLOO to B. WEST WC 4-6-2 / 20.07 B' HURST to LYMINGTON M7 0-4-4T / 20.15 B' HURST to B'MOUTH M7 0-4-4T / Beaulieu Road / Redbridge

the grammar of his fate and retreated, rather crestfallen, towards the door. As he turned to open it the OM – reflecting that perhaps the theatrical nature of the affair had taken control to some extent – called him to attention.

"Non. Terminee c'est trop bon. Un jour suspendee…Un jour sans travail."

The Frenchman scuttled off, followed by his representative who had spent the entire proceedings with goldfish eyes, wondering what on earth was going on. A little later the Inspector came into the office and reported that the Frenchman would not be making an appeal against his day's suspension.

"He did ask what part of France you learnt the language in?"

"It may interest you to know," replied the Old Man with affected haughture, "that our

GOSPORT LOCO : 1954			
GOSPORT 378 : M7 0-4-4T			
Arr	Station	Dep	Train
	Gosport Loco	05.45	
	Gosport Loco	06.10	Pass
06.21	Fareham	07.25	Pass
07.37	Gosport	07.42	Pass
08.51	Alton	09.05	Pass
10.27	Gosport		
	Gosport Loco	11.15	
	Gosport	11.30	Pass
12.49	Alton	13.30	Pass
14.23	Fareham	14.48	Pass
15.47	Alton	16.30	Pass
17.22	Fareham	18.17	Pass
18.28	Gosport		
18.45	Gosport Loco		

something else. The difficulties that Brighton was having with its Bulleids was as good as made public when for a season the train was actually diagrammed to an H2 Atlantic in their latter days. At other times BR Standard 2-6-4 tanks were popular substitutes.

King Arthur 4-6-0's were very uncommon and the only booked example was one of the Eastleigh stud which ran down in the small hours with the Clapham Junction Fish and returned with a late morning parcels train from Portsmouth.

Heavy goods engines such as the S15 4-6-0's were a little more numerous, especially on the through workings to the LBSCR, but the gem of the day was the 08.45 Salisbury – Fratton goods which was relieved by one of Fratton's K 2-6-0's at Eastleigh.

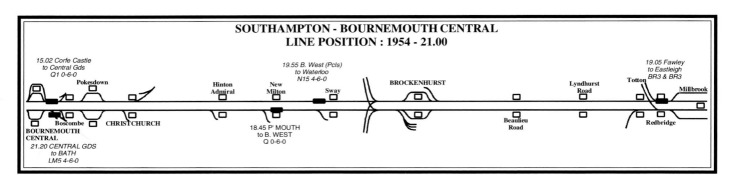

SOUTHAMPTON - BOURNEMOUTH CENTRAL
LINE POSITION : 1954 - 21.00

15.02 Corfe Castle to Central Gds Q1 0-6-0 / Pokesdown / Hinton Admiral / New Milton / 19.55 B. West (Pcls) to Waterloo N15 4-6-0 / Sway / BROCKENHURST / Lyndhurst Road / 19.05 Fawley to Eastleigh BR3 & BR3 Totton / Millbrook / Boscombe / CHRISTCHURCH / BOURNEMOUTH CENTRAL / 21.20 CENTRAL GDS to BATH LM5 4-6-0 / 18.45 P' MOUTH to B. WEST Q 0-6-0 / Beaulieu Road / Redbridge

Frenchman comes from a rural area where they speak French with a very strange dialect. Mine is Queen's French."

"I expect that's what he meant, sir. Funny lot these foreigners, if you ask me. Never trust a Frenchman I was always told. Remember 1588."

"That was the Spanish."

"French – Spanish. They're all the same, sir. I mean, they're not like us real people are they…?"

Fareham's main line ran from Portsmouth to Salisbury via Eastleigh and Chandlers Ford but most of the through passenger traffic ran down the Netley branch to Southampton, re-

joining the direct route at Romsey. The Netley route added four miles to the journey but Southampton was too important to ignore although one or two services which gave connections to and from Waterloo clung to the Eastleigh route.

Large engines were uncommon at Fareham and the normal motive power for the better Portsmouth – Bristol workings was either a U-class 2-6-0 or a Standard 2-6-0. The Three Brighton trains – one each to Plymouth, Bournemouth and Cardiff – were supposed to produce the line's only West Country Pacifics but in practice the availability of the Brighton allocation was so poor that one could put money on the Bournemouth service being hauled by

The heaviest passenger flow on the line was the season ticket trade between Fareham and Portsmouth with five trains being provided between seven and ten each morning: quite a respectable service over what Waterloo regarded as a backwater. The passengers were probably unimpressed but the variety of power on these five trains was about as varied as it could be. The 07.26 (06.28 ex Totton) and the 08.27 (07.39 ex Southampton Terminus) had BR 4MT 2-6-0's, the 08.45 (06.45 ex Andover Junction) was worked by an M7 0-4-4T, the 09.15 (07.47 from Salisbury via Southampton) had a T9 4-4-0 and the 09.26 (08.32 ex Romsey), an LM 2MT 2-6-2T.

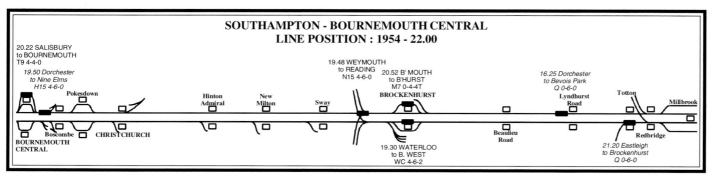

SOUTHAMPTON - BOURNEMOUTH CENTRAL
LINE POSITION : 1954 - 22.00

20.22 SALISBURY to BOURNEMOUTH T9 4-4-0 / 19.50 Dorchester to Nine Elms H15 4-6-0 / Pokesdown / Hinton Admiral / New Milton / Sway / 19.48 WEYMOUTH to READING N15 4-6-0 / 20.52 B' MOUTH to B'HURST M7 0-4-4T / BROCKENHURST / Lyndhurst Road / 16.25 Dorchester to Bevois Park Q 0-6-0 / Totton / Millbrook / Boscombe / CHRISTCHURCH / BOURNEMOUTH CENTRAL / 19.30 WATERLOO to B. WEST WC 4-6-2 / Beaulieu Road / Redbridge / 21.20 Eastleigh to Brockenhurst Q 0-6-0

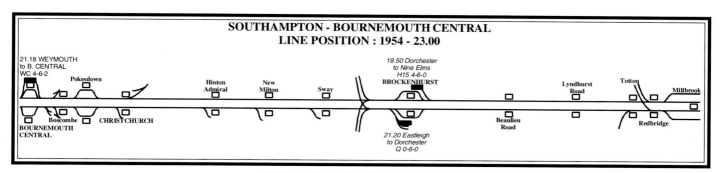

In spite of its heavy traffic commitment, Fareham had no motive power facilities yet Gosport, five miles distant, did and was granted two sets of men and an M7 0-4-4T for sharing the through passenger service to and from Alton. The working was quite an intensive one and covered no less than three trips to and from Alton plus a measure of shunting and local work at Fareham. The one engine was not enough to satisfy the entire needs of the Alton branch and a pair of Guildford engines – a 700 0-6-0 and a T9 4-4-0 - were used to augment the efforts of the Gosport engine. Curiously the 0-6-0 was used for the early passenger service from Alton whilst the T9 found itself relegated to the local goods: a service which required over five hours to cover the twenty-five miles. The 4-4-0 aroused a degree of interest on the Netley line since it filled-in with an evening workman's train between Fareham and Bitterne: Guildford T9 4-4-0's were something of a rarity in the district.

As well as being the Junction for the Southampton, Eastleigh and Alton lines, Fareham was the nearest station of note to the divergence of

LYMINGTON LOCO : 1954			
LYMINGTON 362 : M7 0-4-4T			
Arr	Station	Dep	Train
	Lymington Loco	05.45	
	Lymington Town	06.18	Pass
06.28	Brockenhurst	07.04	Mixed
07.20	Lymington Pier	07.39	Pass
07.53	Brockenhurst	08.41	Pass
08.54	Lymington Pier	09.08	Pass
09.22	Brockenhurst	10.58	Pass
11.12	Lymington Pier	11.25	Pass
11.39	Brockenhurst	12.02	Pass
12.15	Lymington Pier	12.20	Light
12.22	Lymington Loco	13.10	Light
13.12	Lymington Pier	13.22	Pass
13.36	Brockenhurst	13.57	Pass
14.10	Lymington Pier	14.18	Pass
14.31	Brockenhurst	15.06	Pass
15.19	Lymington Pier	15.25	Pass
15.38	Brockenhurst	16.01	On Rear
16.14	Lymington Pier	16.18	Pass
16.31	Brockenhurst	17.00	Pass
17.14	Lymington Pier	17.28	Pass
17.42	Brockenhurst	17.57	Pass
18.10	Lymington Town	18.22	Pass
18.35	Brockenhurst	18.52	Pass
19.05	Lymington Pier	19.27	Pass
19.41	Brockenhurst	20.07	Pass
20.20	Lymington Pier	20.35	Pass
20.48	Brockenhurst	21.00	Pass
21.10	Lymington Town		
21.20	Lymington Loco		

With only five coaches – almost empty for much of the year but full and standing during the summer – an M7 would have had no difficulty in timing the train and as soon as this dawned upon the operators at the Brighton end of the line the booked Pacific became for a time the exception to the rule and almost anything

HAMWORTHY JUNCTION LOCO : 1954			
HAMWORTHY JCN 422 : M7 0-4-4T			
Arr	Station	Dep	Train
	Hamworthy Jn loco	05.45	
	Hamworthy Junction	06.30	Goods
09.52	Swanage	10.45	Goods
11.50	Wareham	13.05	Goods
13.21	Corfe Castle	15.08	Goods
19.06	Hamworthy Jcn		Shunt
20.50	Hamworthy Jn loco		
HAMWORTHY JCN 423 : B4 0-4-0T			
	Hamworthy Jn loco	06.55	
07.00	Hamworthy Jcn	13.00	Yard Pilot
13.03	Hamworthy Jn loco	13.50	
	Hamworthy Jcn	14.00	Goods
14.09	Hamworthy Goods	16.00	Goods
16.09	Hamworthy Jcn	18.24	Goods
18.32	Poole (Yard Pilot)	20.30	Light
20.37	Hamworthy Jn loco		

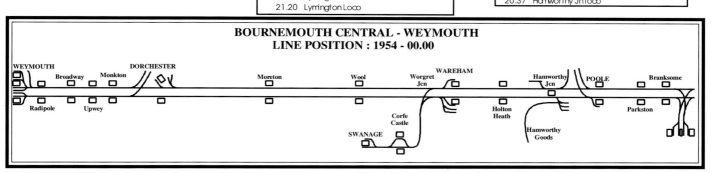

the Portsmouth and the LBSCR west coast lines although, since the electrification east of Portsmouth, there was very little direct running to or from the Brighton direction and was limited to the trio of services which ran from Brighton to Bournemouth, Cardiff and Plymouth. They also gave Fareham its only regular glimpse of Bulleid Pacifics.

Whilst one was grateful to be included on the Pacific map – it meant you were not wholly out on a limb – the Brighton – Bournemouth was a curious working in that it consisted of

only five coaches and yet was booked to one of the biggest engines on the system. It was also one of the few services to pass non-stop through Fareham and this, together with its engine, did nothing to detract from its promise as an express. Naturally one went out of one's way to watch it pass as it crawled at walking speed over the right-angle on the approach to the station before opening up for the long climb to Swanwick; a stretch of line that could be quite perilous for heavy trains in the wrong weather.

that could hold enough water to run between stops was put to work the service. For a time Pacifics, BR4 2-6-4 tanks and Brighton Atlantics vied with each other and it was something of an accolade for the LBSCR when, on summer Saturdays in 1956 – with the train enlarged from five to eight coaches (plus whatever might be added at the last minute) – the working was formally diagrammed to a Marsh H2 Atlantic.

Part of the reason for the absence of Pacifics on the train was because of the poor reliability

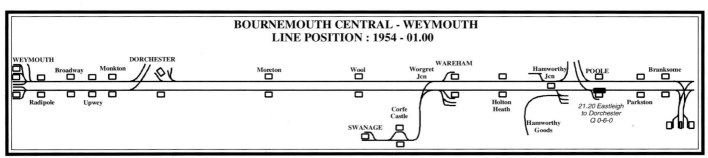

Part 1

	1	2	3	4	5	6	7	8	9	10	11	12	13
Train			10.23			09.40					11.30		11.12
From			B. Park			Brighton					B. Park		F'ham
Class	Vans		Gds	Gds			Gds			ECS	Gds	Pds	Gds
Engine	BR4 2-6-0	N15 4-6-0	M7 0-4-4T	S15 4-6-0	M7 0-4-4T	WC 4-6-2	BR4 2-6-0	N2-6-0	M7 0-4-4T	T9 4-4-0	43xx 2-6-0	N15 4-6-0	N2-6-0
Shed	EL 270	EL 265	EL 304	EL 312	EL 308	BTN 730	EL 272	FRA 370	FRA 371	FRA 365	CH 21	EL 265	FRA 370
PORTSMOUTH H.													
Portsmouth &S	09.33	09.35					10.34		10.45	10.45		11.05	
Fratton		09.40					10.38		10.49	10.50			
Hilsea													
Cosham	09.43					10/42	10.47		10.59			11/15	
Portchester									11.06				
FAREHAM	09.52						10.56		11.11			11.27	
FAREHAM	09.59					10/50	10.58	11.12	11.17			11.38	
Swanwick													
Bursledon													
Hamble								11.28					11.44
Netley	10.06						11.11						
Sholing													
Woolston	10.12						11.17						
Bitterne			10.25										
St Denys	10.18		10/33			11/08	11.23						11/50
Southampton	10C26		10B35	10B23		11C13	11C32				11B30		12B01
Nursling													
Botley									11.29			11.54	
Eastleigh Yard													
EASTLEIGH										11.39		12.06	
EASTLEIGH			10.37	11.10									
Chandlers Ford				11.16									
ROMSEY	10.40		11/04	11.25			11.46				12/00		
Destination	Cardiff			Sarum		B'mouth	Bristol				A. Jcn		

Part 2

	1	2	3	4	5	6	7	8	9	10	11	12	13
Train	12.00	11.30				11.30			11.00		12.24	11.30	
From	S'ton	Gosport				Gosport			Brighton		Elgh	Brighton	
Class			Gds		Gds			Gds		Gds	Gds		Vans
Engine	M7 0-4-4T	M7 0-4-4T	M7 0-4-4T	LM2 2-6-2T	M7 0-4-4T	M7 0-4-4T	T9 4-4-0	700 0-6-0	WC 4-6-2	BR4 2-6-0	M7 0-4-4T	WC 4-6-2	U 2-6-0
Shed	FRA 371	GOS 378	EL 329	EL 306	EL 300	GOS 378	FRA 366	GUI 217	BTN 731	EL 321	EL 329	BTN 732	SAL 450
PORTSMOUTH H.							11.37						
Portsmouth &S			11.19				11.45			12.15			12.45
Fratton			11.23										12.50
Hilsea													
Cosham			11.31				11/56			12/08			
Portchester			11.37								12.26	12/38	
FAREHAM		11.41	11.43				12.05			12.17	12.35	12.46	
FAREHAM			11.45			11.56		12.23		12.24		12.53	
Swanwick			11.54										
Bursledon			11.59					(To					
Hamble			12.03					Alton)					
Netley			12.06										
Sholing			12.11										
Woolston			12.13										
Bitterne			12.17										
St Denys			12.21							12/42		13/11	
Southampton			12C28							12C54		13C21	
Nursling													
Botley					12.25								
Eastleigh Yard			12.24	12.45									
EASTLEIGH													
EASTLEIGH	12.20		12/29										
Chandlers Ford	12.26		12.37								13.10		
ROMSEY	12.35		12.44							13/08	13.25	13.35	
Destination			Romsey	A. Jcn			Alton			Cardiff		Plymouth	

Part 3

	1	2	3	4	5	6	7	8	9	10	11	12	13
Train			13.12	13.12	13.12	13.35							14.40
From			F'ham	F'ham	F'ham	Havant							Gosport
Class	Gds								ECS		Gds		Gds
Engine	BR4 2-6-0	T9 4-4-0	T9 4-4-0	BR4 2-6-0	BR4 2-6-0	BR4 2-6-0	Q1 0-6-0	T9 4-4-0	BR4 2-6-0	M7 0-4-4T	T9 4-4-0	U 2-6-0	BR4 2-6-0
Shed	EL 321	SAL 445	FRA 366	EL 321	EL 321	EL 321	EL 320	FRA 365	EL 270	AJN 246	SAL 485	SAL 485	EL 315
PORTSMOUTH H.													
Portsmouth &S		13.03							14.03	14.15		14.33	
Fratton		13.07							14.07	14.20		14.37	
Hilsea													
Cosham		13.15					13/56		14.15			14/45	
Portchester		13.21							14.21				
FAREHAM		13.27							14.26				14.49
FAREHAM	13.12	13.28	13.29				14/10		14.28		14.48		14/53
Swanwick	13.23		13.38	13.47					14.36				
Bursledon			13.42	13.53	14.05				14.41				
Hamble					14.11	14.30							
Netley			13.48			14.34			14.47				
Sholing			13.53						14.51				
Woolston			13.55						14.54				
Bitterne			13.59						14.58				
St Denys			14.02						15.01			15/11	
Southampton			14C07						15C06			15C20	
Nursling											15.10		
Botley		13.38											
EASTLEIGH		13.47											
EASTLEIGH		13.50											
Eastleigh Yard								14.45					
Chandlers Ford		13.55											
ROMSEY		14.04									15.20	15.34	
Destination		Sarum	B'mouth						S'ton		Alton	Bristol	

C : Southampton Central.   T : Southampton Terminus.   B : Bevois Park.   D : Southampton Docks

ROMSEY (SALISBURY) - PORTSMOUTH : WORKING TIMETABLE 1954

Part 1

Train	08.32	09.05	09.05	09.05		04.10	08.45	09.30		07.25	08.05		09.40
From	Romsey	Alton	Alton	Alton		C'ham	Sarum	A.Jcn		B.Park	Elgh		Elgh
Class				Gds	Gds	Gds	Gds			Gds	Gds	ECS	Gds
Engine	LM2 2-6-2T	M7 0-4-4T	T9 4-4-0	M7 0-4-4T	M7 0-4-4T	43xx 2-6-0	Q1 0-6-0	H15 4-6-0	M7 0-4-4T	Q0 0-6-0	BR4 2-6-0	T9 4-4-0	Q1 0-6-0
Shed	EL 306	GOS 378	FRA 365	GOS 378	EL 304	CH 21	EL 320	EL 311	EL 298	EL 312	EL 315	FRA 366	EL 320
ROMSEY			08.52			09/12	09/37	10.09					
Chandlers Ford								10.19					
EASTLEIGH								10.24					
EASTLEIGH	09.03						10/03						
Eastleigh Yard						09.40			10.08				
Botley	09.14					10/10							
Nursling						09/18							
Southampton			09C25		09B45	10D10							
St Denys			09.33		09/48								
Bitterne			09.36		09.53								
Woolston			09.40										
Sholing			09.43										
Netley			09.48							10.25			
Hamble			09.51										
Bursledon			09.54							10.29			
Swanwick			10.00										
FAREHAM	09.23	09.59	10.06				10.23						
FAREHAM	09.26		10.09	10.16							10.26		10.32
Portchester	09.33		10.16										
Cosham	09.38		10.23										10/51
Hilsea													
Fratton	09.50		10.32									11.00	11.04
Portsmouth & S.	09.53		10.35										
PORTSMOUTH HBR													11.10
Destination		Gosport		Gosport			Fratton	Fratton		F'ham	Gosport		

Part 2

Train		07.25	09.12	09.47		08.10	08.45	11.25	07.25	08.45	12.23	11.27	
From		B.Park	Reading	Sarum		Bristol	Sarum	A.Jcn	B.Park	Sarum	S'ton	Sarum	
Class	ECS	Gds			Gds		Gds		Gds	Gds		Gds	
Engine	T9 4-4-0	Q0 0-6-0	Hall 4-6-0	U 2-6-0	M7 0-4-4T	U 2-6-0	K 2-6-0	LM 2-6-2T	Q0 0-6-0	K 2-6-0	BR4 2-6-0	S15 4-6-0	T9 4-4-0
Shed	FRA 365	EL 312	RDG 50	SAL 485	EL 329	SAL 450	FRA 815	AJN 248	EL 312	FRA 815	EL 271	FEL 116	AJN 248
ROMSEY			10.19	10.35		11.09		12.04				12/20	
Chandlers Ford								12.13					
EASTLEIGH								12.18					
EASTLEIGH			10.49	10/58								12/43	13.08
Eastleigh Yard				11.03		11.45						12.48	
Botley			11.00			12/03							13.16
Nursling													
Southampton				10C40		11C35					12T23		
St Denys				10.47		11.42					12.35		
Bitterne				10.51							12.40		
Woolston				10.56		11.48					12.46		
Sholing				10.58							12.49		
Netley				11.04		11.54					12.56		
Hamble				11.07							12.59		
Bursledon		10.50		11.10							13.03		
Swanwick		10.58		11.16							13.09		
FAREHAM			11.10	11.22		12.06			12.33		13.16		13.25
FAREHAM			11.13	11.24		12.09	12/18				13.19		13.27
Portchester				11.30							13.24		
Cosham			11.24	11.37		12.21	12.33		12.50		13.30		13.37
Hilsea													
Fratton	11.12		11.33	11.47		12.31			13.03		13.39		13.46
Portsmouth & S.	11.17		11.36	11.50		12.34					13.42		13.49
PORTSMOUTH HBR													
Destination		F'ham				Fratton					Chich		

Part 3

Train		10.27	11.27	12.15	13.30		13.06	10.11	10.30	13.50	10.20		
From		Bristol	Sarum	Reading	Alton		Sarum	Ch'ham	Cardiff	B'mouth	Alton		
Class	Pds		Gds								Gds		Gds
Engine	Q0 0-6-0	BR4 2-6-0	S15 4-6-0	BR4 2-6-0	M7 0-4-4T	M7 0-4-4T	T9 4-4-0	78xx 4-6-0	BR4 2-6-0	WC 4-6-2	T9 4-4-0	M7 0-4-4T	T9 4-4-0
Shed	EL 322	EL 270	FEL 116	EL 273	GOS 378	FRA 371	SAL 443	CH 22	EL 272	BTN 730	GUI 187	EL 329	AJN 246
ROMSEY		13/06					13.37	13.15	13.44	13.53		14.05	14.20
Chandlers Ford							13.48					14.15	
EASTLEIGH							13.53					14.20	
EASTLEIGH				13.46									
Eastleigh Yard			13.23										
Botley			13/41	13.57									
Nursling													14.41
Southampton		13C23					13C37		14C12	15C00			
St Denys		13.28					13.50		14.18	15.07			
Bitterne							13.54						
Woolston							13.58			15.13			
Sholing							14.01						
Netley							14.06			15.19			
Hamble							14.09						
Bursledon							14.12						
Swanwick							14.18						
FAREHAM				14.07	14.24		14.24		14.36	15.31	15.34		
FAREHAM	13.30	13/46	14/00	14.10			14.27		14.39	15.33			
Portchester				14.17			14.34						
Cosham		13.55	14/16	14.23			14.40		14.49	15.43			
Hilsea													
Fratton		14.04		14.33			14.49		14.58				
Portsmouth & S.		14.06		14.36			14.51		15.01				
PORTSMOUTH HBR													
Destination	Gosport		Chich				S'ton	S'ton		Brighton			

C : Southampton Central.   T : Southampton Terminus.   B : Bevois Park.   D : Southampton Docks

The Lymington branch was remarkable in that although relatively few Southern classes were allowed to use it, BR Clan Pacifics and LNER B1 4-6-0's were permitted almost without restriction. The latter never came near the line leaving the Summer Saturday through Waterloo trains to give the operators some difficulties since there were very few engines suitable for the long run between London and Brockenhurst that were permitted to go forward to Lymington. T9 and D15 4-4-0's were used for a time whilst in 1954 Nine Elms U1 2-6-0's were employed as far as Brockenhurst where a change of engines took place. Most of the weekday branch trains were worked by M7 0-4-4T's as depicted above with 30028 at Lymington Town in August 1959 and, below, a well-coaled 30481 about to leave Lymington Pier for Brockenhurst in July 1957. One service was noteable for being double-headed by an M7 and a T9 4-4-0.

Although the Swanage branch service consisted very largely of push and pull workings to and from Wareham, the workings called for no less than five separate M7 0-4-4 tanks; three from Bournemouth and one each from Hamworthy Junction and Swanage, the last mentioned being changed daily by a fresh engine from Bournemouth.   The basic train formation consisted of a two car push and pull set strengthened as required by a third vehicle with the addition, several times a day, of a pair of through coaches to or from Waterloo.   M7 0-4-4T's 30105 and 30106 are seen at Corfe Castle with Wareham - Swanage workings during the summer of 1957.

PORTSMOUTH - ROMSEY (SALISBURY) : WORKING TIMETABLE 1954

(first panel)

Train	13.12	13.12				16.36	14.40	14.40			14.20		
From	F'ham	F'ham				S'ton	Gosport	Gosport			W'mouth		
Class	Gds	Gds		Vans			Gds	Gds					ECS
Engine	BR4 2-6-0	BR4 2-6-0	Hal 4-6-0	BR4 2-6-0	T9 4-4-0	78xx 4-6-0	BR4 2-6-0	BR4 2-6-0	T9 4-4-0	M7 0-4-4T	T9 4-4-0	Hal 4-6-0	U 2-6-0
Shed	EL 321	EL 321	RDG 50	EL 273	AJN 248	CH 22	EL 315	EL 315	GUI 187	EL 305	EL 282	OX 208	SAL 450
PORTSMOUTH H.													
Portsmouth &S			14.45	14.50	15.45							16.40	16.45
Fratton			14.49	14.55	15.49							16.45	16.49
Hilsea													
Cosham			14.58		15.57								16.58
Portchester			15.04		16.03								17.05
FAREHAM			15.10		16.09								17.11
FAREHAM			15.14		16.11		16.15			16.29			17.12
Swanwick										16.38			17.21
Bursledon										16.42			17.26
Hamble										16.46			17.30
Netley	15.10									16.49			17.33
Sholing	15.18	15.28											17.38
Woolston										16.54			17.40
Bitterne										16.58			17.44
St Denys		15/39											17,47
Southampton		15B41				16T36							17C56
Nursling													
Botley			15.26		16.22		16.33	16.48					
Eastleigh Yard								17.06					
EASTLEIGH			15.36		16.31								
EASTLEIGH					16.34				17.16		17.42		
Chandlers Ford					16.40				17.22		17.48		
ROMSEY					16.49				17.31		17.57		18.14
Destination			Reading			C'ham			A. Jcn				Sarum

(second panel)

Train		16.47		17.40	17.16								
From		Havant		Gosport	Fawley								
Class	EBV	Gds		Gds		Gds	Gds		ECS				Gds
Engine	U 2-6-0	Q1 0-6-0	BR4 2-6-0	Q0-6-0	T9 4-4-0	BR4 2-6-0	Q1 0-6-0	T9 4-4-0	BR4 2-6-0	N 2-6-0	T9 4-4-0	M7 0-4-4T	43xx 2-6-0
Shed	FRA 363	GUI 213	EL 273	EL 322	EL 280	EL 272	GUI 213	SAL 443	EL 270	FRA 370	GUI 187	EL 298	AJN 250
PORTSMOUTH H.													
Portsmouth &S			17.17			17.45			18.03	18.06			
Fratton			17.21			17.49			18.07	18.11			
Hilsea			17.26						18.12				
Cosham			17.32	17.12		17.58			18.17				
Portchester			17.39						18.24				
FAREHAM		17.27	17.45	18.00		18.07			18.30				
FAREHAM	17.25		17.49			18.10	18.25		18.32		18.48		
Swanwick									18.41				
Bursledon									18.46				
Hamble	17.41								18.49				
Netley						18.23			18.53				
Sholing									18.57				
Woolston						18.30		18.42	19.01				
Bitterne									19.05				
St Denys						18.36		18/50	19.08				
Southampton						18C45		18B52	19T13				19D04
Nursling													19/48
Botley				18.01			18/42						
Eastleigh Yard							19.03						
EASTLEIGH				18.11									
EASTLEIGH					18.30							19.20	
Chandlers Ford					18.36							19.26	
ROMSEY					18.45	18.59						19.35	19/59
Destination			Reading	A. Jcn	Cardiff				S'ton		Alton		C'ham

(third panel)

Train				19.17									
From				Ports									
Class		Gds			Gds		Gds		Gds				
Engine	BR4 2-6-0	C2X 0-6-0	Hal 4-6-0	T9 4-4-0	Q0-6-0	T9 4-4-0	T9 4-4-0	Q1 0-6-0	43xx 2-6-0	H15 4-6-0	T9 4-4-0	T9 4-4-0	U 2-6-0
Shed	EL 271	FRA 814	OX 208	SAL 443	EL 322	BM 403	EL 284	GUI 213	BTL 365	EL 310	EL 286	AJN 248	SAL 482
PORTSMOUTH H.													
Portsmouth &S	18.45		19.17			19.45	20.03				21.38	21.48	23.16
Fratton	18.49		19.21			19.50	20.07			21.05	21.43	21.52	23.21
Hilsea													
Cosham	18.58	19.08	19.30			20.00	20.17			21/16	21.52	22.01	23.30
Portchester	19.04		19.37			20.06					21.58	22.08	23.37
FAREHAM	19.10	19.23	19.44			20.13	20.26				22.04	22.13	23.44
FAREHAM	19.12		19.47		19.55	20.15	20.30			21/34	22.05	22.14	23.47
Swanwick	19.20					20.23					22.13		
Bursledon	19.25					20.28					22.17		
Hamble	19.28												
Netley	19.32					20.34					22.23		
Sholing	19.36					20.39							
Woolston	19.39					20.42					22.29		
Bitterne	19.43					20.47					22.33		
St Denys	19.47					20.50					22.36		
Southampton	19C52					21C01					22T41		
Nursling													
Botley			19.58		20/02		20.43		21/55			22.24	23.58
Eastleigh Yard					20.32			20.50		22.11			
EASTLEIGH			20.07				20.52					22.33	00.07
EASTLEIGH				20.28					20/55	21.44			
Chandlers Ford				20.34						21.50			
ROMSEY				20.43		21.23			21/31	21.59			
Destination	B'mouth		Sarum	Sarum		A. Jcn		Sarum	Sarum		S'ton		

C : Southampton Central. T : Southampton Terminus. B : Bevois Park. D : Southampton Docks

ROMSEY (SALISBURY) - PORTSMOUTH : WORKING TIMETABLE 1954

Train	13.48 Reading	11.00 Plymouth	11.00 Plymouth	16.05 S'ton	*11.18 Srum Gds*	13.00 Cardiff	13.00 Cardiff	16.08 Brock	*15.07 Srum Gds*	16.30 Alton	16.08 Brock	16.12 A. Jcn	
Engine	Hdl 4-6-0	WC 4-6-2	BR4 2-6-0	N 2-6-0	H15 4-6-0	WC 4-6-2	N 2-6-0	T9 4-4-0	43xx 2-6-0	M7 0-4-4T	T9 4-4-0	LM2 2-6-2T	T9 4-4-0
Shed	OX 208	BTN 731	EL 271	FRA 370	*NE 72*	BTN 732	FRA 370	EL 286	*BTL 365*	GOS 378	EL 286	AJN 247	GUI 187
ROMSEY		15/17			*15.38*	16/19			*16.23*			16.50	
Chandlers Ford													
EASTLEIGH													
EASTLEIGH	15.21				*16/06*				*16/45*				
Eastleigh Yard					*16.11*				*16.50*				
Botley	15.33												
Nursling													
Southampton		15C34		16C05		16C36		16C39					
St Denys		15/39		16.17		16/41		16.46					
Bitterne				16.20				16.50					17.15
Woolston				16.25				16.54					17.22
Sholing				16.27				16.57					
Netley				16.32				17.02					17.28
Hamble				16.36				17.06					17.31
Bursledon				16.40				17.10					
Swanwick				16.45				17.16					17.38
FAREHAM	15.43	15.59		16.51		17.01		17.22		17.23			17.45
FAREHAM	15.45	16.03	16.11			17.05	17.11				17.25		
Portchester			16.18								17.32		
Cosham	15.55	16/12	16.24			17/13	17.21				17.38		
Hilsea													
Fratton	16.04		16.33				17.30				17.47		
Portsmouth & S.	16.07		16.36				17.33				17.50		
PORTSMOUTH HBR													
Destination		Brighton				Brighton			S. Dcks				

Train			17.28 S'ton		*16.18 Srum Gds*	13.56 C'ham		17.07 Srum				18.27 S'ton		
Class				ECS	Gds						Gds	Gds		
Engine	LM2 2-6-2T	H15 4-6-0	T9 4-4-0	Hdl 4-6-0	S15 4-6-0	43xx 2-6-0	M7 0-4-4T	T9 4-4-0	T9 4-4-0	*T9 4-4-0*	U 2-6-0	T9 4-4-0	T9 4-4-0
Shed	AJN 247	EL 310	BM 403	OX 208	EL 312	AJN 250	GOS 378	EL 284	AJN 248	AJN 246	FRA 363	FRA 365	AJN 248
Pilot									M7 0-4-4T				
Shed									EL 305				
ROMSEY	17.03				*17.15*	17.23		17.39	17.50	18.05			
Chandlers Ford	17.13								18.00				
EASTLEIGH	17.18								18.06				
EASTLEIGH		17.27			*17/47*					18/35			19.22
Eastleigh Yard					*17.52*					*18.40*			
Botley		17.38											19.33
Nursling													
Southampton			17T28					17C58			18T40		
St Denys			17.36					18.14			18.48		
Bitterne			17.40								18.51		
Woolston			17.44					18.20			18.56		
Sholing			17.47								18.59		
Netley			17.52					18.26		18.44	19.04		
Hamble			17.56								19.07		
Bursledon			18.00					18.31			19.11		
Swanwick			18.05					18.37			19.17		
FAREHAM		17.47	18.12					18.44		*19.10*	19.24		19.43
FAREHAM		17.49	18.13				18.17	18.46			19.25		19.45
Portchester		17.56	18.20								19.32		19.51
Cosham		18.02	18.25					18.57			19.37		19.57
Hilsea													
Fratton		18.11	18.34	18.30				19.06			19.46		20.05
Portsmouth & S.		18.14	18.37	18.52				19.09			19.49		20.07
PORTSMOUTH HBR													
Destination		S'ton					S'ton	Gosport					

Train	16.32 Bristol	18.40 A. Jcn			16.35 Cardiff	19.35 A. Jcn		20.25 Elgh	16.35 Exeter	22.00 S'ton	*15.20 Chel*	19.10 Bristol	16.40 Plymouth
Class			ECS	Gds			Gds	Gds			Gds		
Engine	U 2-6-0	L1 4-4-0	M7 0-4-4T	S15 4-6-0	U 2-6-0	T9 4-4-0	C2X 0-6-0	S15 4-6-0	S15 4-6-0	BR4 2-6-0	BR4 2-6-0	S15 4-6-0	BR4 2-6-0
Shed	FRA 368	EL 309	EL 298	FEL 105	SAL 482	EL 282	FRA 814	FEL 105	FEL 101	EL 270	EL 316	SAL 467	EL 272
ROMSEY	19.28	19.35	19.50		20.09	20.21			20.48		*21/57*	22.34	23/35
Chandlers Ford		19.45				20.31			20.58				
EASTLEIGH		19.50	20.05			20.36			21.03				23.46
EASTLEIGH									21.28		22/20		
Eastleigh Yard				20.25							*22.25*		
Botley				20/42					21.39				
Nursling													
Southampton	19C48				20C28					22T00			
St Denys	19.55				20.37					22.14			
Bitterne	19.59				20.41					22.17			
Woolston	20.03				20.47					22.22			
Sholing	20.06												
Netley	20.11				20.53					22.28			
Hamble													
Bursledon	20.15									22.32			
Swanwick	20.21									22.37			
FAREHAM	20.28			21.01	21.06				21.48	22.44			
FAREHAM	20.30				21.09	21.19	21.32	21.53		22.45			
Portchester	20.37							21.59		22.51			
Cosham	20.44				21.20		21/47	22.07		22.56			
Hilsea													
Fratton	20.54				21.29		*21.48*	22.17		23.05			
Portsmouth & S.	20.57				21.32			22.20		23.08			
PORTSMOUTH HBR													
Destination				Ashford					Ashford				Eastleigh

C : Southampton Central. T : Southampton Terminus. B : Bevois Park. D : Southampton Docks

For most of the year Swanage's trains, goods and passenger, were handled by M7 0-4-4T's and it was only during the summer holiday peak that much in the way of variety was seen. Light Pacifics were diagrammed for the through services to and from Waterloo whilst smaller machines were employed on more local workings. One such appearance was that of T9 4-4-0 30729 in August 1958 when in worked in with a day trip from Salisbury. The engine is seen, above, on the turntable after arrival and, below, preparing to leave with the 18.30 departure.

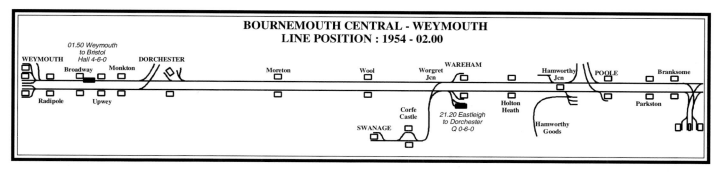

BOURNEMOUTH CENTRAL - WEYMOUTH
LINE POSITION : 1954 - 02.00

of the Brighton West Country engines but it was also due to the fact that whilst Brighton had an allocation of seven light Pacifics, it had - during the busiest periods – six diagrams for the class which, given the difficulties the engines presented to the maintenance staff, was a wholly unreasonable goal. The priority for the Pacifics were the London trains to Kensington Plymouth expresses both called at Fareham to attach portions from Portsmouth: an operation which usually carefully watched since most trains calling at Fareham came and went without any shunting being performed.

The first section to arrive was the four-coach 11.37 from Portsmouth - one of the few workings to start from the Harbour as opposed tween the Brighton Pacific and the Fratton T9 was accentuated by the sight (and sound) of the Alton goods pulling away behind a Guildford 700 0-6-0.

With the train complete, the right away would be given and under a shroud of steam and dense black smoke the Brighton Spam would cough its way out of the station, through

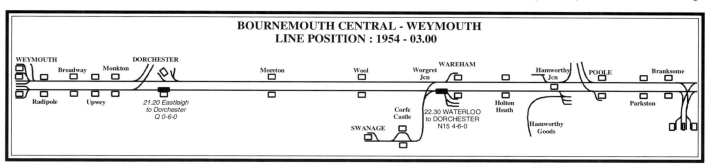

BOURNEMOUTH CENTRAL - WEYMOUTH
LINE POSITION : 1954 - 03.00

or Willesden followed by the Plymouth and Cardiff expresses and only if – which was a big if – there was a West Country remaining, it would get a trip to Bournemouth.

Whilst the Brighton – Bournemouth ran throughout as a fixed formation, the Cardiff and to Portsmouth & Southsea – which ran into the Gosport platform at 12.05, its Fratton T9 running round to pull the stock back as soon as the main portion of the train arrived from Brighton at 12.17. Seven minutes were allowed for the attachment during which time the contrast be-

the junction and – with luck – keep a reasonably steady wheel turning at least until the summit at Swanwick was reached. (This was not usually much of a problem on off-peak days when the combined service loaded to only eight vehicles but at the height of the season when

Train	Arr	Engine	Shed	Dep	Destination	Train	Arr	Engine	Shed	Dep	Destination
Light ex B'mouth loco	04.43	WC 4-6-2	BM 381			12.12 Brockenhurst	12.52	M7 0-4-4T	BM 408	12.53	B'mouth West
03.52 S'disbury	04.50	T9 4-4-0	SAL 444		(Fwd at 04.58)	12.35 B. West	13.02	M7 0-4-4T	BM 409	13.03	Brockenhurst
(03.52 S'disbury)		WC 4-6-2	BM 381	04.58	Weymouth	13.20 B. West	13.52	T9 4-4-0	EL 284	13.54	S'disbury
		T9 4-4-0	SAL 444	05.17	Eastleigh via B'mouth	Light ex Poole	14.37	700 0-6-0	SAL 452		
04.05 S'disbury Gds	05.22	700 0-6-0	SAL 452		(Fwd at 05.50)	14.05 B'mouth	14.44	M7 0-4-4T	BM 409	14.47	B'mouth West
Light ex B'mouth loco	05.44	2 x M7	405/406			14.35 B. West	15.03	M7 0-4-4T	BM 406	15.05	Brockenhurst
(S'disbury Goods)		700 0-6-0	SAL 452	05.50	B'mouth Goods	07.37 S'disbury Goods	15.15	700 0-6-0	SAL 451		
		M7 0-4-4T	BM 405	06.25	Brockenhurst			700 0-6-0	SAL 451	15.25	Light to Hamworthy Jn
		M7 0-4-4T	BM 406	06.45	B'mouth West			700 0-6-0	SAL 452	15.35	S'disbury Goods
06.30 B. Centrd	07.00	M7 0-4-4T		07.02	Brockenhurst	14.45 Hamworthy Jcn Goods	15.37	Q1 0-6-0	EL 319	15.55	Eastleigh
07.14 Broadstone	07.19	T9 4-4-0	EL 284	07.20	S'disbury	12.00 Dorchester Goods	16.43	700 0-6-0	SAL 451		
07.17 Brockenhurst	07.52	M7 0-4-4T	BM 405	07.53	B'mouth West	16.10 Brockenhurst	16.55	M7 0-4-4T	BM 406	16.57	B'mouth West
07.00 Dorchester	07.51	EL 324	Q0-6-0	08.00	Brockenhurst	16.30 B. West	16.56	M7 0-4-4T	BM 409	16.57	Brockenhurst
07.42 B. Centrd	08.12	T9 4-4-0	SAL 444	08.13	S'disbury	16.52 B. West	17.21	T9 4-4-0	SAL 443	17.22	S'disbury
07.15 S'disbury	08.18	T9 4-4-0	SAL 443	08.19	B'mouth West	16.25 Dorchester Goods	18.18	Q0-6-0	EL 317		
08.10 B. West	08.41	M7 0-4-4T	BM 406	08.42	Brockenhurst	17.20 S'disbury	18.20	T9 4-4-0	SAL 445	18.22	B'mouth West
06.40 Brockenhurst Gds	08.54	Q0-6-0	EL 317		(Fwd at 09.32)	18.06 Brockenhurst	18.46	M7 0-4-4T	BM 409	18.47	B'mouth West
08.34 Brockenhurst	09.22	M7 0-4-4T	BM 408	09.25	Poole	17.05 Southampton T.	18.49	EL 265	N15 4-6-0		
08.45 Poole Goods	09.01	Q0-6-0	BM 416	09.11	Brockenhurst			700 0-6-0	SAL 451	18.50	S'disbury Goods
(B'hurst Goods)		Q0-6-0	EL 317	09.32	Dorchester	18.50 B. West	19.17	M7 0-4-4T	BM 406	19.18	Brockenhurst
								EL 265	N15 4-6-0	19.31	Light to B'mouth loco
09.32 Brockenhurst	10.12	EL 324	Q0-6-0			(Dorchester Goods)		Q0-6-0	EL 317	19.56	Eastleigh Goods
09.25 S'disbury	10.21	T9 4-4-0	EL 284	10.23	B'mouth West						
10.32 Poole	10.47	M7 0-4-4T	BM 408	10.49	Brockenhurst	19.43 B. West	20.10	T9 4-4-0	SAL 445	20.11	S'disbury
		EL 324	Q0-6-0	11.12		20.15 Brockenhurst	20.55	M7 0-4-4T	BM 406	20.56	B'mouth West
11.04 Brockenhurst	11.44	M7 0-4-4T	BM 406	11.45	B. Centrd	20.52 B. West	21.18	M7 0-4-4T	BM 409	21.19	Brockenhurst
08.39 Eastleigh Goods		Q1 0-6-0	EL 318	12/40	Poole	23.10 B. West	23.39	M7 0-4-4T	BM 406		
								M7 0-4-4T	BM 406	23.57	Light to B'mouth loco

WIMBORNE : 1954

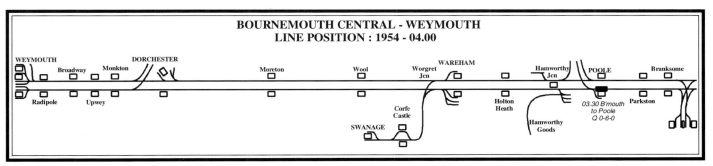

BOURNEMOUTH CENTRAL - WEYMOUTH
LINE POSITION : 1954 - 04.00

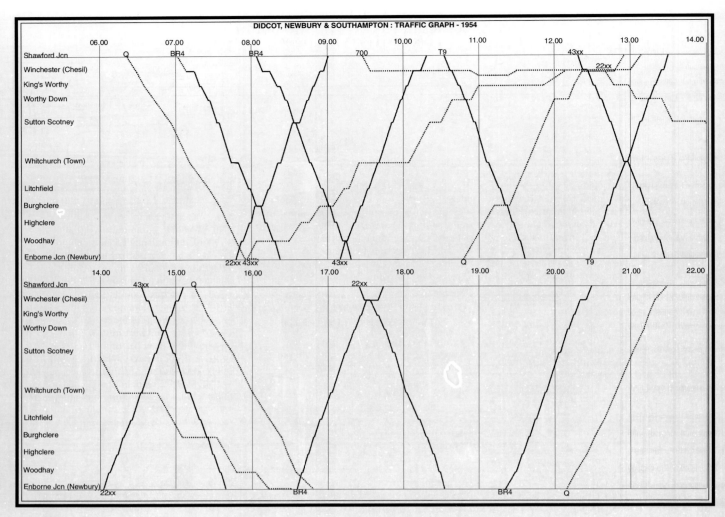

DIDCOT, NEWBURY & SOUTHAMPTON : TRAFFIC GRAPH - 1954

the load increased to eleven coaches, the going could be hard going to say the least).

The T9 would be held back in the up main platform until the 12.15 ex Portsmouth & Southsea arrived in the Gosport platform with its four coaches for Plymouth. The incoming engine, an Eastleigh Standard 2-6-0, would be uncoupled to run off to the yard for the 13.10 Bevois Park goods whilst the T9 shunted to the rear of the set, ready to draw it back as soon as the Brighton section with its Pacific and six coaches came to a stand in the up main platform.

There was always an element of nail biting with the Plymouth and the retribution that might be exacted for any delay that occurred between its arrival in Fareham and getting over the top at Swanwick. No train was deliberately delayed – least of all an express – but one knew that a few minutes dropped, for example, with the Brighton – Cardiff would almost certainly be swamped by greater delays in the Bristol or Newport areas whilst the Plymouth was a Southern train throughout its journey and any problems at Fareham were likely to be scrutinised pretty minutely at Waterloo.

As the T9 propelled the Portsmouth section onto the rear of the Brighton section, the Old Man would be watching the Pacific out of the corner of his eye in order to assess its chances of a

good run up the bank although if the engine showed signs of not being up to the job, there was precious little that could be done. In extremis, the T9 could be used as a pilot up to Swanwick but this was not something that was done lightly since the 4-4-0 was needed at Fareham to dig out the stock for the 13.29 Bournemouth stopping train which it worked as far as Southampton.

On more than one occasion as a Spam coughed and slipped its way round the corner, dropping speed as it tried to get up the first few feet of Swanwick bank, the Old Man would

ask why on earth Brighton couldn't be given a handful of Lord Nelsons or King Arthurs which would romp up to Swanwick as through the line was level. Their use on the three Brighton trains, he believed, would be a far more sensible proposition than some of the all-stations work they performed on the main line between Eastleigh and Bournemouth.

West Country Pacifics were not the only curiosities to be seen at Fareham since, although in the heart of LSWR territory, a pair of Great Western Hall 4-6-0's worked through with the 09.12 and 13.48 Reading - Portsmouth trains, returning at 14.45 and on the 19.17 parcels. Most of the through services between Reading and the South Western ran to and from Southampton Terminus but the 09.12 was a hybrid working of eleven vehicles which split at Eastleigh, the rear six vehicles going forward to Southampton as a parcels train and the engine continuing forward with the Portsmouth section. The 13.48 was a similar type of working except that there was no return working, the engine, an Oxley Hall, returning with the 19.17 Portsmouth – Salisbury as far as Eastleigh and the 21.45 Eastleigh – Crewe Parcels.

The main object of running Great Western services to Portsmouth was to allow Reading (GW) crews to retain route knowledge. During the sum-

WINCHESTER (CHESIL) : 1954					
Train	Arr	Engine	Shed	Dep	Destination
05.35 Bevois Park		Q1 0-6-0	EL 314	06/29	Didcot Goods
06.55 Eastleigh ECS	07.08	BR4 2-6-0	EL 278		
		BR4 2-6-0	EL 278	07.15	Reading
07.32 Southampton (T)	08.08	BR4 2-6-0	EL 278	08.15	Didcot
07.45 Newbury	08.54	22xx 0-6-0	DID 13	08.57	Southampton (T)
08.44 Eastleigh Yard	09.34	700 0-6-0	EL 326		
07.42 Didcot	10.08	43xx 2-6-0	DID 30	10.14	Southampton (T)
10.22 Eastleigh	10.38	T9 4-4-0	EL 279	10.39	Newbury
		700 0-6-0	EL 326	10.53	Winnd Sidings
11.25 Winnd Sdgs	11.30	700 0-6-0	EL 326		
07.52 Newbury Goods	12.08	43xx 2-6-0	RDG 20		
09.50 Didcot Goods	12.47	Q1 0-6-0	EL 314		(Cross Passenger)
11.45 Southampton (T)	12.23	43xx 2-6-0	DID 30	12.26	Didcot
Light ex Eastleigh loco	12.28	22xx 0-6-0	DID 13		
		43xx 2-6-0	RDG 20	12.35	Light to Eastleigh
(09.50 Didcot Goods)		Q1 0-6-0	EL 314	12.48	Eastleigh Yard
		700 0-6-0	EL 326	13.00	Eastleigh Yard
10.50 Didcot	13.25	T9 4-4-0	EL 279	13.27	Eastleigh
		22xx 0-6-0	DID 13	13.45	Didcot Goods
13.56 Southampton (T)	14.37	43xx 2-6-0	RDG 20	14.38	Didcot
12.42 Didcot	15.00	22xx 0-6-0	DID 31	15.02	Southampton (T)
14.48 Eastleigh Yard		Q1 0-6-0	EL 314	15/10	Didcot Goods
16.56 Southampton (T)	17.29	22xx 0-6-0	DID 31	17.34	Didcot
14.56 Oxford	17.31	BR4 2-6-0	EL 277	17.40	Southampton (T)
17.55 Didcot	20.20	BR4 2-6-0	EL 278	20.25	Southampton (T)
19.00 Didcot Goods		Q1 0-6-0	EL 314	21/19	Eastleigh Yard

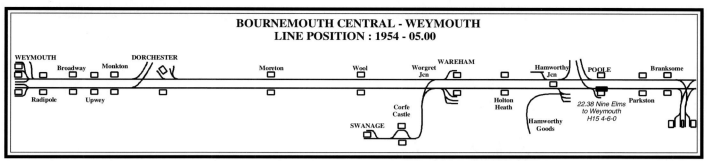

BOURNEMOUTH CENTRAL - WEYMOUTH
LINE POSITION : 1954 - 05.00

Train	Arr	Engine	Shed	Dep	Destination
07.04 Brockenhurst (Mixed)	07.20	M7 0-4-4T	LYM 362	07.39	Brockenhurst
Light ex L. Town	07.52	T9 4-4-0	BM 403		
		T9 4-4-0	BM 403	08.10	Brockenhurst
08.41 Brockenhurst	08.54	M7 0-4-4T	LYM 362	09.08	Brockenhurst
09.39 Brockenhurst	09.53	M7 0-4-4T	BM 406	10.35	Brockenhurst
10.58 Brockenhurst	11.12	M7 0-4-4T	LYM 362	11.25	Brockenhurst
12.02 Brockenhurst	12.15	M7 0-4-4T	LYM 362		
		M7 0-4-4T	LYM 362	12.20	*Light to Lymloco*
Light ex Brockenhurst	13.12	M7 0-4-4T	LYM 362		
		M7 0-4-4T	LYM 362	13.22	Brockenhurst
13.52 Brockenhurst	14.05	M7 0-4-4T	LYM 362		Brockenhurst
15.06 Brockenhurst	15.19	M7 0-4-4T	LYM 362	15.25	Brockenhurst
16.01 Q&M7	16.14	Q&M7	416/362	16.18	Brockenhurst
		Q 0-6-0	BM 416	16.23	*Light to LymTown*
17.00 Brockenhurst	17.14	M7 0-4-4T	LYM 362	17.28	Brockenhurst
17.57 Brockenhurst	18.10	M7 0-4-4T	LYM 362	18.22	Brockenhurst
18.52 Brockenhurst	19.05	M7 0-4-4T	LYM 362		
		M7 0-4-4T	LYM 362	19.10	*Light to Lymloco*
Light ex Lymloco	19.22	M7 0-4-4T	LYM 362		
		M7 0-4-4T	LYM 362	19.27	Brockenhurst
20.07 Brockenhurst	20.20	M7 0-4-4T	LYM 362	20.35	Brockenhurst

mer the Western needed access to the Isle of Wight and by using their own engines and crews were able to organise trains without undue reliance on the South Western. Unfortunately the supply of Hall (and other) 4-6-0's did not always cover the number of trains the GW wanted to operate with the result that the 15.30 Portsmouth – Wolverhampton on Summer Fridays had to be booked to nothing more distinguished than a Reading-based 61xx 2-6-2T.

Located on the boundary between the LSWR and

the war its early morning stopping train from Southampton contained a dining section for Newcastle on Tyne and a pair of through coaches for Glasgow Queen Street, it is probable that very few people would ever have come to hear of the DN&S.

Up to 1952 the line existed as it always had, services being operated by a mix of 22xx 0-6-0 and 43xx 2-6-0's from Reading and Didcot sheds; one of each type being stabled overnight in the shed at Winchester Chesil to kick-off the early

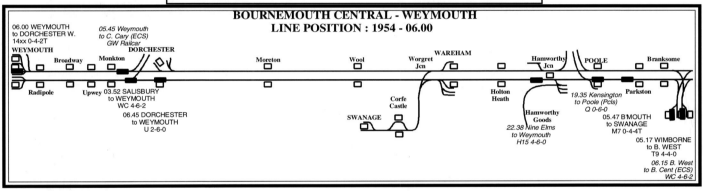

BOURNEMOUTH CENTRAL - WEYMOUTH
LINE POSITION : 1954 - 06.00

LBSCR, one might have expected Brighton engines to have been reasonably numerous at Fareham yet the only examples one saw tended to be a few crumbs from the Fratton table. The pre-war electrification of both routes from London to Portsmouth had pretty well isolated the western extremities of the LBSCR from the rest of the system and the most notable appearance was that of a K class 2-6-0 which ran through at about six in the morning with the 01.45 Brighton – Eastleigh goods, returning six hours later with the 08.45 Salisbury – Fratton. The only other booked Brighton engine was a C2X 0-6-0 which worked to Fareham yard with the 18.40 Cosham goods and went back with the

21.19 to Fratton. During the fruit season it ventured a little deeper into LSWR territory when its working was extended back to Netley.

In a world where very little changed for two decades after the war, the most radical upheavals were probably witnessed by the Didcot, Newbury & Southampton, a rather minor cross-country line that left the LSW a short distance north of Shawford and skirted the outer home counties via Whitchurch, Newbury and Didcot. Worked by the Great Western, many of its trains penetrated south of Chesil – the DN&S Winchester station - to terminate at either Eastleigh or Southampton Terminus. Had it not been for the fact that before

morning workings at the southern end of the line.

With the growth in traffic during the early 1950's and conscious of the limited facilities at Basingstoke and Banbury, the Southern recalled the heavy use made of the DN&S as a diversionary line during the war. With an eye on the expanding activity at Fawley, much of which was connected with West Midlands industry, the Southern proposed that the DN&S be regarded as a joint line for operational purposes with the result that from 1953, Chesil shed closed whilst Eastleigh took over the responsibility for a considerable proportion – but by no means all – of the lines' motive power opera-

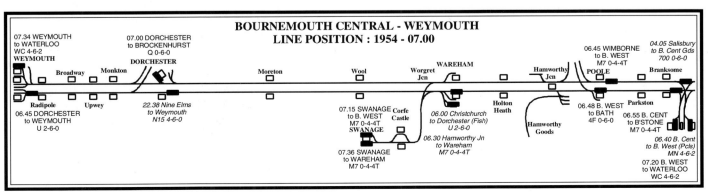

BOURNEMOUTH CENTRAL - WEYMOUTH
LINE POSITION : 1954 - 07.00

Although relegated from express services by King Arthur 4-6-0's around the time of the grouping, the T9 4-4-0's lived a charmed life with over a dozen of the class surviving into the 1960's. Considerable use was made of the class in the Southern division, especially on the stopping trains between Eastleigh and Bournemouth and the local workings from Bournemouth to Salisbury. In the upper view 30706 pulls away from West Moors with a Salisbury - Bournemouth service in 1958 whilst in the lower 30301 approaches Oakley Crossing, Wimborne, with the 10.04 Bournemouth West - Salisbury in June 1959. Generally associated with passenger services, the T9's proved useful on goods work in the Southampton area and regularly worked as many as 60 goods wagons down the main line from Eastleigh to Brockenhurst.

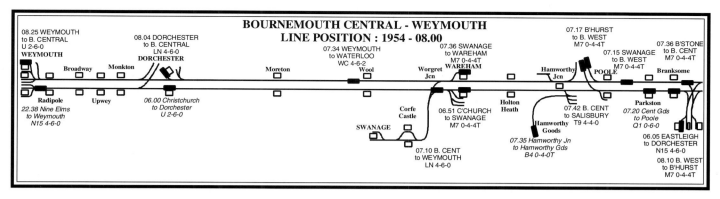

BOURNEMOUTH CENTRAL - WEYMOUTH
LINE POSITION : 1954 - 08.00

tions.

The most significant change was the introduction of two daily fast goods trains in each direction between Eastleigh and Didcot; a highly efficient exercise which was handled by a single Q1 0-6-0 and not much more than two sets of men. It was an extraordinary operation in a world where even the simplest of goods workings was all too often accompanied by a variety of complications. The Q1 rang off Eastleigh shed at 04.54 with a set of relief men (who later took over the 00.55 Feltham – Southampton Docks), ran light to Southampton and pulled away from Bevois Park at 05.35 with 59 empty wagons – the maximum load that could be taken over the DN&S - specially sheeted and roped

and then couple up to the 09.50 return service to Eastleigh.

Given the nature of the line – twenty-five miles of single line between Woodhay and Shawford Junction with gradients that climbed at 1 in 106 from Newbury to Litchfield and then fell at much the same rate to Winchester – the proposal to use it for non-stop goods trains was a novel one and it was probably only the fact that the DN&S had done similar things during the war that saw the scheme adopted. Single-line branches tended to be associated with stopping goods trains and the notion of a hefty great Q1 class with 60 wagons in tow charging hell-for-leather across the sleepier parts of North Hampshire several times a day was one

in districts already saturated with traffic.

The answer lay in the DN&S. If wartime memories were fading, the performance of the Eastleigh – Didcot freights kept its light shining brightly whilst the passenger service was withdrawn in March 1960, nicely (too nicely, some thought) releasing the entire line for freight use. The effect was remarkable and the line that ten years earlier had only had one goods train in each direction was transformed into a major cross-country artery with a basic service of six trains a day conveying a total of one hundred and twenty-nine tanks for Derby, Leicester, Northampton, Denham and Bromford Bridge (two trains). In addition an evening Southampton Docks to Birmingham fitted

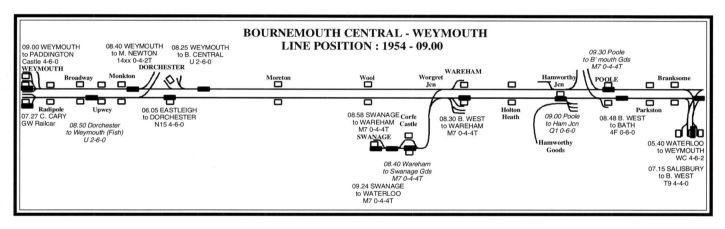

BOURNEMOUTH CENTRAL - WEYMOUTH
LINE POSITION : 1954 - 09.00

over, the main crew taking over as the train paused at Eastleigh at 05.53. The train was then trickled down the slow line to let the 06.04 Eastleigh – Winchester passenger overtake after which the running was hard and fast in order to keep ahead of the 07.32 Southampton – Didcot ('The Glasgow') passenger and to get to Enborne before becoming entangled with the 07.45 down passenger and the pickup.

Haste was also necessary because of the rather ambitious turn-round at Didcot where the Q1 had to leave its train, turn on the angle

that challenged the conventional.

Once it was implemented not only did the scheme work – the passenger service was thin enough to allow some of the through freights a non-stop passage – but it heralded an even brighter future for the DN&S. From 1959 the oil industry at Fawley expanded at an unbelievable rate and began to produce very large quantities of petroleum for the West Midlands which would normally have been routed either via Basingstoke and Reading West or via London; tying up large numbers of engines and crews

freight was run to relieve some of the pressure on Basingstoke and Banbury.

For a time it seemed that the DN&S had its best years ahead of it but the boom proved short-lived. Elsewhere goods traffic was moving from rail to road at an alarming rate – leaving train crews without work at Basingstoke and Reading - whilst the oil companies made no secret of their preference for moving oil round the country by pipeline. The Transport Commission, who had not up to then taken much part in the direction of the regions, had its at-

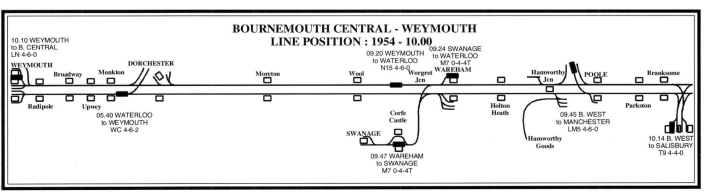

BOURNEMOUTH CENTRAL - WEYMOUTH
LINE POSITION : 1954 - 10.00

The old road - the former main line between Brockenhurst and Hamworthy Junction - did not see much in the way of large engines and on an ordinary weekday the sole appearance of a Bulleid Pacific was confined to the 03.52 Salisbury - Weymouth whose T9 4-4-0 gave way to a West Country 4-6-2 at Wimborne.   The only other large engine to visit the line on a regular basis was an Eastleigh N15 'King Arthur' 4-6-0 which arrived in Wimborne with the 17.05 from Southampton Terminus via Bournemouth.   On Summer Saturdays the incidence of Pacifics increased considerably with the diversion of holiday trains away from the congested Bournemouth area.   The services concerned were the 07.30 Waterloo - Weymouth and 09.15 and 10.54 Waterloo - Swanage expresses plus the 11.34 and 13.23 from Swanage to London; all five workings being booked to West Country Pacifics.   Late in the 1950s a new Pacific-hauled service was added to the list in the shape of the 12.40 Eastleigh - Bournemouth West, a three-coach formation seen at Wimborne in March 1959 (above) and in May of the same year, hauled in both cases by West Country 34044 ' Woolacombe'.

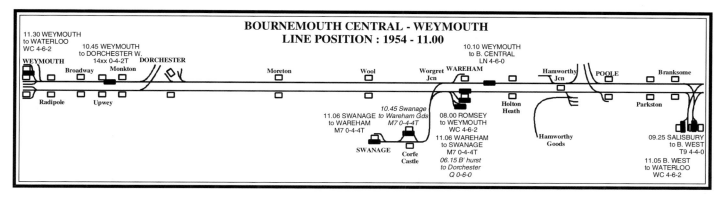

BOURNEMOUTH CENTRAL - WEYMOUTH
LINE POSITION : 1954 - 11.00

tention drawn to the fact that nearly £90m had been lost by the railways in 1961 and was instructed by the government to do something about it. In 1963 the Beeching axe fell and, a year later, took with it the DN&S.

The passenger service over the DN&S had been delightful rather than profitable and, if it served no other role, it allowed the enthusiast to travel between three main lines in the most pleasant way imaginable. In spite of the paucity of trains it also allowed him to sample quite a range of motive power, the service of twelve trains being worked by BR4 2-6-0's and T9 4-4-0's from the Southern end and 22xx 0-6-0's and 43xx 2-6-0's by the Western. For the

Because of this deviation over the Great Western, Eastleigh shed had take care that the booked class of engine was placed into the working. The down train, the 07.15 ex Chesil, changed footplates at Woodhay with the up stopping goods and the Reading men who worked the service forward were none too keen on some of the museum pieces that Eastleigh was apt to find during times of shortages. Equally unenthusiastic were the operators at Paddington after learning that engines uncleared for the Great Western were, by courtesy of the Southern, at large on their main lines.

Although Portsmouth provided the established route to the Isle of Wight, it was by no

also true that for much of the year demand for travel to Yarmouth was far less than it was to Ryde although during on Summer Saturdays the summer the situation changed to the extent that a pair of ten-coach services had to be operated as through trains between Waterloo and Lymington Pier giving the operators a recurring headache as to what motive power to use.

In the past express 4-4-0's, T9 and L12 and S11's, had been the natural choice but as their ranks dissipated the selection narrowed, especially as the remaining T9's were gainfully employed elsewhere on the region. The SECR L and L1 4-4-0's were considered as a substitute – a number were transferred to Eastleigh –

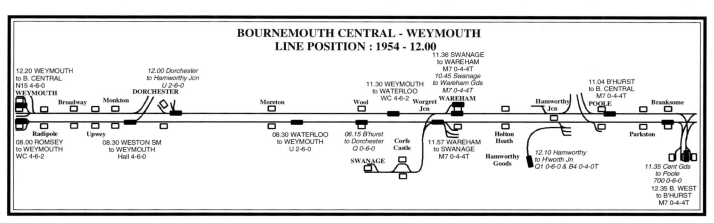

BOURNEMOUTH CENTRAL - WEYMOUTH
LINE POSITION : 1954 - 12.00

most part the trains were self-contained and operated between Southampton Terminus and Didcot but an early train started at Winchester for Reading, running up the Berks and Hants from Newbury. The return service was the 16.32 Didcot to Southampton which started back at Oxford; the stock having formed a Reading – Oxford local in the meantime. Interestingly the engine concerned, an Eastleigh 4MT 2-6-0 did not follow the stock movement but remained on Reading GW shed until mid-afternoon when it worked the 15.45 Reading – Didcot stopping train.

means the only point of entry and the Southern maintained a series of seven (fourteen during the summer) daily sailings between Lymington Pier and Yarmouth for the benefit of passenger wishing to get to the western side of the island.

For most of the year the service was purely local, being operated by a push and pull working between Brockenhurst and Lymington Pier. The absence of through services was in part due to the difficulty involved in finding an engine capable of working at express speeds from Waterloo to Brockenhurst yet light enough for the branch which was highly restricted. It was

but failed to strike the right chord and from 1954 through engine working was abandoned with the main part of the journey being accomplished by Nine Elm's U1 2-6-0's with Q and Q1 0-6-0's taking over from Brockenhurst. (The 2-6-0's were transferred in especially for the Lymington workings and it does not appear to have occurred to anybody that with 0-6-0's working from Brockenhurst, conventional engines such as a Pacific or N15 could have been used from Waterloo).

In spite of the paucity of long-distance trains, the local service to and from Brockenhurst

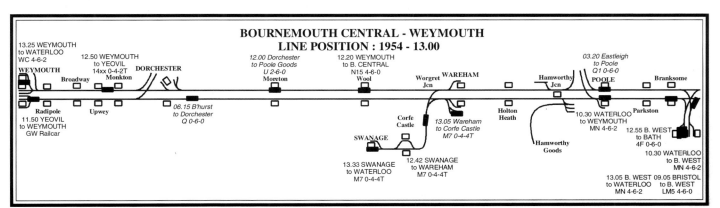

BOURNEMOUTH CENTRAL - WEYMOUTH
LINE POSITION : 1954 - 13.00

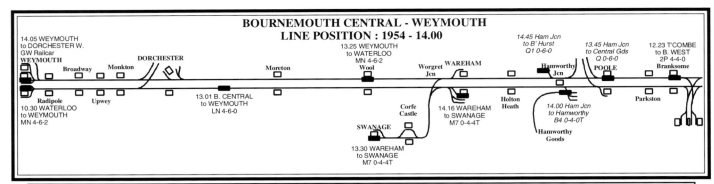

Train	Arr	Station	Dep	Train	Train	Arr	Station	Dep	Train	Train	Arr	Station	Dep	Train
		Bournemouth C	05.47				Swanage	07.38	Waterloo			Bournemouth W.	08.10	
	06.48	Swanage	07.15			08.01	Wareham	08.31			09.18	Brockenhurst	09.39	
	08.12	Bournemouth W.				08.53	Swanage	08.58			09.53	Lymington Pier	10.00	ECS
		Maintenance				09.20	Wareham	09.47			10.02	Lymington Town	10.18	
		Bournemouth W.	16.25			10.09	Swanage	10.30			10.20	Lymington Pier	10.34	
	16.51	Wimborne				10.55	Wareham	11.10			10.38	Brockenhurst	11.04	
		Wimborne	06.15			11.32	Swanage	11.40			12.14	Bournemouth C	17.00	
	06.53	Brockenhurst	07.15			12.02	Wareham	12.25			18.00	Swanage	18.30	
	08.22	Bournemouth W.	08.30			12.49	Swanage	13.33	Waterloo		18.52	Wareham	23.08	ECS
	08.57	Wareham	10.30			13.56	Wareham	14.16			23.37	Bournemouth C	(06.30)	
	10.53	Swanage	11.10			14.38	Swanage	16.23				Bournemouth C	06.30	
	11.35	Wareham	12.00			16.45	Wareham	17.02			07.42	Brockenhurst	07.52	
	12.22	Swanage	12.28			17.24	Swanage	17.38			08.02	Lymington Town	08.05	
	12.50	Wareham	13.05	10.30 Waterloo		18.03	Wareham	18.27			08.15	Brockenhurst	08.34	
	13.27	Swanage	14.45			18.51	Swanage	20.10			09.32	Poole	10.32	
	15.08	Wareham	15.37			20.32	Wareham	21.09	18.30 Waterloo		11.31	Brockenhurst	12.08	
	15.59	Swanage	17.02			21.31	Swanage	22.11			13.19	Bournemouth W.	14.30	
	17.27	Wareham	17.45			22.33	Wareham	23.08	ECS		15.41	Brockenhurst	16.18	
	18.12	Dorchester S.	18.20	ECS		23.37	Bournemouth C	(05.47)			17.26	Bournemouth W.	(08.10)	
	18.40	Wareham	19.30	16.35 Waterloo										
	19.52	Swanage	21.35											
	21.57	Wareham	22.08											
	22.32	Swanage	(07.38)											

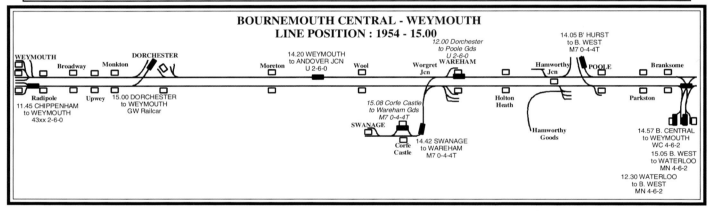

left little to be desired in terms of frequency: so much so that an M7 0-4-4T and two sets of men were permanently stationed at Lymington together with the push and pull set with which most of the services were worked. (A second set was held spare at Brockenhurst to strengthen trains as and when it became necessary).

The level of service was such that the Lymington engine could not cover every work-ing and Bournemouth had to come to the rescue for the 09.39 Brockenhurst – Lymington Pier, the service being worked by the M7 and push & pull of the 08.10 Bournemouth West via West Moors, the engine and stock returning as the 10.35 Lymington/11.04 Brockenhurst to Bournemouth. It was only (!) another M7 but the presence of a second engine added a little colour to what, for regulars on the line, was rather a monotonous working.

School traffic from Lymington occurred at a time when the branch train was in the wrong place and assistance was given by a T9 4-4-0 which came light from Bournemouth shed and made its way up the branch by double-heading the 07.04 from Brockenhurst as far as Lymington Town. The engine used in the work-ing was very often a rarity - keeping the local

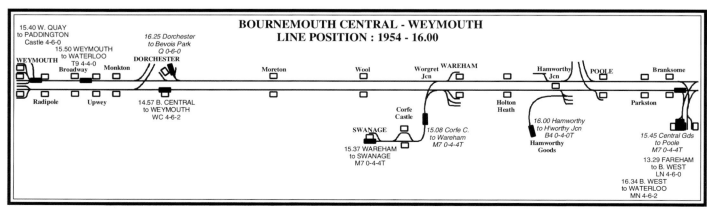

Quite a number of stopping trains operated between Weymouth and Bournemouth in conjunction with the London services and produced an extraordinarily wide array of motive power. No single class had a monopoly of the workings and the engines employed included two U class 2-6-0's, a pair of Lord Nelson 4-6-0's, a West Country Pacific and a King Arthur 4-6-0. In the above view West Country 34105 'Swanage' arrives at Wool in June 1956 with the 12.20 Weymouth to Bournemouth Central. The normal engine for this turn was a King Arthur 4-6-0 but since the duty commenced with the 06.05 Eastleigh - Dorchester, it was often used as a means of returning ex-works Bournemouth Pacifics to their home shed. The three car coaching set worked an interesting, albeit rather unproductive, circuit starting at Salisbury with the early morning newspaper service to Weymouth, returning to Salisbury from Bournemouth with the 07.42 the following morning.

The twenty-strong Q 0-6-0's were rather an enigma so far as the observer was concerned. Lying somewhere between the 700 and Q1 classes, they were too thinly divided between the LBSC and LSW to become really familiar sights at any one location. With only ten examples at work on the South Western, they tended to be rather elusive and one of the best-kept secrets of the South Western was the fact that one of the Eastleigh members of the class had a regular main line passenger booking: the 07.00 Dorchester to Brockenhurst via Wimborne. Seen working a more conventional duty, 30581 enters West Moors with the 08.45 Poole - Brockenhurst goods in August 1958.

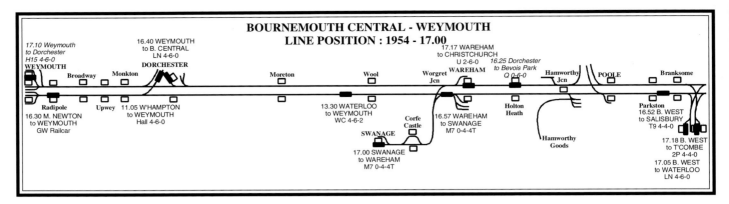

BOURNEMOUTH CENTRAL - WEYMOUTH
LINE POSITION : 1954 - 17.00

trainspotters happy - since the diagram was part of the cyclic arrangement involving Andover Junction, Eastleigh and Bournemouth sheds; a circumstance that militated against the chances of the same engine being used twice in the same week.

Another regular visitor to the line was a Bournemouth Q 0-6-0 which worked the return school train from Brockenhurst and the afternoon goods from Lymington Town during the course of a prolonged circular trip which started with the morning Poole – Brockenhurst goods via Ringwood and concluded with the

For those with time to spare a trip to Bournemouth via Ringwood was another example of the South Western's ability to turn the clock back. The service was almost entirely in the hands of M7 0-4-4T's whilst a change of trains at Wimborne could almost guarantee a T9 4-4-0 over the western section of the line. To further spice the cake an additional change at Broadstone might produce a Midland Railway 2P or 4F: an extraordinary range of variety and all of it, unless one was especially unlucky, pregrouping.

For the less energetic traveller the M7 was

in the mornings, the 0-6-0 subsequently ran light from Brockenhurst to Wimborne and worked the 11.12 local to Bournemouth West.

Although there was no question of the back road ever regaining its status as a main line, a few small steps were taken to ensure that it was not wholly divorced from progress. One – which split the enthusiast from the paying passenger – concerned the replacement during the late 1950's of the traditional LSWR push & pull vehicles in favour of everyday Maunsell corridor stock; the latter being equipped with push & pull apparatus after being made redun-

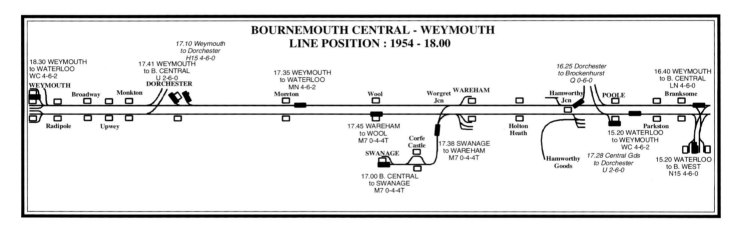

BOURNEMOUTH CENTRAL - WEYMOUTH
LINE POSITION : 1954 - 18.00

18.45 Portsmouth – Bournemouth West from Southampton Central.

As well as being the Junction for Lymington, Brockenhurst was also the point of divergence between the old and new main lines to Bournemouth and although all through traffic had long been transferred to the latter, the former retained a push and pull service which operated at roughly two-hourly intervals taking about seventy minutes to run to Bournemouth West as opposed to three-quarters of an hour by the direct line.

the norm although with an early start at the western end of the line it was possible to break the 0-4-4T mould and secure a run behind a Q 0-6-0; not normally a class of engine associated with passenger work. The engine involved worked down with an overnight goods from Eastleigh to Dorchester goods and returned the following day with the 07.00 passenger to Brockenhurst via Ringwood. The trip also allowed a ride over the little-used Hamworthy Junction – Broadstone section of the old main line. For those who preferred to remain abed

dant by the Kent Coast electrification of 1959. Another was a rumour in early 1957 that the line was once again going to see a daily service to and from London: a circumstance that led to some improbable speculation involving Weymouth services running via Ringwood with their Bournemouth West portions being removed at Broadstone.

The source of the rumours was a request by regular evening travellers from Waterloo for a through service to Wimborne and although the request was granted – the Southern being in

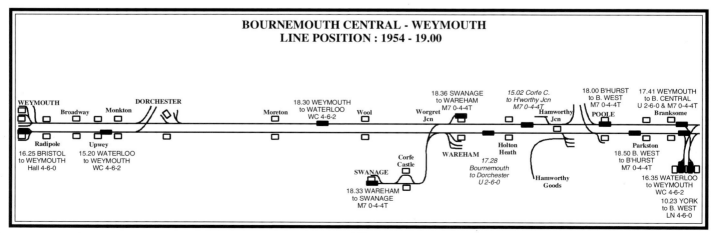

BOURNEMOUTH CENTRAL - WEYMOUTH
LINE POSITION : 1954 - 19.00

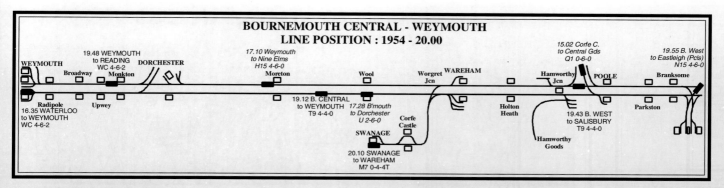

BOURNEMOUTH CENTRAL - WEYMOUTH
LINE POSITION : 1954 - 20.00

an unusually receptive mood that year – the new service had no effect on the West Moors line since the through working was achieved by dividing the 19.30 Waterloo – Bournemouth at Southampton Central and running the rear three vehicles to Wimborne via Bournemouth Central.

Whilst it was pleasing to see the direct London – Wimborne link restored, it was disappointing to find that it did not include the remainder of the back route and a scrutiny of the workings involved revealed that the new arrangement was no more than an amalgam of the 22.13 Brockenhurst – Bournemouth and the 23.10 Bournemouth West – Wimborne.

Curiously, there was no return working and

similarities to those of the Lymington line, the chief difference being that almost any engine in the fleet, if called upon, could work between Wareham and Swanage. Like Lymington, Swanage had a permanent locomotive allocation but it also had the status of through coaches to Waterloo several times a day even though they ran over the branch behind the same M7 0-4-4T's that worked the rest of the service.

Apart from the London trains, which existed for business traffic rather than holiday-makers, the service to Swanage hardly reflected the resort's place as one of the premier watering holes west of Bournemouth and it occurred to one now and again to wonder if greater profit

signs of being *dynamic…*). In any event with a population of only seven thousand, Swanage's potential was hardly boundless.

A handful of branch trains ran beyond Wareham to Bournemouth, more as a means of allowing the branch push and pull set to reach Bournemouth for maintenance every third day than for the handful of office-workers who used it. The working covered the 22.11 Swanage – Wareham and empty to Bournemouth Central, forming the 05.47 Bournemouth Central to Swanage and the 07.15 return after which it received any attention that was due. Later that evening the set returned via a rather circuitous route which involved the 16.25 Bournemouth West – Wimborne and, the following day, the

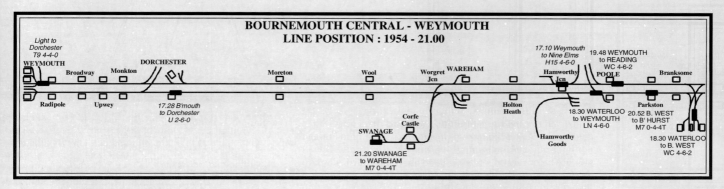

BOURNEMOUTH CENTRAL - WEYMOUTH
LINE POSITION : 1954 - 21.00

the three vehicles that made up the 19,.30 Waterloo – Wimborne continued West with a local train to Bournemouth early the following morning.

In retrospect the level of passenger activity at Brockenhurst was astonishing, especially as the station was reckoned, in operating terms, to be a rather quiet location. M7 push and pull workings from Lymington and the Old Road would run into the up side loop, connect with a main line express before shunting over to the down side to leave after connecting with a down express. It was a procedure repeated with occasional variations of detail many times a day yet it was so ordinary that it went almost unnoticed.

The Swanage branch had many operational

might not be accrued by extending some of the long-distance cross-country workings – such as the York and Birkenhead trains - from Bournemouth West to Swanage, a move which would also have been of benefit to Poole. A couple of coaches twice a day to Broadstone via Hamworthy Junction for attachment to an S&D service would have maintained regular contact with Bath and Bristol but the Southern – like everyone else – preferred to stick to the pattern of workings that had evolved largely on the grounds that since there were relatively few complaints about the status quo, there was little point in taking unnecessary risks by changing it. (Young men *with ideas* on the Southern were pretty firmly put in their place. Waterloo was not too keen on people who showed

06.15 Wimborne – Brockenhurst, 07.15 Brockenhurst – Bournemouth West and the 08.30 to Wareham whereupon it took its place for the next two days in the Swanage workings.

The complexity of carriage workings made them a fascinating business and, alone of the regions, the Southern was the only system to develop a method with made them clear – or as clear as the subject could be made – to the staff on the ground.

To demonstrate the knots that even the local carriages – nothing less than a complete book would do justice to the main line stock - became embroiled in, the workings of the five sets that worked the Swanage branch are set out in an accompanying table. It will be noticed that none were exclusive to the branch but roamed

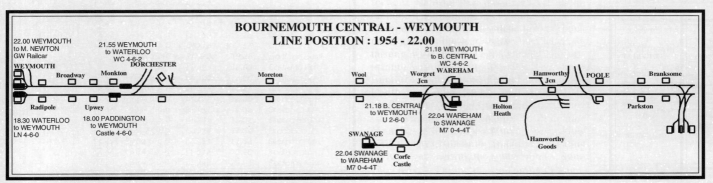

BOURNEMOUTH CENTRAL - WEYMOUTH
LINE POSITION : 1954 - 22.00

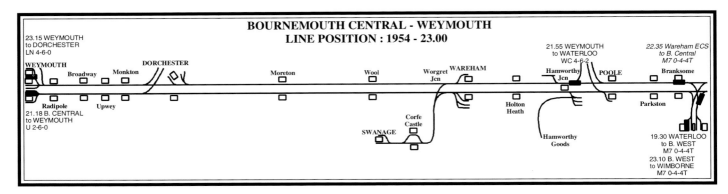

far and wide over the Bournemouth district, including a number of workings over the Lymington line. (It was said, not without truth, that the only man who held the workings in his head was the diagrammer who compiled them and the author can admit that, in his time, he was not the only soul who took a devilish delight in diagramming carriages from one end of a division and back simply for the fun of it).

Of all the South Western byways, perhaps the least was the Salisbury – Wimborne line, a fairly quiet affair with only a handful of trains in each direction; its saving grace being the fact that for most of the year each service was worked by a pregrouping locomotive: T9 4-4-0's on the nine passenger trains and 700 0-6-0's on the four goods services.

Lest it be thought that the line operated on a simple diet of trains shuttling between Bournemouth and Salisbury, the service over this backwater of backwaters contained some unusual imbalances which called for some skill on the part of the diagrammers. In the down direction there were four services from Salisbury – one of which went to Weymouth – whilst there were five up trains, all of which originated in the Bournemouth area and this resulted in some strange workings such as an Eastleigh T9 4-4-0 which performed three trips between Bournemouth and Salisbury and then, to balance matters, some work over the Romsey line. Similarly two of the Salisbury 4-4-0's could be seen at Portsmouth.

One of the reasons for the line remaining as a stronghold of pregrouping power was because very little else was allowed over the route. The L12 and S11 4-4-0's had gone while almost everything that remained, other than the T9 or 700, was heavily restricted. During the summer the supply of suitable engines became – on paper - critical at times; forcing the diagramming office to book the 6-coach 09.05 Salisbury – Weymouth to a West Country Pacific, even though the class was prohibited from exceeding 30 mph between Alderbury Junction and West Moors. Absurd it may have been but, in keeping with the theme of this book, it was yet another example of South Western variety.

Although steam continued to operate the South Western for the greater part of the 1960's, the last few years were ones of unaccustomed standardisation - Rebuilt Bulleid Pacifics and BR 5MT 4-6-0's handled most services - and decline, especially where rolling stock was concerned. The brave new world of the 1960s is encapsulated in the view of 34051 'Winston Churchill' as it winds a Salisbury - Bournemouth working over the old road near Wimborne in 1963. A few years earlier the train would have been formed of a smart set of uniform SR vehicles far removed from the miscellany seen above; the train consisting of a Stanier LMS coach, an LNER Gresley vehicle and a BR standard together with a mixture of parcels vehicles.